D1376376

Ocean and Seabed Acoustics

A Theory of Wave Propagation

George V. Frisk

Woods Hole Oceanographic Institution

P T R Prentice Hall
Englewood Cliffs, New Jersey 07632

Library of Congress Cataloging-In-Publication Data

Frisk, George V.
Ocean and seabed acoustics : a theory of wave propagation / George V. Frisk.
Includes bibliographical references and index.
ISBN 0-13-630112-6
1. Wave-motion, Theory of. 2. Acoustic surface waves.
3. Underwater acoustics. 4. Wave equation--Numerical solutions.
I. Title .
QC157.F75 1994
534' .2--dc20 93-8553
CIP

Editorial / production supervision: *Mary P. Rottino*
Cover design: *Wanda Lubelska Design*
Buyer: *Alexis Heydt*
Acquisitions editor: *Mike Hays*

Prentice-Hall, Inc.
A Paramount Communications Company
Englewood Cliffs, New Jersey 07632

The publisher offers discounts on this book when ordered in bulk quantities.
For more information, contact: Corporate Sales Department, P T R Prentice Hall, 113 Sylvan Avenue, Englewood Cliffs, NJ 07632, Phone: 201-592-2863, FAX: 201-592-2249

Printed in the United States of America
10 9 8 7 6 5 4 3 2 1

ISBN 0-13-630112-6

Prentice-Hall International (UK) Limited, *London*
Prentice-Hall of Australia Pty. Limited, *Sydney*
Prentice-Hall Canada Inc., *Toronto*
Prentice-Hall Hispanoamericana, S.A., *Mexico*
Prentice-Hall of India Private Limited, *New Delhi*
Prentice-Hall of Japan, Inc., *Tokyo*
Simon & Schuster Asia Pte. Ltd., *Singapore*
Editora Prentice-Hall do Brasil, Ltda., *Rio de Janeiro*

To my wife, *Margaret*,

my children, *Daniel*, *Andrew*, and *Maria*,

and my mother, *Lydia*

In loving memory of my father, *Vladimir George Frisk*

You must know that there is nothing higher and stronger and more wholesome and good for life in the future than some good memory, especially a memory of childhood, of home. People talk to you a great deal about your education, but some good, sacred memory, preserved from childhood, is perhaps the best education. If a man carries many such memories with him into life, he is safe to the end of his days, and if one has only one good memory left in one's heart, even that may sometime be the means of saving us.

Alyosha Karamazov in *The Brothers Karamazov*,
by Fyodor Dostoyevsky, 1821-1881

Contents

Preface xi

Acknowledgments xiii

1. Ocean and Seabed Acoustics: A Physicist's Perspective 1

1.1 Introduction 1
1.2 Historical Overview 2
1.3 The Acoustic Wave Equation 5
1.4 The Schrodinger Wave Equation 6
1.5 The Electromagnetic Wave Equation 7
1.6 The Elastic Wave Equation 11
Problems 13

2. Elementary Solutions and Basic Acoustic Quantities 17

2.1 Introduction 17
2.2 The One-Dimensional Wave Equation 17
2.3 The Three-Dimensional Wave Equation 19
2.4 The Two-Dimensional Wave Equation 21
2.5 Velocity Potential 23
2.6 Specific Acoustic Impedance 25
2.6.1 Plane Wave Impedance 25
2.6.2 Spherical Wave Impedance 25
2.7 Energy Flux Density and Intensity 26
2.7.1 Plane Wave Intensity 27
2.7.2 Spherical Wave Intensity 27
Problems 28

3. Plane Wave Reflection from Planar Boundaries 31

3.1 Introduction 31
3.2 Boundary Conditions 32
3.2.1 Soft Boundary 32
3.2.2 Hard Boundary 32
3.2.3 Impedance Boundary 33
3.2.4 Sommerfeld Radiation Condition 33
3.3 Plane Wave Reflection from a Horizontally Stratified Medium 34
3.3.1 Reflection from a Soft Boundary 37
3.3.2 Reflection from a Hard Boundary 37
3.3.3 Reflection from a Homogeneous Fluid Half-Space 37
Reflection from a Lower Velocity, Less Dense Half-Space (The Water-Air Interface) 41
Reflection from a Lower Velocity, More Dense Half-Space (The Water-Bottom Interface: Model I) 43
Reflection from a Higher Velocity, More Dense Half-Space (The Water-Bottom Interface: Model II) 45
3.3.4 Reflection from a Homogeneous Fluid Layer Overlying an Arbitrary Horizontally Stratified Medium 51
Problems 53

4. Acoustic Sources and Green's Functions 59

4.1 Introduction 59
4.2 Wave Equations with Source Terms 60
4.3 Properties of the Green's Function 61
4.4 General Solution of the Boundary Value Problem with Sources 64
4.5 A Recipe for Solving Problems Using Green's Functions 67
4.6 Construction of the Green's Function: The Method of Images 68
4.6.1 The Method of Images for a Plane with Dirichlet Conditions 68
4.6.2 The Method of Images for a Plane with Neumann Conditions 70
4.6.3 The Method of Images for a Quadrant with Dirichlet Conditions 72
4.6.4 The Lloyd Mirror Effect 73

4.6.5 A Point Source in a Homogeneous Fluid Layer with Impenetrable Boundaries 80
4.7 Construction of the Green's Function: The Endpoint Method 83
4.8 A Point Source in a Homogeneous Fluid Half-Space Overlying an Arbitrary Horizontally Stratified Medium 85
4.8.1 Asymptotic Analysis for Reflection of a Spherical Wave from a Homogeneous Fluid Half-Space 89
Singularities of the Integrand 93
Reflection from a Lower Velocity Half-Space ($n>1$) 93
Reflection from a Higher Velocity Half-Space ($n<1$) 96
Problems 105

5. The Method of Normal Modes 110

5.1 Introduction 110
5.2 A Point Source in a Horizontally Stratified, Fluid Medium 111
5.2.1 Normal Modes for a Homogeneous Fluid Layer with a Soft Top and Hard Bottom 116
Nodes of the Modal Depth Eigenfunctions 118
Modal Phase Velocity 119
Modal Cutoff Frequency 120
Modal Group Velocity 120
Modal Intensity, Interference Wavelength, and Cycle Distance 123
The Modal Continuum 126
5.3 Eigenvalue Equation for a Homogeneous Fluid Layer Bounded by Arbitrary Horizontally Stratified Media 129
5.4 Normal Modes for a Homogeneous Fluid Layer Bounded Above by a Pressure-Release Surface and Below by a Lower Velocity, Homogeneous Fluid Half-Space 130
5.4.1 An Improper Sturm-Liouville Problem 130
5.4.2 A Proper Sturm-Liouville Problem 131
5.4.3 Improper Modes 136
An Equivalent Eigenvalue Equation 137
Poles of the Integrand 138
A Leaky Mode Decomposition 141
5.5 Normal Modes for a Homogeneous Fluid Layer Bounded Above by a Pressure-Release Surface and Below by a Higher Velocity, Homogeneous Fluid Half-Space 145

5.5.1 Proper Modes for the Pekeris Waveguide 148
Nodes of the Modal Depth Eigenfunctions 149
Modal Cutoff Frequency 150
Modal Phase Velocity 152
Modal Group Velocity 152
Modal Intensity, Interference Wavelength, and Cycle Distance 155
5.5.2 Improper Modes for the Pekeris Waveguide 157
A Leaky Mode Decomposition 158
5.5.3 The Total Field in the Pekeris Waveguide 162
Problems 163

6. The Hankel Transform: A Unified Approach to Wave Propagation in Horizontally Stratified Media 168

6.1 Introduction 168
6.2 A Point Source in a Horizontally Stratified, Fluid Layer Bounded by Arbitrary Horizontally Stratified Media 169
6.3 A Point Source in a Homogeneous Fluid Layer Bounded by Arbitrary Horizontally Stratified Media 172
6.3.1 A Geometrical Acoustics Decomposition 173
Zeroth Order Model of the Ocean Acoustic Waveguide 174
The General Case of Penetrable Boundaries 176
An Alternative Geometrical Acoustics Decomposition for the Pekeris Waveguide 181
6.3.2 A Normal Mode Decomposition 186
The Pekeris Waveguide 187
Problems 198

7. Approximate Methods for Inhomogeneous Media: Ray and WKB Theory 202

7.1 Introduction 202
7.2 The Refracting Ocean Environment 203
7.3 Two-Dimensional Plane Wave Ray/WKB Theory in an Unbounded, Horizontally Stratified Medium with No Sources 206
7.4 Three-Dimensional Ray Theory for an Arbitrary Unbounded Medium with No Sources 212

7.5 Three-Dimensional Ray Theory for a Point Source in an Unbounded, Horizontally Stratified Medium 214
7.5.1 An Alternative Derivation of Snell's Law 215
7.5.2 Ray Range 217
Linear Gradient Example 218
7.5.3 Ray Phase and Travel Time 219
Linear Gradient Example 220
7.5.4 Ray Acoustic Intensity 221
Linear Gradient Example 225
7.6 A Point Source in a Medium with a Velocity Profile Having a Single Minimum 226
7.6.1 The Ray Series Solution 226
7.6.2 Asymptotic Evaluation of the Hankel Transform Representation in the WKB Approximation 231
7.6.3 Long-Range SOFAR Propagation for a Symmetric Bilinear Profile 239
Ray Range, Initial Angle, and Turning Depth 242
Ray Phase and Travel Time 244
Ray Amplitude 247
Time-Dependent Behavior of a Delta Function Pulse 248
7.6.4 Normal Modes in the WKB Approximation 253
7.7 Deficiencies of Classical Ray Theory 258
7.7.1 Intensity Predictions 259
Shadow Zones 260
Perfect Foci, Caustics, and Convergence Zones 261
7.7.2 Reflection Coefficients at Discontinuous Interfaces and Turning Points 264
Problems 265

Appendix A. Adiabatic Mode/WKB Theory for a Slowly Varying, Range-Dependent Environment 269

Appendix B. Sound Propagation in a Wedge with Impenetrable Boundaries 275

References 281

Index 291

Preface

A journey of a thousand miles must begin with a single step.
Lao-Tzu, c. 604-531 B.C.

This project began one morning more than fifteen years ago when I was driving to work with the late Earl Hays, my boss and Chairman of the Ocean Engineering Department at Woods Hole Oceanographic Institution. He asked me if I had ever taught a course before, and I naively said, "No," but expressed my enthusiasm at the prospect of doing so. Thus began my involvement as principal instructor in Course 13.861, *Ocean and Seabed Acoustics I*, a part of the Massachusetts Institute of Technology/Woods Hole Oceanographic Institution Joint Graduate Program in Oceanography and Applied Ocean Science and Engineering. This one-semester course is offered in alternate years, and, as part of a sequence of courses in underwater acoustics, is intended to provide students with an introduction to theoretical ocean and seabed acoustics. Early on, I realized that although a number of fine books existed in the field, none of them presented the material in the manner that I desired. As a result, I developed my own set of notes and homework problems in which I strived to convey the notion that ocean and seabed acoustics is a part of the mainstream of theoretical physics. In particular, I hoped to show that the ocean environment provides an excellent setting in which to display general principles of wave propagation which are applicable to other areas of wave physics as well.

With this approach in mind, I begin the book with a discussion of the acoustic wave equation and its relationship to wave equations which arise in other fields of physics. The development then proceeds with a derivation of elementary solutions to the wave equation in free space and progressively addresses problems of increasing complexity, culminating in a discussion of acoustic wave propagation due to a point source in an inhomogeneous waveguide with lossy boundaries. Theoretical methods in wave propagation physics are developed in a logical sequence and then applied to canonical examples which deal primarily with horizontally stratified (range-independent) fluid media. The formalism presented is general, however, and can readily accommodate elastic boundaries, for example, through the incorporation of appropriate reflection coefficients. Furthermore, two important problems in wave propagation in laterally varying media are solved in the appendices.

My initial plan for the book had included three additional chapters dealing with solution techniques for range-dependent media, wave phenomena in elastic media, and rough surface scattering. However, time constraints and single authorship have forced me to leave these subjects for a possible expanded future edition. It is a consolation to me that some of the topics I had in mind are addressed in recent books by Ogilvy [1991] on rough surface scattering and Jensen et al. [1993] on computational ocean acoustics.

This book can be used as a textbook as well as a reference book for the practicing wave physicist. The development is driven primarily by the pedagogical considerations of logic, clarity, and completeness. The material is presented at a level which is appropriate for a student having a background in complex variables, Fourier transforms, and differential equations, as well as some exposure to wave physics. However, no previous experience in acoustics is required. The homework problems have been thoroughly thought out and are an integral part of the presentation. A solutions manual is currently being prepared and will be available from the publisher. For the student, the development is most valuable if it is followed in logical sequence, for this approach displays the full beauty and richness of the subject matter. On the other hand, individual chapters and sections can be studied in isolation or out of sequence, depending on the nature of the course, the inclination of the professor, and the backgrounds of the students. The scope of the book has gone well beyond my original one-semester course and, if used in its entirety, could possibly occupy two semesters.

In conclusion, there are two significant aspects of this project that I wish to mention. First, I learned that writing is basically a peaceful activity, despite the mental turmoil the author sometimes experiences. I found the solitude associated with the writing process to be particularly appealing. Finally, the work presented here is an application of fundamental ideas developed by great minds, primarily during the nineteenth century. Any successes in the development should therefore be credited largely to those geniuses, and any failures should be considered errors in my interpretation of their great work.

GEORGE VLADIMIR FRISK

July 1993

Senior Scientist and Chair
Department of Applied Ocean Physics and Engineering
Woods Hole Oceanographic Institution
Woods Hole, Massachusetts 02543

Acknowledgments

Reading maketh a full man, conference
a ready man, and writing an exact man.
Francis Bacon, 1561-1626

I began writing this book in 1989 and was supported by a J. Seward Johnson Chair in Oceanography and the Mellon Independent Study Award Program at the Woods Hole Oceanographic Institution. I am grateful to the Institution and members of the previous WHOI administration - John Steele, Derek Spencer, Charles Hollister, and Robert Spindel - for granting me these awards. I thank members of the subsequent and current administrations - Craig Dorman, Robert Gagosian, John Farrington, and Albert J. Williams, III - for their continued support of my academic activities.

A substantial portion of this book was written during a sabbatical year (1989-1990) at the Applied Research Laboratory, Pennsylvania State University, and I am grateful to Ray Hettche and Richard Stern for providing this opportunity. I also thank Diana McCammon, Suzanne McDaniel, Scott Sommerfeldt, and David Swanson for their hospitality and helpful comments.

The theoretical development presented in the book was strongly influenced by my graduate career in physics at the Catholic University of America. I am thankful for the course notes (published and unpublished) of Thomas Eisler [1969], Charles Montrose, and Herbert Überall, my doctoral thesis adviser. My dissertation research [Frisk and Überall, 1976] left me with a lasting impression concerning the importance of asymptotic methods and limiting processes in improving our understanding of wave physics problems. As a result, these techniques manifest themselves frequently as very useful analysis and interpretation tools in the book. Over the years, my interactions with Norman Bleistein, John DeSanto, Leopold Felsen, and David Stickler have continued to shape my thinking about wave propagation problems.

Early in the project, Thomas Gabrielson volunteered to proofread chapters as they were generated, particularly with respect to their technical content. For his invaluable feedback, I am forever grateful. I am also thankful for the support and comments of David

Bradley, Kazuhiko Ohta, James Preisig, Dajun Tang, and D. Keith Wilson.

In addition to using this material in MIT/WHOI Course 13.861, I have presented portions of the development in a course at the Colorado School of Mines and an Acoustical Society of America Short Course on Ocean and Seabed Acoustics. In all, more than 150 students have been exposed to my presentation of this topic. Their feedback has been enormously helpful in the continuing effort to improve the accuracy and content of the text.

I am grateful to Alan Oppenheim for recommending my work to Prentice-Hall and helping to establish my relationship with them. I thank Michael Hays, Executive Editor, for his patience and cooperation.

With the exception of the cover, I generated this book entirely using a Macintosh IICX computer, a Laserwriter IINT printer, and three excellent pieces of Software: (1) Microsoft Word 4.00A for the text, (2) MathType 2.05 (Design Science, Inc.) for the equations, and (3) SuperPaint 2.0 (Silicon Beach Software) for the figures. The power, freedom, and flexibility which this arrangement offered me in terms of production, formatting, and editing was remarkable. I thank my sister, Helen Buzyna, for her helpful suggestions concerning the cover design.

Finally, a project of this magnitude has many high and low points, all of which my family had to endure for an extended period of time. I am eternally grateful to them for their patience, encouragement, and love.

G. V. F.

Chapter I. Ocean and Seabed Acoustics: A Physicist's Perspective

There is no new thing under the sun.
Ecclesiastes 1:9

The more things change, the more they remain the same.
Alphonse Karr, 1849

1.1 Introduction

Ocean and seabed acoustics is a remarkable science with significant practical applications. For the physicist specializing in wave propagation, the ocean provides a medium in which virtually every conceivable basic problem in wave propagation can be studied. The ocean is an inhomogeneous medium with an acoustic index of refraction that varies with depth as a result of changes in temperature and hydrostatic pressure and, to a lesser extent, in salinity. On a finer spatial scale, thermal microstructure introduces a random component to the sound velocity structure. Wave propagation through inhomogeneous and random media therefore constitutes an important part of ocean acoustic research. The ocean surface and bottom are, in general, rough boundaries, thus requiring the study of acoustic scattering from rough surfaces. The ocean bottom is a multilayered structure composed of sediments and rocks which, in general, supports both compressional and shear waves. Therefore, the interaction of acoustic waves with multilayered media containing velocity gradients is also an active area of research in underwater acoustics. In recent years, the inverse problem of determining oceanographic properties from acoustic measurements has become increasingly important and has given rise to the term "acoustical oceanography," a variant of "ocean acoustics" that emphasizes the oceanographic implications of acoustic experiments.

Acoustic waves, in contrast to electromagnetic waves, which are strongly absorbed by water, can propagate over hundreds, even thousands, of miles through the ocean under the proper conditions. As a result, sound waves and sonar assume the major role in the ocean that electromagnetic waves and radar play in the atmosphere. Sound is used for navigation, communication, underwater detection, and as a tool for studying the oceanography of the water column and the structure of the seabed. Active and passive sonar systems are the most important methods used by the world's navies to detect, classify, and track submarines in antisubmarine warfare.

Because of its fundamental scientific nature on the one hand and its practical character on the other, ocean and seabed acoustics embraces a broad spectrum of activities ranging from crude estimates of sonar performance for engineering purposes to esoteric calculations of wave propagation through inhomogeneous media. In this book, we will concentrate on the development of the fundamental principles of wave physics which underlie ocean and seabed acoustics and show that it is part of a great tradition of theoretical physics which also includes quantum mechanics, electromagnetics, and seismics.

1.2 Historical Overview

The history of ocean and seabed acoustics and the history of acoustics in general are to a large extent one and the same thing prior to the twentieth century. Although Lindsay [1972] and Pierce [1981] discuss insights into the science of acoustics dating back to Marcus Vitruvius Pollio (ca 25 B.C.) and Aristotle (384-322 B.C.), the classical foundation of modern acoustics was laid by a distinguished list of nineteenth century physicists and mathematicians which included Simeon Denis Poisson, Pierre Simon Laplace, Jean Baptiste Joseph Fourier, George Green, George Gabriel Stokes, Gustav Robert Kirchhoff, and Hermann von Helmholtz. The culmination of this intense activity in mathematical physics manifested itself in the acoustics arena with the publication of *The Theory of Sound* by Lord Rayleigh (John William Strutt) in 1877-78. In this monumental, two-volume work, Rayleigh not only described his own fundamental contributions, but also synthesized the state of the art in acoustics as a whole. Rayleigh's advances in the study of mechanical vibration and sound propagation through fluid media have had a profound influence on the science of acoustics [Lindsay, 1966]. Specifically, in ocean and seabed acoustics, his solutions of basic problems in waveguide propagation, scattering from bounded objects, and scattering from rough surfaces are regarded as benchmarks. As a result, to this day, Rayleigh is considered the acoustician nonpareil, the physicist who set the standard in acoustics and who represents the conscience of every practicing acoustical scientist.

One notable research activity specifically related to ocean acoustics which occurred in the nineteenth century was the measurement of the speed of sound in water. In 1826, Daniel Colladon and Charles Sturm conducted an experiment in Lake Geneva, Switzerland in which they measured a sound speed of 1435 m/s, a remarkably accurate value [Lindsay, 1966, 1972]. It is of interest to note that Sturm was the well-known mathematician associated with the Sturm-Liouville theory of differential equations [Arfken, 1985], which we will use extensively in this book. This versatility, particularly with regard to theoretical *vis-a-vis* experimental work, was typical of the nineteenth century scientist.

At the turn of the century, the emergence of quantum mechanics and the special theory of relativity permanently altered the course of theoretical physics [Schiff, 1968; Jackson, 1975]. The novel experimental observations, the mathematically sophisticated theories, and the philosophical implications associated with these new disciplines made them the fields of choice for the twentieth century physicist. The excitement of these new developments in modern physics was so intense that they eclipsed the classical areas of theoretical physics which had been so heavily pursued in the nineteenth century. This turn of events, when combined with Rayleigh's treatise, gave the appearance that acoustics was dead as a fundamental science and was only of interest for engineering applications. The irony of this situation was that many of the mathematical methods used in the new modern physics were borrowed from the rich classical tradition of the nineteenth century. In fact, this deep classical foundation in wave physics enabled the rapid development of a suitable wave equation to explain the quantum-mechanical observations. The intimate connection between the Schrodinger wave equation and the classical wave equation was then exploited to the fullest. For example, the method of partial wave expansion, which became a mainstay of quantum-mechanical potential scattering theory, was originally developed by Rayleigh to solve the problem of acoustic scattering by spheres [Rayleigh, 1877-78; Schiff, 1968; Merzbacher, 1970]. The Rayleigh-Schrodinger perturbation technique and the Rayleigh-Ritz variational method, widely used approximate approaches for estimating the eigenstates and eigenvalues of quantum-mechanical systems, originated with Rayleigh in applications to mechanically vibrating systems. The WKB method for determining approximate solutions to wave equations was named for its quantum-mechanical proponents, Wentzel, Kramers, and Brillouin, and yet it had been developed earlier by Liouville, Rayleigh, and Jeffreys. As quantum mechanics and relativity rapidly progressed in the twentieth century, finally evolving into a theory of relativistic quantum mechanics, they assumed the lead in the advancement of wave-mechanical techniques. Whereas at the beginning of the century, the modern physicist looked to the classical tradition for assistance in developing the new wave mechanics, by the middle of the century, the classical wave physicist turned to the enormous wealth of knowledge in modern physics for guidance. A typical example is the application of Feynman diagrams to the solution of problems in wave propagation in random media and scattering from randomly rough surfaces [Feynman and Hibbs, 1965; DeSanto, 1979].

In the midst of this excitement in the world of modern physics, ocean and seabed acoustics got off to a slow start. The *Titanic* tragedy triggered patent applications for underwater acoustic echo ranging by L. R. Richardson in Great Britain in 1912 and R. A. Fessenden in the United States in 1913 [Clay and Medwin, 1977; Burdic, 1984]. The first applications of underwater sound to submarine detection were made in 1915-16 during World War I by Constantin Chilowsky and Pierre Langevin in France and Robert W. Boyle in Great Britain. Subsequent advances were made in the use of sound to determine water depth (echo sounding) and to find schools of fish. But it was only in World War II that ocean and seabed acoustics came into its own as a science and as a powerful tool for detecting submarines. An intense effort to study underwater sound phenomena was initiated by the United States government under the auspices of the National Defense Research Committee of the Office of Scientific Research and Development. The Sonar Analysis Group, directed by Lyman Spitzer, Jr., was formed and included distinguished scientists from institutions such as the Woods Hole Oceanographic Institution, Scripps Institution of Oceanography, Columbia University, and the Massachusetts Institute of Technology. Their efforts culminated in the publication of *Physics of Sound in the Sea* [1946], a remarkable summary of the state of the art in underwater sound which is still a useful reference today. The most significant finding of this period was the discovery of the SOFAR (SOund Fixing And Ranging) channel by W. Maurice Ewing and John L. Worzel [1948] in the United States and Leonid M. Brekhovskikh and co-workers [1980, 1991] in the Soviet Union. This presence of a sound velocity minimum with depth enables sound to be channeled by the oceanic waveguide and to propagate for thousands of kilometers under the appropriate conditions. This effect virtually guaranteed the position of acoustics as the single most important remote sensing tool in the ocean.

In the postwar period, the funding base for ocean and seabed acoustics research was solidified in the United States with the formation of the Office of Naval Research in 1946. Significant international efforts developed as well, and it became clear that the field was here to stay. The rapidly expanding knowledge of the complex ocean environment repeatedly challenged the acoustical wave theorist and experimentalist, and acoustics and oceanography became increasingly intertwined [Clay and Medwin, 1977; Flatte, 1979]. Although ocean and seabed acoustics never achieved the glamour of high-energy physics or cosmology, the intellectual excitement generated when a sound wave meets the ocean and its boundaries rivals the best of modern physics problems.

1.3 The Acoustic Wave Equation

The equations of linear acoustics are obtained by linearizing the basic equations of inviscid, compressible fluid mechanics without heat conduction [DeSanto, 1979; Boyles, 1984; Landau and Lifshitz, 1987]. The details of this derivation are left as an exercise for the reader (cf. Prob. 1.1). The key result of this procedure is the *time-dependent acoustic wave equation* with density and sound velocity stratification and no sources

$$\rho_0(\mathbf{r})\nabla\bullet\left[\frac{1}{\rho_0(\mathbf{r})}\nabla P(\mathbf{r},t)\right]-\frac{1}{c^2(\mathbf{r})}\frac{\partial^2 P(\mathbf{r},t)}{\partial t^2}=0\ . \tag{1.1}$$

Here $P(\mathbf{r},t)$ is the excess acoustic pressure about the ambient pressure as a function of position $\mathbf{r}$ and time t. The background density $\rho_0(\mathbf{r})$ and sound velocity $c(\mathbf{r})$ are arbitrary functions of position. In many cases of interest in acoustics, we consider solutions to Eq. (1.1) in regions of constant density, in which case it becomes the *standard wave equation*

$$\nabla^2 P(\mathbf{r},t)-\frac{1}{c^2(\mathbf{r})}\frac{\partial^2 P(\mathbf{r},t)}{\partial t^2}=0\ . \tag{1.2}$$

Equation (1.2) is a fundamental equation of acoustics, as well as other areas of wave physics, as we shall see. When combined with suitable source terms and boundary and initial conditions, it constitutes the focal point of most acoustics research. When we apply the Fourier transform operator $F.T.\{\bullet\}$ to Eq. (1.2),

$$F.T.\{\bullet\}=\frac{1}{\sqrt{2\pi}}\int_{-\infty}^{\infty}\{\bullet\}e^{i\omega t}\,dt\ , \tag{1.3}$$

where $\omega=2\pi f$ is the angular frequency (f is the frequency), we obtain the *Helmholtz equation* or the *reduced wave equation*

$$\left[\nabla^2 + k^2(\mathbf{r})\right]p(\mathbf{r},\omega) = 0 \quad , \tag{1.4}$$

where $k(\mathbf{r}) = \omega/c(\mathbf{r})$ is the total acoustic wavenumber. Here $p(\mathbf{r},\omega)$ and $P(\mathbf{r},t)$ are conjugate Fourier transform pairs

$$p(\mathbf{r},\omega) = \frac{1}{\sqrt{2\pi}}\int_{-\infty}^{\infty} P(\mathbf{r},t)e^{i\omega t}\,dt \quad , \tag{1.5}$$

$$P(\mathbf{r},t) = \frac{1}{\sqrt{2\pi}}\int_{-\infty}^{\infty} p(\mathbf{r},\omega)e^{-i\omega t}\,d\omega \quad . \tag{1.6}$$

Solutions $p(\mathbf{r},\omega)$ of the Helmholtz equation, corresponding to time-harmonic solutions of the standard wave equation,

$$P(\mathbf{r},t) = p(\mathbf{r},\omega)e^{-i\omega t} \quad , \tag{1.7}$$

are of considerable interest both because of the importance of narrowband signals and because time-dependent solutions can be synthesized from time-harmonic solutions via the Fourier transform [cf. Eqs. (1.5) and (1.6)]. In this book, we will therefore emphasize the Helmholtz equation and its ramifications.

1.4 The Schrodinger Wave Equation

In nonrelativistic quantum mechanics [Schiff, 1968; Merzbacher, 1970], the motion of a particle of mass m in a force field $\mathbf{F}(\mathbf{r},t)$ characterized by the potential energy $V(\mathbf{r},t)$,

$$\mathbf{F}(\mathbf{r},t) = -\nabla V(\mathbf{r},t) \quad , \tag{1.8}$$

is described by the *time-dependent Schrodinger equation*

$$i\hbar\frac{\partial\Psi(\mathbf{r},t)}{\partial t} = \left[-\frac{\hbar^2}{2m}\nabla^2 + V(\mathbf{r},t)\right]\Psi(\mathbf{r},t) \quad , \tag{1.9}$$

where $\hbar = h/2\pi$ and h is Planck's constant. The magnitude of the wave function squared $|\Psi(\mathbf{r},t)|^2$ is interpreted as the position probability density of finding the particle about the point $\mathbf{r}$ at the time t. If we assume a time-harmonic solution of the form

$$\Psi(\mathbf{r},t) = \psi(\mathbf{r},\omega)e^{-i\omega t} \tag{1.10}$$

and a time-independent potential $V(\mathbf{r})$, then Eq. (1.9) becomes the *time-independent Schrodinger equation*

$$\left\{\nabla^2 + \frac{2m}{\hbar^2}[E - V(\mathbf{r})]\right\}\psi(\mathbf{r},\omega) = 0 \ , \tag{1.11}$$

where $E = \hbar\omega$ is the total energy of the particle. Although the time-dependent Schrodinger equation is only indirectly related to the standard wave equation, the time-independent version is clearly in the form of a Helmholtz equation with

$$k^2(\mathbf{r}) = \frac{2m}{\hbar^2}[E - V(\mathbf{r})] \ . \tag{1.12}$$

This intimate relationship between the Helmholtz and Schrodinger equations is the primary reason that acoustics and quantum mechanics have had such close ties historically.

1.5 The Electromagnetic Wave Equation

With the grand synthesis of classical electromagnetic theory by James Clerk Maxwell in 1865, it appeared that the fields of electromagnetism and optics were on the same course as acoustics as the twentieth century approached. But the nagging, inconsistent assumption of an ether as the medium for the propagation of electromagnetic waves became the *raison d'etre* for the special theory of relativity. As a result, electromagnetism played a far more visible role in the development of modern physics than did acoustics [Jackson, 1975]. In addition, the invention of radio just prior to the turn of the century and the development of radar during World War II led to substantial progress in the understanding of classical electromagnetic wave propagation. In the postwar period, advances such as quantum electrodynamics and the laser ensured the place of electromagnetism in the mainstream of modern physics.

Classical electromagnetic theory is completely described by *Maxwell's equations* [Tyras, 1969; Jackson, 1975] which, for a nonconducting, dielectric medium with no sources or currents, are (in rationalized MKS units)

Gauss' Law:

$$\nabla \bullet \mathbf{D}(\mathbf{r},t) = 0 \ , \tag{1.13}$$

No Magnetic Monopoles:

$$\nabla \bullet \mathbf{B}(\mathbf{r},t) = 0 \ , \tag{1.14}$$

Faraday's Law:

$$\nabla \times \mathbf{E}(\mathbf{r},t) + \frac{\partial \mathbf{B}(\mathbf{r},t)}{\partial t} = 0 \ , \tag{1.15}$$

Ampere's Law:

$$\nabla \times \mathbf{H}(\mathbf{r},t) - \frac{\partial \mathbf{D}(\mathbf{r},t)}{\partial t} = 0 \ . \tag{1.16}$$

In these equations, **E** and **H** are the electric and magnetic fields, respectively, **D** is the dielectric displacement, and **B** is the magnetic induction. These quantities are related through the constitutive relations

$$\mathbf{D}(\mathbf{r},t) = \varepsilon(\mathbf{r})\mathbf{E}(\mathbf{r},t) \ , \ \ \mathbf{H}(\mathbf{r},t) = \mathbf{B}(\mathbf{r},t)/\mu(\mathbf{r}) \ , \tag{1.17}$$

where the dielectric constant $\varepsilon(\mathbf{r})/\varepsilon_0$ and the magnetic permeability $\mu(\mathbf{r})/\mu_0$ are functions of position, and ε_0 and μ_0 are the values in vacuum of ε and μ.

Taking the curl of Eq. (1.15), using the identity

$$\nabla \times \nabla \times \mathbf{E} = \nabla(\nabla \bullet \mathbf{E}) - \nabla^2 \mathbf{E} \ , \tag{1.18}$$

and Eqs. (1.13) and (1.16), we find the result

$$\left[\nabla^2 - \mu(\mathbf{r})\varepsilon(\mathbf{r})\frac{\partial^2}{\partial t^2}\right]\mathbf{E}(\mathbf{r},t) + \nabla\left[\frac{\nabla\varepsilon(\mathbf{r}) \bullet \mathbf{E}(\mathbf{r},t)}{\varepsilon(\mathbf{r})}\right] = \nabla\mu(\mathbf{r}) \times \frac{\partial \mathbf{H}(\mathbf{r},t)}{\partial t}$$

which, using Eq. (1.15), becomes [suppressing the $(\mathbf{r},t)$ dependence in $\mathbf{E}(\mathbf{r},t)$]

$$\left[\nabla^2 - \mu(\mathbf{r})\varepsilon(\mathbf{r})\frac{\partial^2}{\partial t^2}\right]\mathbf{E} + \nabla\left[\frac{\nabla\varepsilon(\mathbf{r}) \bullet \mathbf{E}}{\varepsilon(\mathbf{r})}\right] = -\frac{\nabla\mu(\mathbf{r})}{\mu(\mathbf{r})} \times \nabla \times \mathbf{E} . \tag{1.19}$$

Equation (1.19) reduces to a standard wave equation only when μ and ε are constants

$$\nabla^2\mathbf{E}(\mathbf{r},t) - \frac{1}{c^2}\frac{\partial^2\mathbf{E}(\mathbf{r},t)}{\partial t^2} = 0 \quad , \tag{1.20}$$

where the wave velocity $c = 1/\sqrt{\mu\varepsilon}$. In a similar manner, if we take the curl of Eq. (1.16) and use Eqs. (1.14) and (1.15), we obtain

$$\left[\nabla^2 - \mu(\mathbf{r})\varepsilon(\mathbf{r})\frac{\partial^2}{\partial t^2}\right]\mathbf{H}(\mathbf{r},t) + \nabla\left[\frac{\nabla\mu(\mathbf{r})\bullet\mathbf{H}(\mathbf{r},t)}{\mu(\mathbf{r})}\right] = -\nabla\varepsilon(\mathbf{r})\times\frac{\partial\mathbf{E}(\mathbf{r},t)}{\partial t}$$

which, using Eq. (1.16), becomes

$$\left[\nabla^2 - \mu(\mathbf{r})\varepsilon(\mathbf{r})\frac{\partial^2}{\partial t^2}\right]\mathbf{H} + \nabla\left[\frac{\nabla\mu(\mathbf{r})\bullet\mathbf{H}}{\mu(\mathbf{r})}\right] = -\frac{\nabla\varepsilon(\mathbf{r})}{\varepsilon(\mathbf{r})}\times\nabla\times\mathbf{H}. \tag{1.21}$$

Equation (1.21) also reduces to a standard wave equation only when μ and ε are constants

$$\nabla^2\mathbf{H}(\mathbf{r},t) - \frac{1}{c^2}\frac{\partial^2\mathbf{H}(\mathbf{r},t)}{\partial t^2} = 0 \quad . \tag{1.22}$$

Thus, Eqs. (1.19) and (1.21) are uncoupled, *non-standard wave equations* for the electric and magnetic fields which reduce to standard wave equations only for homogeneous, isotropic media.

It is sometimes convenient to define vector and scalar potentials from which the electromagnetic field quantities can be obtained [Stratton, 1941; Jackson, 1975]. However, the utility of this procedure is again restricted to the case where μ and ε are constants. Equation (1.14) implies that $\mathbf{B}(\mathbf{r},t)$ can be written in terms of a vector potential $\mathbf{A}(\mathbf{r},t)$,

$$\mathbf{B}(\mathbf{r},t) = \nabla\times\mathbf{A}(\mathbf{r},t) \quad , \tag{1.23}$$

while Eq. (1.15) suggests that $\mathbf{E}(\mathbf{r},t)$ can be expressed as

$$\mathbf{E}(\mathbf{r},t) = -\nabla\Phi(\mathbf{r},t) - \frac{\partial\mathbf{A}(\mathbf{r},t)}{\partial t} \quad , \tag{1.24}$$

where $\mathbf{\Phi}(\mathbf{r},t)$ is a scalar potential. Substituting Eq. (1.24) into Eq. (1.13), we obtain

$$\nabla^2\Phi(\mathbf{r},t)+\frac{\partial[\nabla\bullet\mathbf{A}(\mathbf{r},t)]}{\partial t}=0 \ , \tag{1.25}$$

while substituting Eqs. (1.23) and (1.24) into Eq. (1.16), we find that

$$\left[\nabla^2-\mu\varepsilon\frac{\partial^2}{\partial t^2}\right]\mathbf{A}(\mathbf{r},t)-\nabla\left[\nabla\bullet\mathbf{A}(\mathbf{r},t)+\mu\varepsilon\frac{\partial\Phi(\mathbf{r},t)}{\partial t}\right]=0. \tag{1.26}$$

The arbitrariness associated with the choice of potentials enables us to invoke the *Lorentz condition,*

$$\nabla\bullet\mathbf{A}(\mathbf{r},t)+\mu\varepsilon\frac{\partial\Phi(\mathbf{r},t)}{\partial t}=0 \ , \tag{1.27}$$

whereby Eqs. (1.25) and (1.26) become standard wave equations,

$$\nabla^2\Phi(\mathbf{r},t)-\frac{1}{c^2}\frac{\partial^2\Phi(\mathbf{r},t)}{\partial t^2}=0 \ , \tag{1.28}$$

$$\nabla^2\mathbf{A}(\mathbf{r},t)-\frac{1}{c^2}\frac{\partial^2\mathbf{A}(\mathbf{r},t)}{\partial t^2}=0 \ , \tag{1.29}$$

which, when combined with Eq. (1.27), form a set of equations completely equivalent to Maxwell's equations. The Lorentz condition is associated with the class of *gauge transformations*, under which the fields are invariant.

Thus, we see that the electromagnetic fields and potentials satisfy the standard wave equation, in general, only for homogeneous, isotropic media, a situation to be contrasted with the acoustic field, which satisfies the standard wave equation for media with arbitrarily varying sound velocity. However, the importance of the standard wave equation in the electromagnetic case should not be underestimated for several reasons: (a) there are many realistic examples of interest which can be described by homogeneous, isotropic media; (b) there are special cases of horizontally stratified media and specific field polarizations for which the standard wave equation holds for continuously varying dielectric properties (cf. Prob.

1.2); (c) arbitrarily varying media can sometimes be approximated by a judicious choice of a suitable number of homogeneous, isotropic regions.

1.6 The Elastic Wave Equation

The fields of elastic wave propagation and acoustics were built on the same nineteenth century intellectual foundation [Achenbach, 1975]. Both relied heavily on developments in the theory of elasticity and mechanical vibration. At the turn of the century, a distinct thread associated with geophysical applications emerged, and the science of seismology was born [Aki and Richards, 1980]. Here, the list of notable contributors again includes Rayleigh, as well as A. E. H. Love and Horace Lamb. The field progressed rapidly in the twentieth century, motivated by the applications of earthquake prediction and, in the postwar era, the detection of underground nuclear explosions. Seismology is considered to be a cornerstone of geophysics, but like acoustics, falls outside the realm of modern physics.

The particle displacement $\mathbf{u}(\mathbf{r},t)$ for an isotropic, perfectly elastic solid with density $\rho(\mathbf{r})$, Lame parameters $\lambda(\mathbf{r})$ and $\mu(\mathbf{r})$, and no sources is given by [Grant and West, 1965]

$$\begin{aligned}\rho(\mathbf{r})\frac{\partial^2\mathbf{u}(\mathbf{r},t)}{\partial t^2} &= [\lambda(\mathbf{r})+\mu(\mathbf{r})]\nabla\theta(\mathbf{r},t)+\theta(\mathbf{r},t)\nabla\lambda(\mathbf{r})\\ &+\mu(\mathbf{r})\nabla^2\mathbf{u}(\mathbf{r},t)+[\nabla\mu(\mathbf{r})\bullet\nabla]\mathbf{u}(\mathbf{r},t)\\ &+\nabla[\nabla\mu(\mathbf{r})\bullet\mathbf{u}(\mathbf{r},t)] \ , \end{aligned} \tag{1.30}$$

where $\theta(\mathbf{r},t)=\nabla\bullet\mathbf{u}(\mathbf{r},t)$ is the dilatation. If we assume that $\rho(\mathbf{r})$ is constant and take the divergence of Eq. (1.30), we obtain

$$\begin{aligned}\rho\frac{\partial^2\theta(\mathbf{r},t)}{\partial t^2} &= [\lambda(\mathbf{r})+2\mu(\mathbf{r})]\nabla^2\theta(\mathbf{r},t)+2\nabla[\lambda(\mathbf{r})+2\mu(\mathbf{r})]\bullet\nabla\theta(\mathbf{r},t)\\ &-2\nabla\mu(\mathbf{r})\bullet[\nabla\times\xi(\mathbf{r},t)] \ , \end{aligned} \tag{1.31}$$

where $\xi(\mathbf{r},t)=\nabla\times\mathbf{u}(\mathbf{r},t)$ is the rotation. If we take the curl of Eq. (1.30), we obtain

$$\rho \frac{\partial^2 \xi(\mathbf{r},t)}{\partial t^2} = \mu(\mathbf{r})\nabla^2\xi(\mathbf{r},t) + [\nabla\mu(\mathbf{r}) \bullet \nabla]\xi(\mathbf{r},t)$$
$$- \nabla\mu(\mathbf{r}) \times [\nabla \times \xi(\mathbf{r},t)] + 2\nabla\mu(\mathbf{r}) \times \nabla\theta(\mathbf{r},t). \quad (1.32)$$

From Eqs. (1.31) and (1.32), we see that, in general, the dilatational and rotational motions are coupled and become uncoupled only when λ and μ are constants. In that case, we obtain standard wave equations for the compressional wave,

$$\nabla^2\theta(\mathbf{r},t) - \frac{1}{c_p^2}\frac{\partial^2\theta(\mathbf{r},t)}{\partial t^2} = 0 \ , \quad (1.33)$$

with $c_p = \sqrt{(\lambda + 2\mu)/\rho}$, and for the shear wave,

$$\nabla^2\xi(\mathbf{r},t) - \frac{1}{c_s^2}\frac{\partial^2\xi(\mathbf{r},t)}{\partial t^2} = 0 \ , \quad (1.34)$$

with $c_s = \sqrt{\mu/\rho}$.

Finally, we can decompose the displacement into the sum of contributions from a scalar potential $\Phi(\mathbf{r},t)$ and a vector potential $\mathbf{A}(\mathbf{r},t)$ using Helmholtz's theorem [Grant and West, 1965; Arfken, 1985]

$$\mathbf{u}(\mathbf{r},t) = \nabla\Phi(\mathbf{r},t) - \nabla \times \mathbf{A}(\mathbf{r},t) \ . \quad (1.35)$$

The potentials are related to the dilatation and rotation through the equations

$$\theta(\mathbf{r},t) = \nabla^2\Phi(\mathbf{r},t) \ , \quad (1.36)$$

$$\xi(\mathbf{r},t) = \nabla^2\mathbf{A}(\mathbf{r},t) \ , \quad (1.37)$$

and it can be shown that the potentials also satisfy standard wave equations in the case of homogeneous media (cf. Prob. 1.3)

$$\nabla^2\Phi(\mathbf{r},t) - \frac{1}{c_p^2}\frac{\partial^2\Phi(\mathbf{r},t)}{\partial t^2} = 0 \ , \quad (1.38)$$

$$\nabla^2 \mathbf{A}(\mathbf{r},t) - \frac{1}{c_s^2}\frac{\partial^2 \mathbf{A}(\mathbf{r},t)}{\partial t^2} = 0 \ . \tag{1.39}$$

Thus, the elastic case presents us with a situation similar to the one encountered with electromagnetic waves. The dilatational and rotational fields and potentials satisfy the standard wave equation only for homogeneous, isotropic, elastic solids. However, as in the electromagnetic case, propagation through homogeneous media is of considerable importance both in applications to realistic situations and in approximations to propagation through inhomogeneous media. In the latter case, the continuous coupling of the fields is replaced by coupling through the boundary conditions imposed at the discontinuous interfaces separating the homogeneous regions. There is also a special case of a one-dimensional, continuously varying medium for which the dilatational and rotational motions uncouple and give rise to non-standard wave equations (cf. Prob. 1.4).

PROBLEMS

1.1 The basic equations of linear acoustics are obtained by linearizing the following equations of inviscid, compressible fluid mechanics without heat conduction (P' is pressure, ρ' is density, S' is entropy, and $\mathbf{v}'$ is particle velocity):

Euler's Equation (momentum balance):

$$\rho'\left[\frac{\partial \mathbf{v}'}{\partial t} + \mathbf{v}' \bullet \nabla \mathbf{v}'\right] = -\nabla P' \ , \tag{1.40}$$

Equation of Continuity (mass conservation):

$$\frac{\partial \rho'}{\partial t} + \nabla \bullet \rho' \mathbf{v}' = 0 \ , \tag{1.41}$$

Adiabatic Condition (no heat transfer):

$$\frac{\partial S'}{\partial t} + \mathbf{v}' \bullet \nabla S' = 0 \ , \tag{1.42}$$

Equation of State:

$$P' = P'(\rho', S') \ . \tag{1.43}$$

The linearization is implemented by writing each physical quantity as the sum of a steady-state, time-independent value and a small fluctuating value:

$$
\begin{aligned}
P'(\mathbf{r},t) &= P_0(\mathbf{r}) + P(\mathbf{r},t) \ , \\
\rho'(\mathbf{r},t) &= \rho_0(\mathbf{r}) + \rho(\mathbf{r},t) \ , \\
S'(\mathbf{r},t) &= S_0(\mathbf{r}) + S(\mathbf{r},t) \ , \\
\mathbf{v}'(\mathbf{r},t) &= \qquad\qquad \mathbf{v}(\mathbf{r},t) \ . \qquad \text{(no mean flow)}
\end{aligned}
$$

By using this procedure and keeping only terms to first order in the fluctuating quantities, find the equations corresponding to Eqs. (1.40)-(1.42) [let us call them Eqs. (1.40a)-(1.42a)] for the acoustic part of the field. We expand Eq. (1.43) about the equilibrium values

$$
P' = P_0 + \left(\frac{\partial P_0}{\partial \rho_0}\right)_{S'} (\rho' - \rho_0) + \left(\frac{\partial P_0}{\partial S_0}\right)_{\rho'} (S' - S_0) + \ldots \ ,
$$

or

$$
P = \left(\frac{\partial P_0}{\partial \rho_0}\right)_{S'} \rho + \left(\frac{\partial P_0}{\partial S_0}\right)_{\rho'} S \ . \tag{1.43a}
$$

Eliminate the entropy S between Eqs. (1.42a) and (1.43a) by differentiating Eq. (1.43a) with respect to t and using the fact that for a fluid in hydrostatic equilibrium with no external forces

$$
\nabla P_0 = \left(\frac{\partial P_0}{\partial \rho_0}\right)_{S'} \nabla \rho_0 + \left(\frac{\partial P_0}{\partial S_0}\right)_{\rho'} \nabla S_0 = 0 \ .
$$

Combine the resulting equation with Eq. (1.41a) to obtain

$$
\frac{\partial P}{\partial t} + \rho_0 \left(\frac{\partial P_0}{\partial \rho_0}\right)_{S'} \nabla \bullet \mathbf{v} = 0 \ . \tag{1.44}
$$

Equations (1.40a) and (1.44) are the two basic field equations of acoustics, analogous to Maxwell's equations in electromagnetic theory. Show that P satisfies the time-dependent wave equation for density and sound velocity stratification [cf. Eq. (1.1)]

$$\rho_0 \nabla \bullet \left[\frac{1}{\rho_0} \nabla P \right] - \frac{1}{c^2} \frac{\partial^2 P}{\partial t^2} = 0 \ ,$$

where the sound velocity c is given by

$$c^2 = \left(\frac{\partial P_0}{\partial \rho_0} \right)_{S'} \ .$$

1.2 The *horizontally stratified case*, where the medium properties are a function of only one spatial coordinate, is extremely important both in theory and in practical applications.

(a) Consider a horizontally stratified dielectric medium where $\varepsilon = \varepsilon(z)$ and μ is constant. Assume that the electric field is horizontally polarized

$$\mathbf{E} = \mathbf{j} E_y(x, z) \ ,$$

where the x-z plane is in the plane of the paper, and $\mathbf{j}$ is a unit vector in the y direction. Show that E_y satisfies a standard wave equation.

(b) Consider a horizontally stratified dielectric medium where $\mu = \mu(z)$ and ε is constant. Assume that the electric field is vertically polarized, so that the magnetic field is given by

$$\mathbf{H} = \mathbf{j} H_y(x, z) \ .$$

Show that H_y satisfies a standard wave equation.

1.3 Show that the elastic scalar and vector potentials satisfy standard wave equations for homogeneous media.

1.4 Consider the elastic equation of motion for a one-dimensional, continuously varying medium where

$$\rho = \rho(z), \ \ \lambda = \lambda(z), \ \ \mu = \mu(z) \ .$$

The total displacement $\mathbf{u}(z,t)$ can be decomposed into a shear component $\mathbf{u}_s(z,t)$ and a compressional component $\mathbf{u}_p(z,t)$, where

$$\mathbf{u}_s(z,t) = u_x(z,t)\mathbf{i} + u_y(z,t)\mathbf{j} \ ,$$
$$\mathbf{u}_p(z,t) = u_z(z,t)\mathbf{k} \ ,$$

and **i**, **j**, and **k** are unit vectors in the x, y, and z directions, respectively. Show that the shear and compressional components satisfy *uncoupled*, non-standard wave equations

$$\rho(z)\frac{\partial^2 \mathbf{u}_s(z,t)}{\partial t^2} = \frac{\partial}{\partial z}\left[\mu(z)\frac{\partial \mathbf{u}_s(z,t)}{\partial z}\right] \ ,$$

$$\rho(z)\frac{\partial^2 \mathbf{u}_p(z,t)}{\partial t^2} = \frac{\partial}{\partial z}\left\{[\lambda(z)+2\mu(z)]\frac{\partial \mathbf{u}_p(z,t)}{\partial z}\right\} \ ,$$

and that these reduce to standard wave equations when λ and μ are constants. Note that in the derivation of Eq. (1.30), second and higher order derivatives of the Lame parameters are neglected.

1.5 For harmonic time dependence, the time-dependent acoustic wave equation with density and sound velocity stratification becomes

$$\rho_0(\mathbf{r})\nabla \bullet \left[\frac{1}{\rho_0(\mathbf{r})}\nabla p(\mathbf{r},\omega)\right] + k^2(\mathbf{r})p(\mathbf{r},\omega) = 0 \ , \qquad (1.45)$$

where $k(\mathbf{r}) = \omega/c(\mathbf{r})$. Show that, by making the transformation

$$\mathrm{P}(\mathbf{r},\omega) = \frac{p(\mathbf{r},\omega)}{\sqrt{\rho_0(\mathbf{r})}} \ ,$$

Eq. (1.45) can be put into the form of a Helmholtz equation

$$\left[\nabla^2 + K^2(\mathbf{r})\right]\mathrm{P}(\mathbf{r},\omega) = 0 \ ,$$

where

$$K^2(\mathbf{r}) = k^2(\mathbf{r}) + \frac{1}{2\rho_0(\mathbf{r})}\nabla^2\rho_0(\mathbf{r}) - \frac{3}{4}\left[\frac{1}{\rho_0(\mathbf{r})}\nabla\rho_0(\mathbf{r})\right]^2 \ .$$

Chapter 2. Elementary Solutions and Basic Acoustic Quantities

Take care of the sense and the sounds will take care of themselves.
Lewis Carroll, 1865

2.1 Introduction

A number of the fundamental concepts associated with wave propagation emerge when we examine elementary solutions to the acoustic wave equation in homogeneous media with no boundaries or sources. One-dimensional solutions are of particular interest because of their (sometimes deceptive) simplicity. Two- and three-dimensional solutions suggest the notions of cylindrical and spherical waves. In all three cases, the importance of plane-wave, Fourier decompositions of the wave fields is clearly evident. Elementary solutions also provide illustrative examples of basic acoustic quantities such as impedance and intensity.

2.2 The One-Dimensional Wave Equation

In a one-dimensional, homogeneous medium with sound velocity c, Eq. (1.2) becomes

$$\frac{\partial^2 P(x,t)}{\partial x^2} - \frac{1}{c^2}\frac{\partial^2 P(x,t)}{\partial t^2} = 0 \ . \tag{2.1}$$

Equation (2.1) has the general solution

$$P(x,t) = f(x-ct) + g(x+ct) \ , \tag{2.2}$$

where f and g are arbitrary functions corresponding to disturbances which travel, without changing shape, in the directions of increasing and decreasing x, respectively. These disturbances are defined as *waves* traveling with *phase velocity* c. Although waves in inhomogeneous, lossy, or dispersive media may change shape as they propagate, the notion of a disturbance carrying energy through a medium without permanently deforming it remains a fundamental concept in wave physics.

The functions f and g can be constructed by using Fourier decomposition. The solution $P(x,t)$ is expressed as [cf. Eq. (1.6)]

$$P(x,t)=\frac{c}{\sqrt{2\pi}}\int_{-\infty}^{\infty}p(x,k)e^{-ikct}\,dk \quad , \tag{2.3}$$

where we have used the relation $\omega = ck$ and changed the integration variable to k for convenience in the subsequent development. The function $p(x,k)$ satisfies the one-dimensional Helmholtz equation

$$\frac{d^2p(x,k)}{dx^2}+k^2p(x,k)=0 \quad . \tag{2.4}$$

Equation (2.4) is a linear, second-order, ordinary differential equation, and therefore has a solution which is a superposition of two linearly independent solutions

$$p(x,k)=A(k)e^{ikx}+B(k)e^{-ikx} \quad , \tag{2.5}$$

where $A(k)$ and $B(k)$ are constants determined by the boundary and initial conditions. The solution $p(x,k)$ can also be constructed from the trigonometric functions $\cos kx$ and $\sin kx$ or some combination of exponential and trigonometric functions, as long as they are linearly independent. Normally the form of the solution is chosen for convenience in dealing with the problem at hand. When the exponential solutions are combined with the harmonic time dependence, the results

$$e^{ik(x-ct)},\ e^{-ik(x+ct)} \quad ,$$

are interpreted as *propagating plane waves* because the surfaces of constant phase, moving with velocity c, are planes (actually points in one dimension) separated by the wavelength $\lambda = 2\pi/k$. The trigonometric functions yield solutions

$$\cos kx\, e^{-ikct},\ \sin kx\, e^{-ikct} \quad ,$$

which are interpreted as *standing waves* because there are points in the wave field called *nodes* which vanish for all times. These interpretations are based on the convention that measurable and physically meaningful quantities are obtained by taking the real part of quantities expressed by complex representations. Finally, combining Eqs. (2.3) and (2.5), we obtain the complete one-dimensional solution

$$P(x,t) = \frac{c}{\sqrt{2\pi}} \int_{-\infty}^{\infty} \left[A(k)e^{ikx} + B(k)e^{-ikx} \right] e^{-ikct}\, dk \ . \tag{2.6}$$

2.3 The Three-Dimensional Wave Equation

We consider the three-dimensional wave equation next because of its mathematical similarity to the one-dimensional case. In a three-dimensional medium with sound velocity c, Eq. (1.2) becomes (in Cartesian coordinates)

$$\left(\frac{\partial^2}{\partial x^2} + \frac{\partial^2}{\partial y^2} + \frac{\partial^2}{\partial z^2} \right) P(\mathbf{r},t) - \frac{1}{c^2} \frac{\partial^2 P(\mathbf{r},t)}{\partial t^2} = 0 \ , \tag{2.7}$$

where $\mathbf{r} = (x, y, z)$, and the Helmholtz equation is

$$\left(\frac{\partial^2}{\partial x^2} + \frac{\partial^2}{\partial y^2} + \frac{\partial^2}{\partial z^2} + k^2 \right) p(\mathbf{r},k) = 0 \ , \tag{2.8}$$

with

$$P(\mathbf{r},t) = \frac{c}{\sqrt{2\pi}} \int_{-\infty}^{\infty} p(\mathbf{r},k) e^{-ikct}\, dk \ . \tag{2.9}$$

Equation (2.8) has the solution [$A(k)$ and $B(k)$ are arbitrary constants]

$$p(\mathbf{r},k) = A(k)e^{i\mathbf{k}\cdot\mathbf{r}} + B(k)e^{-i\mathbf{k}\cdot\mathbf{r}} \ , \tag{2.10}$$

where the solutions

$$e^{i(\mathbf{k}\cdot\mathbf{r} - \omega t)}, \ e^{-i(\mathbf{k}\cdot\mathbf{r} + \omega t)} \ ,$$

are interpreted as *three-dimensional propagating plane waves* with *wave vector* $\mathbf{k} = (k_x, k_y, k_z)$ perpendicular to the *wavefronts*, or planes of constant phase, and satisfying the relation

$$|\mathbf{k}| = k = \sqrt{k_x^2 + k_y^2 + k_z^2} \quad . \tag{2.11}$$

Analogous to the one-dimensional case, the solutions

$$\cos \mathbf{k} \bullet \mathbf{r}\, e^{-ikct}, \ \sin \mathbf{k} \bullet \mathbf{r}\, e^{-ikct} \quad ,$$

correspond to *three-dimensional standing waves.* The full three-dimensional solution is then given by

$$P(\mathbf{r},t) = \frac{c}{\sqrt{2\pi}} \int_{-\infty}^{\infty} \left[A(k) e^{i\mathbf{k} \bullet \mathbf{r}} + B(k) e^{-i\mathbf{k} \bullet \mathbf{r}} \right] e^{-ikct}\, dk \quad . \tag{2.12}$$

It is illuminating to solve Eq. (1.2) in spherical coordinates, assuming spherical symmetry, so that we obtain

$$\frac{1}{r^2} \frac{\partial}{\partial r} \left[r^2 \frac{\partial P(r,t)}{\partial r} \right] - \frac{1}{c^2} \frac{\partial^2 P(r,t)}{\partial t^2} = 0 \quad , \tag{2.13}$$

where $r = \sqrt{x^2 + y^2 + z^2}$ is the radial spherical coordinate. Equation (2.13) can be rewritten as

$$\frac{\partial^2 [rP(r,t)]}{\partial r^2} - \frac{1}{c^2} \frac{\partial^2 [rP(r,t)]}{\partial t^2} = 0 \quad , \tag{2.14}$$

which admits solutions of the form

$$P(r,t) = \frac{f(r - ct)}{r} + \frac{g(r + ct)}{r} \quad , \tag{2.15}$$

where f and g are arbitrary functions. The first and second terms in Eq. (2.15) correspond to *spherical waves* diverging from and converging toward the origin, respectively. Of particular interest is the first term in the harmonic solution

$$P(r,t) = A(k) \frac{e^{ik(r-ct)}}{r} + B(k) \frac{e^{-ik(r+ct)}}{r} \quad , \tag{2.16}$$

because it corresponds to a spherical wave emanating from a *point source* at the origin.

2.4 The Two-Dimensional Wave Equation

In two dimensions, Eqs. (2.7) and (2.8) become

$$\left(\frac{\partial^2}{\partial x^2}+\frac{\partial^2}{\partial y^2}\right)P(\mathbf{r},t)-\frac{1}{c^2}\frac{\partial^2 P(\mathbf{r},t)}{\partial t^2}=0 \ , \tag{2.17}$$

$$\left(\frac{\partial^2}{\partial x^2}+\frac{\partial^2}{\partial y^2}+k^2\right)p(\mathbf{r},k)=0 \ , \tag{2.18}$$

where $\mathbf{r}=(x,y)$, and $P(\mathbf{r},t)$ and $p(\mathbf{r},k)$ satisfy Eq. (2.9). Then the full two-dimensional solution in Cartesian coordinates is given by Eq. (2.12) with $\mathbf{k}=(k_x,k_y)$ and

$$|\mathbf{k}|=k=\sqrt{k_x^2+k_y^2} \ . \tag{2.19}$$

In cylindrical coordinates, assuming cylindrical symmetry, we find that Eq. (1.2) becomes

$$\frac{1}{r}\frac{\partial}{\partial r}\left[r\frac{\partial P(r,t)}{\partial r}\right]-\frac{1}{c^2}\frac{\partial^2 P(r,t)}{\partial t^2}=0 \ , \tag{2.20}$$

where $r=\sqrt{x^2+y^2}$ is the radial cylindrical coordinate. Interestingly, in general, Eq. (2.20) does *not* permit solutions containing the functional forms $f(r-ct)$ and $g(r+ct)$. Instead, when we examine the corresponding Helmholtz equation

$$\frac{1}{r}\frac{d}{dr}\left[r\frac{dp(r,k)}{dr}\right]+k^2p(r,k)=0 \ , \tag{2.21}$$

we see that it can be rewritten as

$$\frac{d^2p(r,k)}{dr^2}+\frac{1}{r}\frac{dp(r,k)}{dr}+k^2p(r,k)=0 \ , \tag{2.22}$$

which has the form of *Bessel's equation* [Abramowitz and Stegun, 1964],

$$\frac{d^2R(x)}{dx^2}+\frac{1}{x}\frac{dR(x)}{dx}+\left(1-\frac{\nu^2}{x^2}\right)R(x)=0 \ , \tag{2.23}$$

for $\nu=0$. In general, both x and ν may be complex variables, although in this discussion we shall assume they are real. The linearly independent solutions $R(x)$ of Eq. (2.23), analogous to the trigonometric solutions of Eq. (2.4), are the *Bessel function* $J_\nu(x)$ and *Neumann function* $N_\nu(x)$ of *order* ν. Another pair of linearly independent solutions, analogous to the exponential functions, are the *Hankel functions of the first and second kind*, $H_\nu^{(1)}(x)$ and $H_\nu^{(2)}(x)$, respectively. These four solutions are related to one another by

$$H_\nu^{(1)}(x)=J_\nu(x)+iN_\nu(x) \ , \tag{2.24}$$

$$H_\nu^{(2)}(x)=J_\nu(x)-iN_\nu(x) \ , \tag{2.25}$$

where, as in the case of Eq. (2.4), we select the pair of linearly independent solutions which are most convenient for solving the problem under consideration. Their physical interpretation is clarified when we examine their *asymptotic forms* for fixed ν and $x>>1$,

$$J_\nu(x)\sim\sqrt{\frac{2}{\pi x}}\cos\left(x-\frac{\nu\pi}{2}-\frac{\pi}{4}\right) \ , \tag{2.26}$$

$$N_\nu(x)\sim\sqrt{\frac{2}{\pi x}}\sin\left(x-\frac{\nu\pi}{2}-\frac{\pi}{4}\right) \ , \tag{2.27}$$

$$H_\nu^{(1)}(x)\sim\sqrt{\frac{2}{\pi x}}e^{i\left(x-\frac{\nu\pi}{2}-\frac{\pi}{4}\right)} \ , \tag{2.28}$$

$$H_\nu^{(2)}(x)\sim\sqrt{\frac{2}{\pi x}}e^{-i\left(x-\frac{\nu\pi}{2}-\frac{\pi}{4}\right)} \ . \tag{2.29}$$

Thus, we see that, when combined with the harmonic time dependence, Eqs. (2.26) and (2.27) correspond to standing wave solutions, while Eqs. (2.28) and (2.29) correspond to propagating wave solutions. The solution of Eq. (2.20) can therefore be written as

$$P(r,t)=\left[A(k)H_0^{(1)}(kr)+B(k)H_0^{(2)}(kr)\right]e^{-ikct} \quad , \tag{2.30}$$

which in the *far field* ($kr>>1$), using Eqs. (2.28) and (2.29), becomes

$$P(r,t)\sim\sqrt{\frac{2}{\pi k}}\left[A(k)\frac{e^{ik(r-ct)}}{\sqrt{r}}+B(k)\frac{e^{-ik(r+ct)}}{\sqrt{r}}\right] \quad , \tag{2.31}$$

where the first and second terms correspond to outwardly and inwardly propagating *cylindrical waves*, respectively. The first term in Eq. (2.30) is associated with the radiation due to a *line source* at the origin.

Finally, for completeness, we examine the behavior of the solutions to Bessel's equation for fixed ν and small arguments $x<<1$,

$$J_\nu(x)\approx\frac{1}{\Gamma(\nu+1)}\left(\frac{x}{2}\right)^\nu, \quad \nu\neq-1,-2,-3,\ldots \quad , \tag{2.32}$$

$$N_0(x)\approx\frac{2}{\pi}\ln x \quad , \tag{2.33}$$

$$N_\nu(x)\approx-\frac{\Gamma(\nu)}{\pi}\left(\frac{2}{x}\right)^\nu, \quad \nu>0 \quad , \tag{2.34}$$

$$-iH_\nu^{(1)}(x)\approx iH_\nu^{(2)}(x)\approx N_\nu(x) \quad . \tag{2.35}$$

In these equations, $\Gamma(\nu)$ is the *Gamma function*, which for integer n is equal to the *factorial function*

$$\Gamma(n+1)=n!=n(n-1)(n-2)\cdots 1 \quad . \tag{2.36}$$

2.5 Velocity Potential

In acoustics, we typically assume that the fluid flow is *irrotational*, i.e.,

$$\nabla \times \mathbf{v}(\mathbf{r},t) = 0 \ , \tag{2.37}$$

so that the particle velocity $\mathbf{v}(\mathbf{r},t)$ can be expressed in terms of a *velocity potential* $\Phi(\mathbf{r},t)$

$$\mathbf{v}(\mathbf{r},t) = -\nabla\Phi(\mathbf{r},t) \ . \tag{2.38}$$

Substituting Eq. (2.38) into the momentum balance Eq. (1.40a), we obtain

$$\rho_0(\mathbf{r})\nabla\left[\frac{\partial\Phi(\mathbf{r},t)}{\partial t}\right] = \nabla P(\mathbf{r},t) \ , \tag{2.39}$$

which, for constant density, implies that

$$P(\mathbf{r},t) = \rho_0 \frac{\partial\Phi(\mathbf{r},t)}{\partial t} \ . \tag{2.40}$$

By substituting Eqs. (2.38) and (2.40) into Eq. (1.44), we can show that the velocity potential satisfies the standard wave equation

$$\nabla^2\Phi(\mathbf{r},t) - \frac{1}{c^2(\mathbf{r})}\frac{\partial^2\Phi(\mathbf{r},t)}{\partial t^2} = 0 \ . \tag{2.41}$$

For harmonic time dependence

$$\Phi(\mathbf{r},t) = \phi(\mathbf{r},\omega)e^{-i\omega t} \ , \tag{2.42}$$

Eq. (2.40) becomes

$$P(\mathbf{r},t) = -i\omega\rho_0\Phi(\mathbf{r},t) \ . \tag{2.43}$$

Furthermore, it is clear that $\phi(\mathbf{r},\omega)$ satisfies the Helmholtz equation

$$\left[\nabla^2 + k^2(\mathbf{r})\right]\phi(\mathbf{r},\omega) = 0 \ . \tag{2.44}$$

2.6 Specific Acoustic Impedance

The acoustic pressure is analogous to voltage in electric circuit theory, while the particle speed V, defined as the vector magnitude of the particle velocity

$$V(\mathbf{r},t) = |\mathbf{v}(\mathbf{r},t)| = \sqrt{\mathbf{v}(\mathbf{r},t) \bullet \mathbf{v}(\mathbf{r},t)} \quad , \tag{2.45}$$

is analogous to electric current. Pursuing this analogy, we find it useful to define the *specific acoustic impedance* [Kinsler et al., 1982; Burdic, 1984]

$$Z = \frac{P(\mathbf{r},t)}{V(\mathbf{r},t)} \quad . \tag{2.46}$$

2.6.1 Plane Wave Impedance

For a three-dimensional plane wave, we have

$$P(\mathbf{r},t) = A(k)e^{i(\mathbf{k}\bullet\mathbf{r}-\omega t)} \quad , \tag{2.47}$$

$$\mathbf{v}(\mathbf{r},t) = \frac{\mathbf{k}}{\omega\rho_0}P(\mathbf{r},t) \quad , \tag{2.48}$$

and therefore $Z = \rho_0 c$. The quantity $\rho_0 c$ is also called the *characteristic impedance*, which is an acoustic property of the medium, independent of the particular type of wave propagating through it.

2.6.2 Spherical Wave Impedance

For a spherical wave, we have

$$P(r,t) = A(k)\frac{e^{i(kr-\omega t)}}{r} \quad , \tag{2.50}$$

$$\mathbf{v}(r,t) = \frac{\hat{\mathbf{r}}}{i\omega\rho_0}\frac{\partial P(r,t)}{\partial r} = \frac{i\hat{\mathbf{r}}}{\omega\rho_0}\left(\frac{1}{r} - ik\right)P(r,t) \quad , \tag{2.51}$$

where $\hat{\mathbf{r}}$ is the unit vector in the radial direction. In the *near field* ($kr << 1$), the particle speed is given by

$$V(r,t) \approx \frac{iP(r,t)}{\omega \rho_0 r} \ , \tag{2.52}$$

while in the far field ($kr >> 1$), its behavior is

$$V(r,t) \sim \frac{kP(r,t)}{\omega \rho_0} \ . \tag{2.53}$$

The specific acoustic impedance for the spherical wave is therefore

$$Z = \rho_0 c \left[\frac{1 - i/(kr)}{1 + 1/(kr)^2} \right] \ , \tag{2.54}$$

which can be rewritten in terms of its real and imaginary parts as

$$Z = \rho_0 c \frac{(kr)^2}{(kr)^2 + 1} - i \rho_0 c \frac{kr}{(kr)^2 + 1} \ . \tag{2.55}$$

The first and second terms in Eq. (2.55) are called the *resistive* and *reactive* components of the impedance, respectively. For $kr << 1$, the reactive term dominates, while for $kr >> 1$, the resistive term dominates with

$$\lim_{kr \to \infty} Z = \rho_0 c \ . \tag{2.56}$$

The fact that the spherical wave impedance approaches the plane wave impedance in Eq. (2.56) is not surprising because a spherical wavefront has the appearance of a planar wavefront for large kr.

2.7 Energy Flux Density and Intensity

The *energy flux density* $\mathbf{S}$ of the acoustic wave field is defined as

$$\mathbf{S}(\mathbf{r},t) = [\mathrm{Re}\, P(\mathbf{r},t)] \cdot [\mathrm{Re}\, \mathbf{v}(\mathbf{r},t)] \ , \tag{2.57}$$

and is a measure of the energy transported by the field. It is analogous to the *Poynting vector* in electromagnetic theory. For harmonic time dependence [cf. Eq. (1.7)], it can be shown that the *time-averaged energy flux density* **I** is given by (cf. Prob. 2.7)

$$\mathbf{I} = \langle \mathbf{S}(\mathbf{r},t) \rangle = \frac{i}{4\omega\rho_0}\left[p(\mathbf{r},\omega)\nabla p^*(\mathbf{r},\omega) - p^*(\mathbf{r},\omega)\nabla p(\mathbf{r},\omega)\right]. \quad (2.58)$$

The intensity I is the vector magnitude of **I**.

2.7.1 Plane Wave Intensity

For a plane wave of the form in Eq. (2.47),

$$\begin{aligned} p(\mathbf{r},k) &= A(k)e^{i\mathbf{k}\cdot\mathbf{r}}, & \nabla p(\mathbf{r},k) &= i\mathbf{k}p(\mathbf{r},k), \\ p^*(\mathbf{r},k) &= A^*(k)e^{-i\mathbf{k}\cdot\mathbf{r}}, & \nabla p^*(\mathbf{r},k) &= -i\mathbf{k}p^*(\mathbf{r},k) \ , \end{aligned} \quad (2.59)$$

so that

$$\mathbf{I} = \frac{\mathbf{k}}{2\omega\rho_0}|p(\mathbf{r},k)|^2 = \frac{\mathbf{k}}{2\omega\rho_0}|A(k)|^2 \ . \quad (2.60)$$

As expected, the acoustic energy flows in the direction of wave propagation **k**. It is straightforward to show that the plane wave intensity is given by [cf. Prob. (2.8)]

$$I = \frac{|p(\mathbf{r},k)|^2}{2\rho_0 c} = \frac{|A(k)|^2}{2\rho_0 c} = \frac{\langle P^2(\mathbf{r},t)\rangle}{\rho_0 c} \ , \quad (2.61)$$

where $\langle P^2(\mathbf{r},t)\rangle$ is the *mean-square pressure*.

2.7.2 Spherical Wave Intensity

For a spherical wave of the form in Eq. (2.50),

$$p(r,k) = A(k)\frac{e^{ikr}}{r}, \qquad \nabla p(r,k) = -\frac{p(r,k)}{r}(1-ikr)\hat{\mathbf{r}},$$

$$p^*(r,k) = A^*(k)\frac{e^{-ikr}}{r}, \quad \nabla p^*(r,k) = -\frac{p^*(r,k)}{r}(1+ikr)\hat{\mathbf{r}}, \quad (2.62)$$

so that

$$\mathbf{I} = \frac{\hat{\mathbf{r}}k}{2\omega\rho_0}|p(\mathbf{r},k)|^2 = \frac{\hat{\mathbf{r}}k}{2\omega\rho_0}\frac{|A(k)|^2}{r^2} , \quad (2.63)$$

and

$$I = \frac{|p(\mathbf{r},k)|^2}{2\rho_0 c} = \frac{1}{2\rho_0 c}\frac{|A(k)|^2}{r^2} = \frac{\langle P^2(\mathbf{r},t)\rangle}{\rho_0 c} . \quad (2.64)$$

Thus, we see that due to the $1/r$ *spherical spreading* factor in the spherical wave field, the intensity falls off as $1/r^2$.

PROBLEMS

2.1 Show by direct substitution that the solution in Eq. (2.2) satisfies the one-dimensional wave equation [cf. Eq. (2.1)] for a homogeneous medium.

2.2 Prove that a solution of the form $\Psi(x,t) = f(x-ct)$, where $c = E/(\hbar k)$ and $E = \hbar^2 k^2/(2m) + V$, satisfies the one-dimensional, time-dependent Schrodinger equation [cf. Eq. (1.9)] for a constant potential V.

Hint: Use the Fourier representation for $\Psi(x,t)$.

2.3 Suppose that at $t = 0$, the one-dimensional wave field in a homogeneous medium has the form

$$f(x) = e^{-\alpha x^2} , \quad (2.65)$$

where α is a constant and we assume $g = 0$ in Eq. (2.2). Determine the wavenumber content $A(k)$ of this wave function and verify that at time t it has the form

$$f(x - ct) = e^{-\alpha(x-ct)^2} \quad . \tag{2.66}$$

Hint:
$$\int_{-\infty}^{\infty} e^{-\alpha x^2}\, dx = \sqrt{\frac{\pi}{\alpha}} \quad . \tag{2.67}$$

2.4 Consider the following standing-wave solution to the wave equation in spherical coordinates in a homogeneous medium:

$$P(r,t) = A\frac{\sin kr}{r} e^{-i\omega t} \quad . \tag{2.68}$$

A is a constant, complex amplitude.

(a) Calculate the acoustic particle speed and its near- and far-field behavior.

(b) Calculate the specific acoustic impedance and its near- and far-field behavior. [In parts (a) and (b), keep the first two terms in the small argument expansions for the trigonometric functions.]

(c) Calculate the intensity of this wave and provide a physical interpretation of the result.

2.5 Consider the following solution to the wave equation in spherical coordinates in a homogeneous medium:

$$P(r,t) = \frac{1}{r}\left[e^{i(kr - \pi/4)} + \cos kr\right] e^{-i\omega t} \quad . \tag{2.69}$$

(a) Calculate the acoustic particle speed and its near- and far-field behavior to leading order.

(b) Calculate the far-field behavior of the specific acoustic impedance to leading order.

(c) Calculate the intensity of this wave.

2.6 Consider the following solution to the wave equation in cylindrical coordinates in a homogeneous medium:

$$P(r,t) = AH_0^{(1)}(kr)e^{-i\omega t} \quad . \tag{2.70}$$

A is a constant, complex amplitude.

(a) Calculate the acoustic particle speed and its near- and far-field behavior.

(b) Calculate the specific acoustic impedance and its near- and far-field behavior.

(c) Calculate the intensity and its near- and far-field behavior and provide physical interpretations of the results.

2.7 Show that, for harmonic time dependence, the time-averaged energy flux density is given by Eq. (2.58).

2.8 Show that for time-harmonic wave fields,

$$\langle P^2(\mathbf{r},t)\rangle = \frac{|p(\mathbf{r},\omega)|^2}{2} \quad . \tag{2.70}$$

Note that when we take a product of quantities which have complex representations, we must take the real parts of these quantities *before* we form the product.

Chapter 3. Plane Wave Reflection from Planar Boundaries

There are two ways of spreading light: to be
The candle or the mirror that reflects it.
Edith Wharton, c. 1900

3.1 Introduction

Plane waves of infinite extent are figments of the theoretical physicist's imagination. Realistic sources produce wavefronts with finite curvature, and homogeneous half-spaces of the type we will discuss in this chapter do not exist in nature. Yet there are no concepts more fundamental in wave physics than those of plane wave propagation and interaction with boundaries. In addition to contributing directly to our understanding of wave phenomena, the study of plane wave propagation and reflection provides us with a means by which we can analyze more complicated problems. For example, as we shall see in subsequent chapters, the behavior of wave fields generated by real sources (e.g., spherical waves) can be synthesized from the plane wave interaction. In this chapter, we will concentrate on the problem of plane wave reflection from planar boundaries, in which case the analysis is simplified because the geometries of the incident wave field and the reflecting boundary are the same. However, even then, the development is nontrivial and leads to ideas which we will use repeatedly later. First, we discuss the notion of boundary conditions, which provides a mathematical way in which the effects of boundaries can be incorporated into a problem. Then we consider reflection from horizontally stratified media and introduce the ideas of reflection coefficient and specular reflection. The bulk of the chapter is devoted to examining the reflection from a homogeneous fluid half-space. This problem naturally leads to the concepts of Snell's law, total transmission, total internal reflection, interface waves, and inhomogeneous waves. We conclude with the derivation of the reflection coefficient for a homogeneous fluid layer overlying an arbitrary horizontally stratified medium.

3.2 Boundary Conditions

In order to obtain unique, physically meaningful solutions to the time-dependent acoustic wave equation, we must, in general, impose constraints on the temporal and spatial behavior of the wave field. The temporal restrictions are called *initial conditions*, which specify the nature of the field for specific times at all points in space (cf. Prob. 2.3). For the narrowband, time-harmonic fields typically of interest to us, we essentially impose the condition that they have existed and will continue to exist for all time. Although this assumption may appear to be unrealistic, in fact, the notion of harmonic time dependence contributes enormously to our theoretical and practical understanding of wave phenomena. Initial conditions associated with broadband pulse propagation will therefore play a minor role in our development. On the other hand, *boundary conditions*, which specify the nature of the field at specific spatial locations for all times, will significantly influence our discussion.

3.2.1 Soft Boundary

For a *soft*, or *pressure-release*, boundary S, the pressure vanishes at all positions on the boundary:

$$P(\mathbf{r},t) = 0 \quad \text{on } S. \tag{3.1}$$

Equation (3.1) is also known as a *Dirichlet* boundary condition, and for harmonic time dependence and constant density (cf. Sec. 2.5) is equivalent to the condition of vanishing velocity potential:

$$\Phi(\mathbf{r},t) = 0 \quad \text{on } S. \tag{3.2}$$

3.2.2 Hard Boundary

For a *hard* boundary S, the derivative of the pressure normal to the boundary vanishes at all positions on S:

$$\frac{\partial P(\mathbf{r},t)}{\partial n} = \nabla P(\mathbf{r},t) \bullet \hat{\mathbf{n}} = 0 \quad \text{on } S, \tag{3.3}$$

where $\hat{\mathbf{n}}$ is the unit normal to the boundary. Equation (3.3) is also known as a *Neumann* boundary condition, and is related to the *rigid* boundary condition, which states that the normal component of the particle velocity V_n vanishes on S:

$$V_n(\mathbf{r},t) = \mathbf{v}(\mathbf{r},t) \bullet \hat{\mathbf{n}} = -\nabla\Phi(\mathbf{r},t) \bullet \hat{\mathbf{n}} = -\frac{\partial\Phi(\mathbf{r},t)}{\partial n} = 0 \quad \text{on } S. \quad (3.4)$$

For harmonic time dependence and constant density, the hard and rigid boundary conditions are equivalent to one another.

3.2.3 Impedance Boundary

The *impedance, mixed,* or *Cauchy* boundary condition specifies the ratio of pressure and normal particle velocity to be a constant ξ independent of position and time:

$$\frac{P(\mathbf{r},t)}{V_n(\mathbf{r},t)} = \xi \quad \text{on } S. \quad (3.5)$$

It is clear that the soft and rigid boundary conditions are limiting cases of the impedance condition for ξ approaching 0 and ∞, respectively. For harmonic time dependence and constant density, Eq. (3.5) becomes

$$\frac{P(\mathbf{r},t)}{V_n(\mathbf{r},t)} = \frac{i\omega\rho_0 P(\mathbf{r},t)}{\partial P(\mathbf{r},t)/\partial n} = \frac{i\omega\rho_0 p(\mathbf{r},\omega)}{\partial p(\mathbf{r},\omega)/\partial n} = \xi \quad \text{on } S. \quad (3.6)$$

3.2.4 Sommerfeld Radiation Condition

The *Sommerfeld radiation condition* [Sommerfeld, 1949] quantifies the notion, stemming from our physical intuition, that sources confined to a finite spatial domain produce outgoing, radiating wave fields at infinity. Specifically, for homogeneous media, the condition for n dimensions is expressed as

$$\lim_{r\to\infty} r^{\frac{n-1}{2}} \left[\frac{\partial p(r,k)}{\partial r} - ikp(r,k) \right] = 0 \ , \quad (3.7)$$

where r is the appropriate spatial coordinate. Equation (3.7) leads to the following asymptotic behavior:

One Dimension:

$$p(x,k) \sim A(k)e^{ik|x|} \quad , \tag{3.8}$$

Two Dimensions:

$$p(r,k) \sim A(k)\frac{e^{ikr}}{\sqrt{r}} \quad , \tag{3.9}$$

Three Dimensions:

$$p(r,k) \sim A(k)\frac{e^{ikr}}{r} \quad . \tag{3.10}$$

In Eq. (3.9), r is the cylindrical radial coordinate, while in Eq. (3.10), r is the spherical radial coordinate.

3.3 Plane Wave Reflection from a Horizontally Stratified Medium

We consider a plane wave of unit amplitude incident from a homogeneous medium with density ρ and sound speed c upon a horizontally stratified medium, as shown in Fig. 3.1. We recall that horizontal stratification means that the medium properties depend on only one spatial coordinate, in this case z. For convenience, we assume that the incident wave lies in the x-z plane and suppress the harmonic time dependence. The total field can be decomposed into the sum of an incident field $p_i(\mathbf{r})$ and a reflected field $p_r(\mathbf{r})$ with wave vectors $\mathbf{k}$ and $\mathbf{k}'$, respectively:

$$p(\mathbf{r}) = p_i(\mathbf{r}) + p_r(\mathbf{r}) = e^{i\mathbf{k}\cdot\mathbf{r}} + Re^{i\mathbf{k}'\cdot\mathbf{r}} \quad , \tag{3.11}$$

where

$$\begin{aligned} \mathbf{k} &= (k_x, k_z) = (k\sin\theta,\ -k\cos\theta), \\ \mathbf{k}' &= (k_x', k_z') = (k\sin\theta',\ k\cos\theta') \ . \end{aligned} \tag{3.12}$$

The amplitude R of the reflected field is called the *plane wave reflection coefficient*, and its detailed behavior is determined by the boundary conditions. While the incident and reflected wave vectors must have the same magnitude,

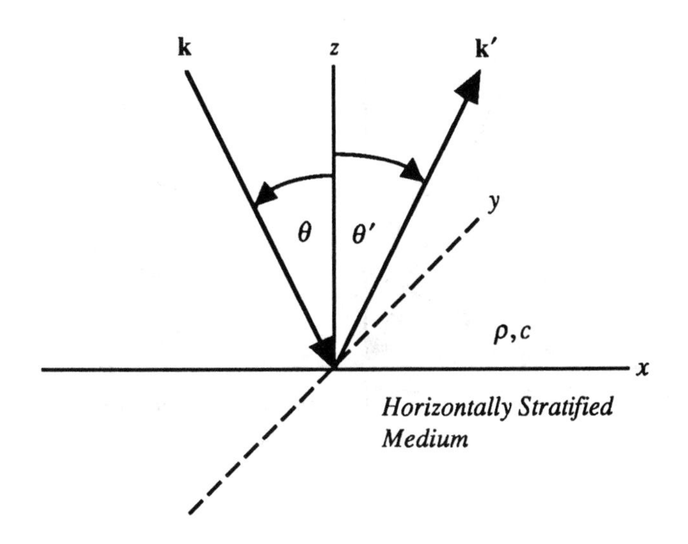

Figure 3.1 Plane wave incident upon a horizontally stratified medium.

$$|\mathbf{k}| = |\mathbf{k}'| = k = \frac{\omega}{c} \quad , \tag{3.13}$$

it is not at all clear, *a priori*, that the angle of the reflected wave must equal the angle of the incident wave. In order to show that the two angles must, in fact, be equal, we impose the physical requirement that the wavefronts of the incident and reflected fields must be continuous at the interface. From Fig. 3.2(a), we see that the projection λ_i of the incident wavefronts onto the boundary is given by

$$\lambda_i = \frac{\lambda}{\sin\theta} \quad , \tag{3.14}$$

where λ is the wavelength of the incident field. Similarly, the projection λ_r of the reflected wavefronts onto the boundary is [cf. Fig. 3.2(b)]

$$\lambda_r = \frac{\lambda}{\sin\theta'} \quad . \tag{3.15}$$

Continuity of the incident and reflected wavefronts at the interface requires that $\lambda_i = \lambda_r$, and therefore

$$\sin\theta = \sin\theta' \quad , \tag{3.16}$$

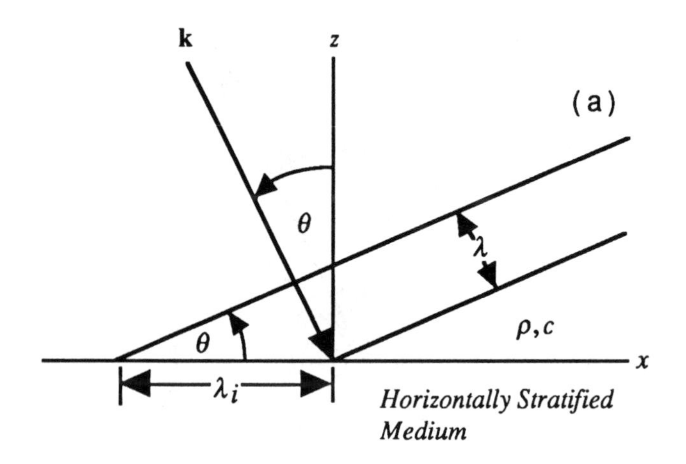

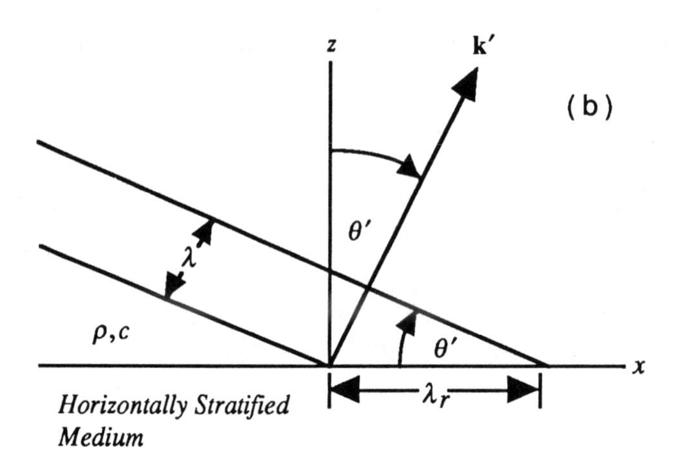

Figure 3.2 Projections of the (a) incident and (b) reflected wavefronts onto the planar boundary.

which implies that $\theta = \theta'$, or *the angle of incidence equals the angle of reflection.* Upon interaction with the boundary, the incident wave undergoes the process of *specular reflection.* Equation (3.11) for the total field therefore becomes

$$p(x,z) = e^{ik(x\sin\theta - z\cos\theta)} + Re^{ik(x\sin\theta + z\cos\theta)} \quad , \tag{3.17}$$

and the reflection coefficient can be expressed as the ratio of the reflected and incident fields evaluated on the boundary:

$$R = |R|e^{i\varphi} = \left.\frac{p_r(x,z)}{p_i(x,z)}\right|_{z=0} , \tag{3.18}$$

where $|R|$ and φ are the reflection coefficient magnitude and phase, respectively. We also note that, by conservation of energy (cf. Prob. 3.2), $|R| \le 1$ for $0 \le \theta \le \pi/2$.

3.3.1 Reflection from a Soft Boundary

For a soft boundary, $p(x,z) = 0$ at $z = 0$, Eq. (3.17) becomes

$$0 = e^{ikx\sin\theta} + Re^{ikx\sin\theta} , \tag{3.19}$$

and therefore $R = -1$.

3.3.2 Reflection from a Hard Boundary

For a hard boundary at $z = 0$,

$$\frac{\partial p(x,z)}{\partial n} = -\frac{\partial p(x,z)}{\partial z} = 0 \text{ at } z = 0, \tag{3.20}$$

where we have chosen the unit normal to be outward from the upper medium. Equation (3.17) therefore becomes

$$0 = -ik\cos\theta e^{ikx\sin\theta} + ikR\cos\theta e^{ikx\sin\theta} , \tag{3.21}$$

which has the solution $R = 1$. We note that for both the soft and hard boundaries, all of the incident acoustic energy is reflected since $|R| = 1$, and there is no acoustic wave in the lower medium. The soft and hard interfaces are therefore also called *impenetrable boundaries*.

3.3.3 Reflection from a Homogeneous Fluid Half-Space

We suppose that a plane wave is incident upon a homogeneous fluid half-space with density ρ_1 and sound speed c_1, as shown in Fig. 3.3. The pressure field in the upper medium is then given by Eq. (3.17), while the field in the lower medium is given by

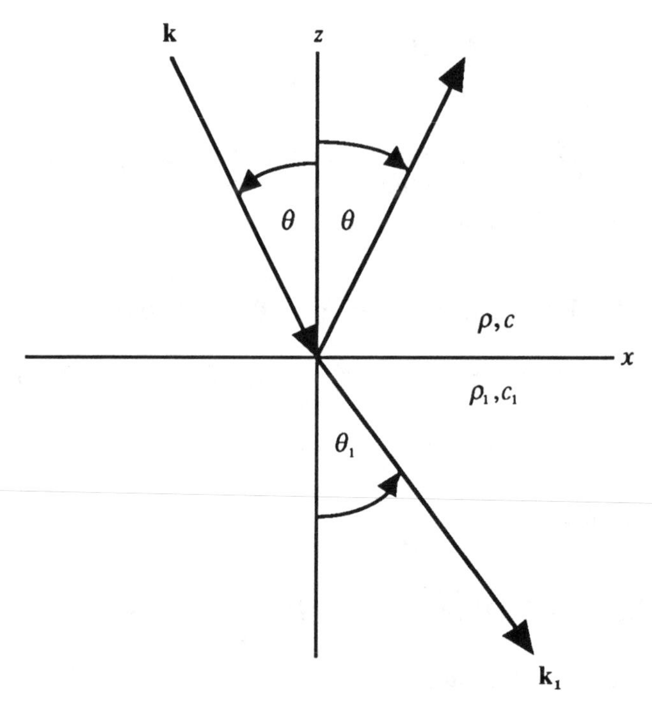

Figure 3.3 Plane wave incident upon a homogeneous fluid half-space.

$$p_1(x,z) = Te^{ik_1(x\sin\theta_1 - z\cos\theta_1)} \quad , \tag{3.22}$$

where $k_1 = \omega/c_1 = 2\pi/\lambda_1$, and T is defined as the *transmission coefficient*. We note that the plane wave in the lower medium propagating in the direction of increasing z is eliminated by virtue of the radiation condition, i.e., there are no sources at $z = -\infty$. We then impose the physical constraints, known as *continuity conditions*, that the pressure and normal component of particle velocity are continuous across the boundary, i.e.,

$$p(x,z)\big|_{z=0} = p_1(x,z)\big|_{z=0} \quad , \tag{3.23}$$

$$\frac{1}{\rho}\frac{\partial p(x,z)}{\partial z}\bigg|_{z=0} = \frac{1}{\rho_1}\frac{\partial p_1(x,z)}{\partial z}\bigg|_{z=0} \quad . \tag{3.24}$$

Equation (3.23) leads to the result

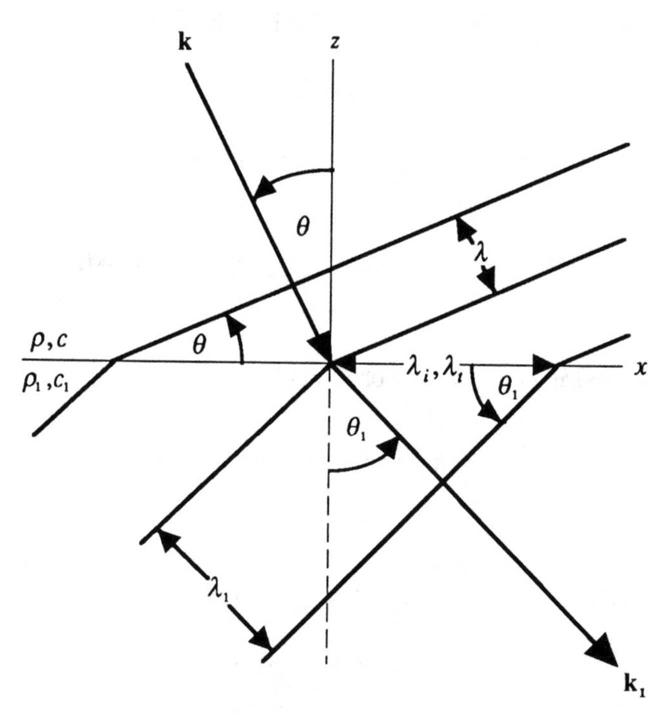

Figure 3.4 Projections of the incident and transmitted wavefronts onto the planar boundary.

$$(1+R)e^{ikx\sin\theta} = Te^{ik_1 x\sin\theta_1} \quad , \tag{3.25}$$

or

$$(1+R) = Te^{i(k_1\sin\theta_1 - k\sin\theta)x} \quad . \tag{3.26}$$

At this point, it is useful to impose the requirement that the incident and transmitted wavefronts be continuous across the boundary, as shown in Fig. 3.4. Then the projection λ_t of the transmitted wavefront onto the boundary,

$$\lambda_t = \frac{\lambda_1}{\sin\theta_1} \quad , \tag{3.27}$$

must equal λ_i, which leads to *Snell's law*:

$$k\sin\theta = k_1\sin\theta_1 \quad . \tag{3.28}$$

Using Snell's law, we simplify Eq. (3.26), which becomes

$$T = 1 + R \ . \tag{3.29}$$

Applying Eq. (3.24) to the fields in the upper and lower media, we find that

$$\frac{-ik\cos\theta}{\rho}(1-R)e^{ikx\sin\theta} = \frac{-ik_1 T\cos\theta_1}{\rho_1}e^{ik_1 x\sin\theta_1} \ , \tag{3.30}$$

which, again using Snell's law, reduces to

$$(1-R)\cos\theta = \frac{n}{m}T\cos\theta_1 \ , \tag{3.31}$$

where the *index of refraction* $n = k_1/k = c/c_1$ and $m = \rho_1/\rho$. Substituting Eq. (3.29) into Eq. (3.31), we find that

$$R = \frac{m\cos\theta - n\cos\theta_1}{m\cos\theta + n\cos\theta_1} \ , \tag{3.32}$$

or, using Snell's law,

$$R = \frac{m\cos\theta - \sqrt{n^2 - \sin^2\theta}}{m\cos\theta + \sqrt{n^2 - \sin^2\theta}} \ . \tag{3.33}$$

Equation (3.33) is frequently referred to as the *Rayleigh reflection coefficient*, although it was originally derived by George Green [Rayleigh, 1877-78; Lindsay, 1972].

The Rayleigh reflection coefficient can also be expressed in terms of the *normal specific acoustic impedance* ζ, which is defined for a plane wave as

$$\zeta = \frac{P(\mathbf{r},t)}{V_n(\mathbf{r},t)} \ \text{on } S. \tag{3.34}$$

We note that Eq. (3.34) deals with the pressure and normal particle velocity associated with an individual plane wave constituent of the field, as opposed to Eq. (3.5), which specifies the impedance for the *total* field. For an incident plane wave, the normal specific acoustic impedance is given by

$$\zeta = \left.\frac{i\omega\rho p_i(x,z)}{\partial p_i(x,z)/\partial n}\right|_{z=0} = \left.\frac{-i\omega\rho p_i(x,z)}{\partial p_i(x,z)/\partial z}\right|_{z=0} = \frac{\rho c}{\cos\theta} , \tag{3.35}$$

while for the transmitted wave, it is

$$\zeta_1 = \left.\frac{i\omega\rho_1 p_1(x,z)}{\partial p_1(x,z)/\partial n}\right|_{z=0} = \left.\frac{-i\omega\rho_1 p_1(x,z)}{\partial p_1(x,z)/\partial z}\right|_{z=0} = \frac{\rho_1 c_1}{\cos\theta_1} , \tag{3.36}$$

where the unit normal is outward from the upper medium. Equation (3.32) then simply becomes

$$R = \frac{\zeta_1 - \zeta}{\zeta_1 + \zeta} , \tag{3.37}$$

which, for normal incidence ($\theta = \theta_1 = 0$), reduces to

$$R = \frac{\rho_1 c_1 - \rho c}{\rho_1 c_1 + \rho c} = \frac{m-n}{m+n} . \tag{3.38}$$

Equation (3.38) shows that at normal incidence, only the characteristic impedances, i.e., the material properties of the two media, influence the reflection coefficient. On the other hand, Eq. (3.37) shows that at oblique incidence, the angular dependencies must also be taken into account, and, as we shall see, play a significant role in the reflection process.

Finally, it is clear from Eq. (3.33) that the Rayleigh reflection coefficient is frequency-independent and that $R = -1$ when $\theta = \pi/2$ for any values of the material properties of the two media. In fact, the latter limiting behavior at grazing incidence holds true for reflection from an arbitrary horizontally stratified medium.

Reflection from a Lower Velocity, Less Dense Half-Space (The Water-Air Interface)

We consider reflection from a lower velocity, less dense medium, e.g., the water-air interface in the ocean acoustic case. Then

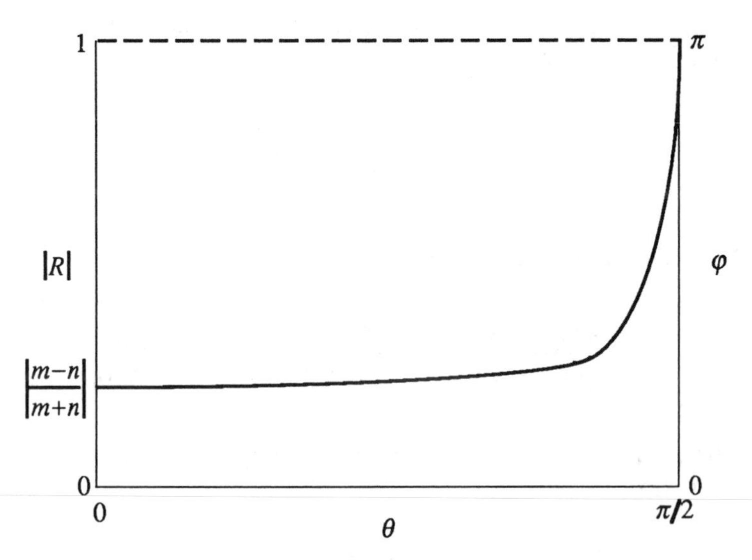

Figure 3.5 Reflection coefficient magnitude (—) and phase (----) for reflection from a lower velocity, less dense half-space (n>1, m<1).

$n > 1$ and $m < 1$, and the reflection coefficient can be expanded in terms of the small parameter ζ_1/ζ,

$$\frac{\zeta_1}{\zeta} = \frac{m \cos\theta}{n \cos\theta_1} < 1 \quad , \tag{3.39}$$

since $\cos\theta_1 > \cos\theta$ for $n > 1$. Using the expansion [Arfken, 1985]

$$\frac{1}{1-z} = \sum_{n=0}^{\infty} z^n \text{ for } |z| < 1 \quad , \tag{3.40}$$

we find that

$$R \approx -1 + 2\frac{\zeta_1}{\zeta} - 2\frac{\zeta_1^2}{\zeta^2} + \ldots \quad . \tag{3.41}$$

Equation (3.41) shows that R is real and can be expanded around its pressure-release value for any incident angle. The general character of the reflection coefficient in this case is shown in Fig. 3.5.

For the water-air interface,

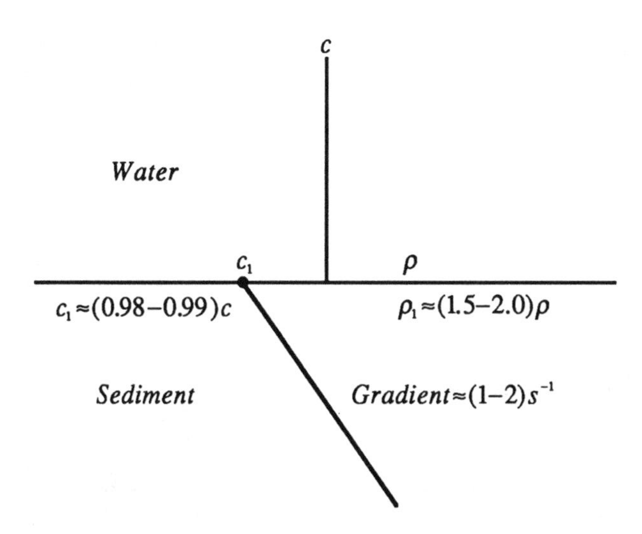

Figure 3.6 The water-bottom interface for silty-clay sediments.

$$\rho = 1.0 \text{ g/cm}^3,\ c = 1500 \text{ m/s},$$
$$\rho_1 = 1.3\times10^{-3} \text{ g/cm}^3,\ c_1 = 332 \text{ m/s}\ , \tag{3.42}$$

and we obtain

$$\frac{\zeta_1}{\zeta} \le 2.88\times10^{-4}\ . \tag{3.43}$$

We therefore approximate the water-air interface as a pressure-release surface for all incident angles.

Reflection from a Lower Velocity, More Dense Half-Space (The Water-Bottom Interface: Model I)

We suppose that the lower half-space consists of a lower velocity, more dense medium with the specific relationship among the parameters $m > n > 1$. In this case, there is a greater discontinuity in density than in sound velocity at the boundary, a situation which occurs, for example, at the water-bottom interface for silty-clay marine sediments [Hamilton, 1980]. Here, a substantial increase in density of the water-saturated surficial sediments is accompanied by a slight decrease in sound velocity (cf. Fig. 3.6). This low-velocity zone may persist for up to 30 m in the sediment column before the positive sound

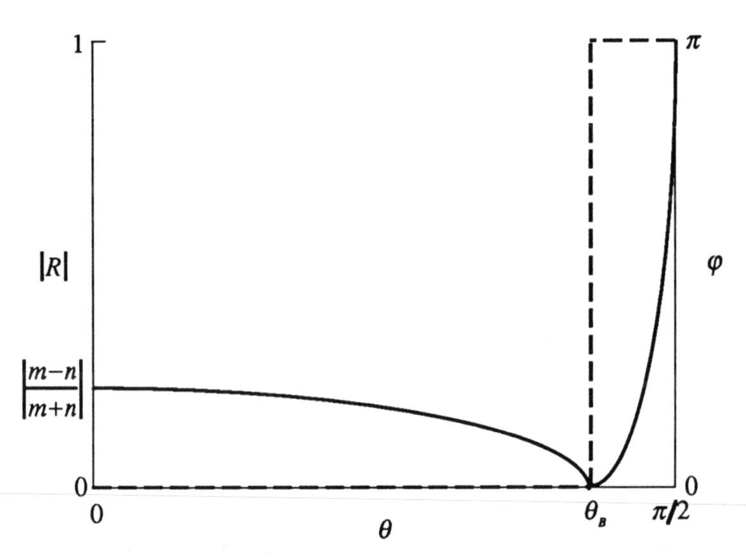

Figure 3.7 Reflection coefficient magnitude (—) and phase (----) for reflection from a lower velocity, more dense half-space ($m>n>1$).

velocity gradient, associated with the increasing compaction of the sediments, causes the sound velocity in the bottom to equal and ultimately exceed the sound velocity in the water.

An interesting effect occurs in this case, namely, the phenomenon of *total transmission*. When $\theta = \theta_B$, where

$$\theta_B = \sin^{-1}\sqrt{\frac{m^2 - n^2}{m^2 - 1}} \quad , \tag{3.44}$$

then $\zeta = \zeta_1$ and $R = 0$, and there is total transmission into the lower medium. The real angle θ_B, by analogy with the electromagnetic case [Jackson, 1975], is sometimes called the *acoustic Brewster angle*; it is also referred to as the *angle of intromission*. A real Brewster angle can also occur for the parameters $m < n < 1$, a case which is not of physical interest to us.

For $\theta < \theta_B$, we can expand R in the parameter $\zeta/\zeta_1 < 1$,

$$R \approx 1 - 2\frac{\zeta}{\zeta_1} + 2\frac{\zeta^2}{\zeta_1^2} + \ldots \quad , \tag{3.45}$$

while if $\theta > \theta_B$, then $\zeta_1/\zeta < 1$, and we can use Eq. (3.41). Again, R is real and, in this case, undergoes a π phase change when the angle of incidence passes through the Brewster angle. The general nature of the reflection coefficient is shown in Fig. 3.7.

Typical silty-clay bottom parameters are

$$\rho_1 = 1.5 \text{ g/cm}^3, \; c_1 = 0.985c = 1478 \text{ m/s} \;, \tag{3.46}$$

for which $\theta_B = 81.0°$. Since n/m=0.68 and m/n=1.48, treating the boundary as a hard interface in the region $\theta < \theta_B$ or a soft interface in the region $\theta > \theta_B$ are crude approximations, and, in general, the detailed behavior of the reflection coefficient must be computed.

Reflection from a Higher Velocity, More Dense Half-Space (The Water-Bottom Interface: Model II)

The case where the lower half-space consists of a higher velocity, more dense medium with $n < 1$ and $m > 1$ is of considerable interest in ocean acoustics because of its frequent occurrence in the seabed. In sandy sediments, for example, there are pronounced increases in both the sound velocity and density at the water-bottom interface [Hamilton, 1980]. Under these conditions, as the angle of the incident wave increases, the angle of the transmitted wave increases until the incident angle reaches the *critical angle* θ_c,

$$\theta = \theta_c = \sin^{-1} n \;, \tag{3.47}$$

for which the transmitted wave lies parallel to the boundary ($\theta_1 = \pi/2$) and $R = 1$ [cf. Eq. (3.33)]. This phenomenon, and the corresponding behavior for $\theta > \theta_c$ for which $|R| = 1$, is known as *total internal reflection* because the incident wave is totally reflected by the boundary. We will now consider the detailed behavior of the fields for the three angular regions $\theta < \theta_c$, $\theta = \theta_c$, and $\theta > \theta_c$.

$\underline{\theta < \theta_c}$: In this region, $\zeta/\zeta_1 < 1$, and the expansion in Eq. (3.45) holds true. The reflection coefficient is real with zero phase.

$\underline{\theta = \theta_c}$: Since $R = 1$ in this case, the total field in the upper medium is given by

$$p(x,z) = 2e^{ikx\sin\theta_c} \cos(kz\cos\theta_c) \;, \tag{3.48}$$

and has a standing wave, cosine envelope in the z direction. Nulls occur in the field at the positions $z = z_n$ such that

$$\cos(kz_n \cos\theta_c) = 0 \quad , \tag{3.49}$$

and therefore

$$kz_n \cos\theta_c = (n + 1/2)\pi, \quad n = 0,1,2,\ldots \quad , \tag{3.50}$$

so that

$$z_n = \frac{(n+1/2)}{\cos\theta_c}\frac{\lambda}{2}, \quad n = 0,1,2,\ldots \quad . \tag{3.51}$$

The corresponding null spacing is

$$z_{n+1} - z_n = \frac{\lambda}{2\cos\theta_c} \quad . \tag{3.52}$$

When $\theta = \theta_c$, the incident field is coupled directly into the boundary and produces a bulk wave in the lower medium which propagates along the boundary, as shown in Fig. 3.8. The transmitted wave has the form [cf. Eq. (3.29)]

$$p_1(x,z) = Te^{ik_1(x\sin\theta_1 - z\cos\theta_1)}\Big|_{\theta_1=\pi/2} = 2e^{ik_1 x} \quad , \tag{3.53}$$

which propagates with velocity c_1 and has a wavefront projection onto the boundary of

$$\lambda_t = \frac{\lambda}{\sin\theta_c} = \lambda_1 \quad , \tag{3.54}$$

since $\sin\theta_1 = 1$. This wave and its manifestation for the case of a source at finite distance from the boundary is variously called a *lateral*, *interface*, *boundary*, *trace*, or *surface wave*; it is also sometimes referred to as a *refracted arrival.* When we calculate the time-averaged energy flux density for the transmitted wave [cf. Eq. (2.58)], we find that

$$\begin{aligned} &p_1(x) = 2e^{ik_1 x}, \quad \nabla p_1(x) = \mathbf{i}ik_1 p_1(x), \\ &p_1^*(x) = 2e^{-ik_1 x}, \quad \nabla p_1^*(x) = -\mathbf{i}ik_1 p_1^*(x) \quad , \end{aligned} \tag{3.55}$$

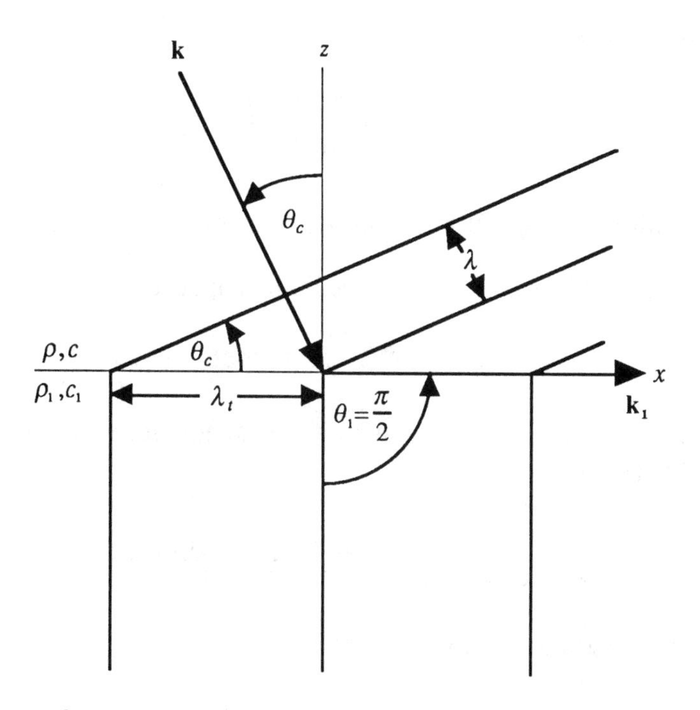

Figure 3.8 Geometry of the incident and transmitted waves at the critical angle.

and therefore

$$\mathbf{I} = \frac{2\mathbf{i}}{\rho_1 c_1} \quad . \tag{3.56}$$

Thus, as expected, the energy in the lateral wave is propagated only in the horizontal, not in the vertical, direction.

$\underline{\theta > \theta_c}$: In this instance, the reflection coefficient can be written as (taking positive square roots)

$$R = \frac{m\cos\theta - i\sqrt{\sin^2\theta - n^2}}{m\cos\theta + i\sqrt{\sin^2\theta - n^2}} \quad , \tag{3.57}$$

and therefore

$$R = e^{i\varphi} \text{ with } \varphi = -2\tan^{-1}\left[\frac{\sqrt{\sin^2\theta - n^2}}{m\cos\theta}\right] \quad . \tag{3.58}$$

Here, we have used the identity

$$\tan\vartheta = \frac{2\tan(\vartheta/2)}{1-\tan^2(\vartheta/2)} \quad . \tag{3.59}$$

The reflection coefficient therefore has unit magnitude and changing phase.

The total field in the upper medium then becomes

$$p(x,z) = 2e^{i\varphi/2}e^{ikx\sin\theta}\cos(kz\cos\theta+\varphi/2) \quad , \tag{3.60}$$

an expression which again contains a cosine envelope, but now with nulls at the positions

$$z_n = \frac{(n+1/2)}{\cos\theta}\frac{\lambda}{2} - \frac{\varphi\lambda}{4\pi\cos\theta}, \quad n=0,1,2,\ldots \quad , \tag{3.61}$$

and a corresponding null spacing

$$z_{n+1} - z_n = \frac{\lambda}{2\cos\theta} \quad . \tag{3.62}$$

When $\theta > \theta_c$, then $\sin\theta > n$, which, from Snell's law, implies that $\sin\theta_1 > 1$. This can only occur for a complex angle

$$\theta_1 = \theta_{1r} + i\theta_{1i} = \pi/2 + i\theta_{1i} \quad , \tag{3.63}$$

since

$$\sin(\theta_{1r} + i\theta_{1i}) = \sin\theta_{1r}\cosh\theta_{1i} + i\cos\theta_{1r}\sinh\theta_{1i} \quad . \tag{3.64}$$

We also find, using Snell's law, that

$$\cos\theta_1 = i\sqrt{\sin^2\theta_1 - 1} = \frac{i}{n}\sqrt{\sin^2\theta - n^2} \quad , \tag{3.65}$$

and therefore

$$p_1(x,z) = Te^{ik_1(x\sin\theta_1 - z\cos\theta_1)}\Big|_{\theta_1=\pi/2+i\theta_{1i}} = \left(1+e^{i\varphi}\right)e^{ikx\sin\theta}e^{kz\sqrt{\sin^2\theta-n^2}} \; . \tag{3.66}$$

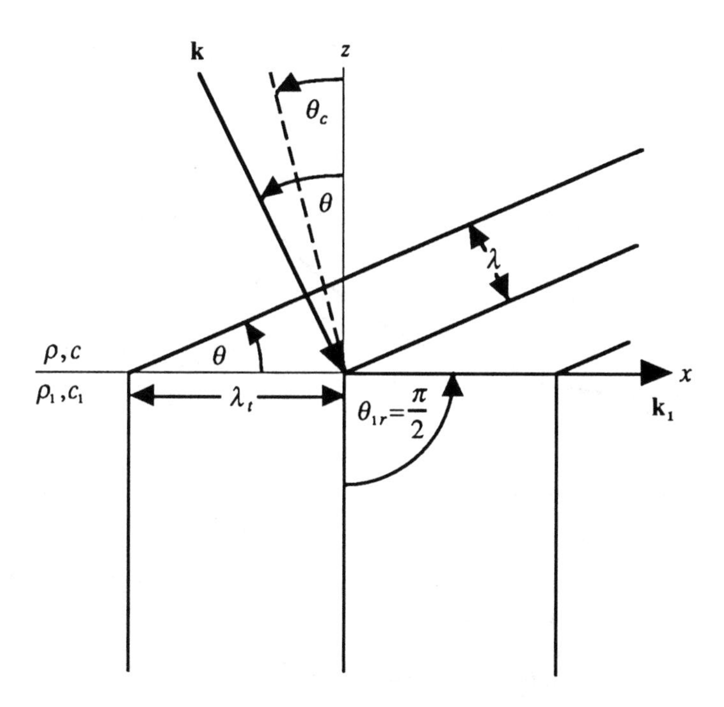

Figure 3.9 Geometry of the incident and transmitted waves when the incident angle exceeds the critical angle.

Equation (3.66) describes an *inhomogeneous plane wave* because it is propagating in one coordinate (the x direction) and exponentially decaying in the other (the negative z direction). This lateral wave now propagates along the boundary with a velocity c_1' such that

$$c_1' = \frac{\omega}{k \sin\theta} = \frac{\omega}{k_1 \sin\theta_1} = \frac{c_1}{\sin\theta_1} < c_1 \ , \qquad (3.67)$$

since $\sin\theta_1 > 1$. In addition, its wavefront projection onto the boundary is

$$\lambda_t = \frac{\lambda}{\sin\theta} = \frac{\lambda_1}{\sin\theta_1} < \lambda_1 \ , \qquad (3.68)$$

as shown in Fig. 3.9. Thus, both the velocity and effective wavelength of the lateral wave are reduced by the factor $1/\sin\theta_1$ when compared with the case $\theta = \theta_c$.

Calculating the time-averaged energy flux density for the transmitted wave, we find that

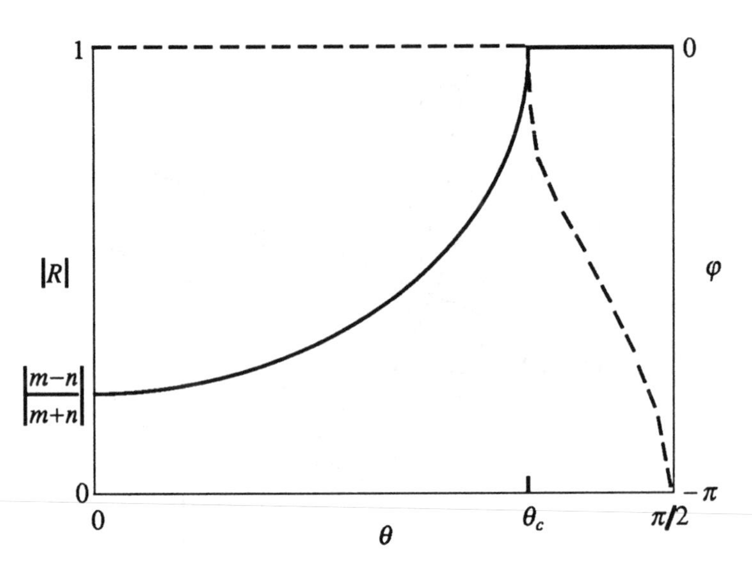

Figure 3.10 Reflection coefficient magnitude (—) and phase (----) for reflection from a higher velocity, more dense half-space ($n<1$, $m>1$).

$$p_1^*(x,z)=\left(1+e^{-i\varphi}\right)e^{-ikx\sin\theta}e^{kz\sqrt{\sin^2\theta-n^2}},$$
$$\nabla p_1(x,z)=\left[\mathbf{i}ik\sin\theta+\mathbf{j}k\sqrt{\sin^2\theta-n^2}\right]p_1(x,z),$$
$$\nabla p_1^*(x,z)=\left[-\mathbf{i}ik\sin\theta+\mathbf{j}k\sqrt{\sin^2\theta-n^2}\right]p_1^*(x,z) \quad , \tag{3.69}$$

and therefore

$$\mathbf{I}=\frac{\mathbf{i}}{\rho_1 c}(1+\cos\varphi)\sin\theta e^{2kz\sqrt{\sin^2\theta-n^2}} \quad . \tag{3.70}$$

Thus, although there is an exponentially decaying (*evanescent*) field in the lower medium, there is no net energy flow across the boundary; energy propagates only parallel to the boundary.

The general behavior of the reflection coefficient for the higher velocity, more dense half-space for all incident angles is shown in Fig. 3.10. Typical sandy bottom parameters are

$$\rho_1=1.8\ \mathrm{g/cm^3},\ c_1=1800\ \mathrm{m/s} \quad , \tag{3.71}$$

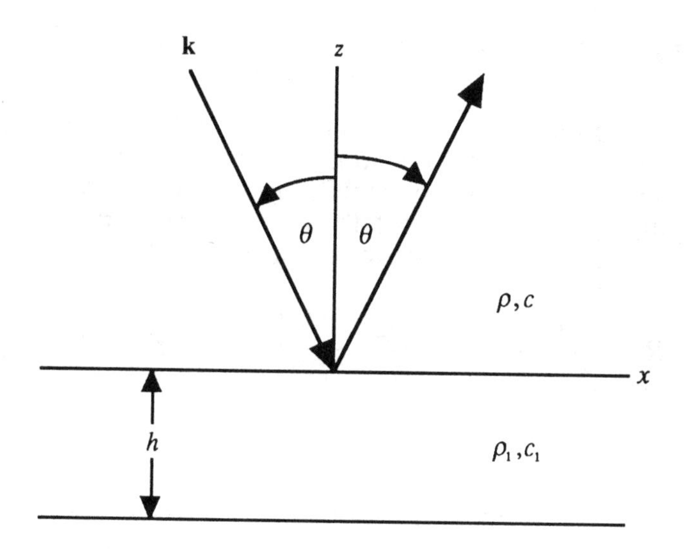

Figure 3.11 Plane wave incident upon a homogeneous fluid layer overlying an arbitrary horizontally stratified medium.

and therefore $\theta_c = 56.4°$. Just as in the case of the *Water-Bottom Interface: Model I*, we can see that approximating *Model II* by a hard or soft boundary is only a crude approximation, at best.

3.3.4 Reflection from a Homogeneous Fluid Layer Overlying an Arbitrary Horizontally Stratified Medium

A problem which contributes significantly to our ability to deal with multilayered media involves the reflection of a plane wave from a homogeneous fluid layer, with density ρ_1 and sound velocity c_1, overlying an arbitrary horizontally stratified medium (cf. Fig. 3.11). Not only is the solution to this problem very useful, but the technique for obtaining the solution is clever and informative in its exploitation of the definition of the reflection coefficient.

The field in the upper medium is given by Eq. (3.17), while the field in the layer of thickness h consists of down- and up-going waves, $p_{1i}(x,z)$ and $p_{1r}(x,z)$, which are incident upon and reflected from the horizontally stratified region:

$$p_1(x,z) = p_{1_i}(x,z) + p_{1_r}(x,z) = e^{ik_1 x \sin\theta_1}\left[Ae^{-ik_1 z\cos\theta_1} + Be^{ik_1 z\cos\theta_1}\right]. \quad (3.72)$$

Here, the coefficients A and B contain the effects of all multiple reflections within the layer. Equation (3.72) can be rewritten in terms of one of these coefficients and the reflection coefficient R' for a plane wave impinging on the horizontally stratified medium from a half-space with the acoustic properties of the layer,

$$R' = \left.\frac{p_{1_r}(x,z)}{p_{1_i}(x,z)}\right|_{z=-h} = \left.\frac{Be^{ik_1 z\cos\theta_1}}{Ae^{-ik_1 z\cos\theta_1}}\right|_{z=-h} = \frac{B}{A}e^{-2ik_1 h\cos\theta_1} , \quad (3.73)$$

so that

$$p_1(x,z) = Ae^{ik_1 x\sin\theta_1}\left[e^{-ik_1 z\cos\theta_1} + R'e^{2ik_1 h\cos\theta_1}e^{ik_1 z\cos\theta_1}\right] . \quad (3.74)$$

Imposing continuity of pressure and normal particle velocity at $z=0$, we obtain

$$1+R = A\left[1+R'e^{2ik_1 h\cos\theta_1}\right] , \quad (3.75)$$

and

$$\frac{k\cos\theta}{\rho}[-1+R] = A\frac{k_1\cos\theta_1}{\rho_1}\left[-1+R'e^{2ik_1 h\cos\theta_1}\right] . \quad (3.76)$$

Solving Eqs. (3.75) and (3.76) for R, we find that, after some manipulation,

$$R = \frac{R_{01} + R'e^{2ik_1 h\cos\theta_1}}{1+R_{01}R'e^{2ik_1 h\cos\theta_1}} , \quad (3.77)$$

where R_{01} is the Rayleigh reflection coefficient for a wave incident from the upper medium upon a half-space with the acoustic properties of the layer [cf. Eq. (3.33)]. In addition to its dependence on angle, R is now a function of frequency as well because of the phase factor $\exp(2ik_1 h\cos\theta_1)$. This term reflects the introduction of the length scale associated with the layer thickness.

Equation (3.77) is useful because we can simply substitute the reflection coefficient for the specific horizontally stratified case of interest to us. For example, if the lower medium is a half-space with density ρ_2 and sound velocity c_2, then $R' = R_{12}$ and

$$R = \frac{R_{01} + R_{12}e^{2ik_1 h\cos\theta_1}}{1 + R_{01}R_{12}e^{2ik_1 h\cos\theta_1}}, \tag{3.78}$$

where R_{12} is the Rayleigh reflection coefficient

$$R_{12} = \frac{m_1\cos\theta_1 - \sqrt{n_1^2 - \sin^2\theta_1}}{m_1\cos\theta_1 + \sqrt{n_1^2 - \sin^2\theta_1}}, \tag{3.79}$$

with $n_1 = k_2/k_1$ and $m_1 = \rho_2/\rho_1$. Furthermore, Eq. (3.77) can be used to construct the reflection coefficient for more complicated, multilayered media, for example, a stack of homogeneous layers (cf. Prob. 3.8).

PROBLEMS

3.1 A plane wave in a medium with density ρ and sound speed c is incident at an angle θ on a plane boundary (cf. Fig. 3.1) characterized by a complex impedance $\xi = \alpha + i\beta$, as defined in Eq. (3.6). This boundary can also be characterized by its plane wave reflection coefficient R, which has magnitude $|R|$ and phase φ.

(a) Express α and β in terms of ρ, c, θ, $|R|$, and φ.

(b) Suppose the lower medium has density ρ_1 and sound speed $c_1 > c$. Evaluate α and β for angles of incidence which are (i) less than the critical angle and (ii) greater than the critical angle.

3.2 The concept of *conservation of energy* for the wave field is deceptively simple in theory and subtle to implement in practice. For example, in the case of plane wave reflection from a horizontally stratified medium, energy is conserved in the following sense: *The sum of the magnitudes of the time-averaged energy flux densities of the reflected and transmitted waves normal to the boundary must equal the magnitude of the time-averaged energy flux density of the incident wave normal to the boundary.*

Consider the reflection of a plane wave at normal incidence from a homogeneous fluid half-space.

(a) Express the conservation of energy in terms of the reflection and transmission coefficients and the parameters m and n.

(b) Verify your result in part (a) by substituting the actual expressions for the reflection and transmission coefficients as functions of m and n.

3.3 *Bottom loss, BL,* is defined as

$$BL = -20\log_{10}|R| \quad , \tag{3.80}$$

and is expressed in units of decibels (dB). Since it is related only to the magnitude of the reflection coefficient, bottom loss is a measure of the energy lost into the bottom; however, it does not include the effect of the bottom on *coherent propagation*, which requires phase information as well.

Sketch the bottom losses associated with the cases shown in Figs. 3.7 and 3.10 using the parameters in Eqs. (3.46) and (3.71).

3.4 In realistic media, some of the acoustic energy is converted into heat as the sound wave propagates through the material. This effect can be due to a variety of mechanisms which include, for example, chemical relaxation in seawater and frictional and viscous losses in bottom sediments [Clay and Medwin, 1977; Stoll, 1985, 1989]. In order to accommodate this behavior in the Helmholtz equation, we make the ansatz

$$k(\mathbf{r}) = \frac{\omega}{c(\mathbf{r})} \rightarrow k(\mathbf{r}) = \frac{\omega}{c(\mathbf{r})} + i\alpha'(\mathbf{r},\omega) \quad , \tag{3.81}$$

so that, for example, a previously unattenuated plane wave is now attenuated with distance:

$$p(x) = e^{ikx} \rightarrow p(x) = e^{i\frac{\omega}{c}x - \alpha' x} \quad . \tag{3.82}$$

We can introduce loss into the loss-free solution to a particular problem simply by making the transformation in Eq. (3.81) as long as

$$\alpha'(\mathbf{r},\omega) << \frac{\omega}{c(\mathbf{r})} \quad , \tag{3.83}$$

thereby keeping the original wave equation unaltered. The quantity α' and its related decibel equivalent α are called the *intrinsic absorption* or *intrinsic attenuation* of the medium:

$$\alpha = \alpha' \cdot 20 \log_{10} e = 8.686 \alpha' \quad . \tag{3.84}$$

The term *intrinsic* emphasizes the fact that we are referring to a heat conversion process as opposed to other loss mechanisms such as scattering.

(a) What is the effect on the reflection coefficient of the introduction of a small amount of attenuation in the lower medium for the case depicted in Fig. 3.10? Sketch the reflection coefficient magnitude in this case.

(b) State the manner in which energy is conserved in part (a).

(c) A common, though crude, approximation assumes a linear dependence of absorption on frequency for the ocean and the seabed,

$$\alpha \approx \kappa f \quad , \tag{3.85}$$

where $\omega = 2\pi f$, and the coefficient $\kappa \approx 10^{-4}$ dB / m / kHz for the ocean volume and $\kappa \approx 10^{-1}$ dB / m / kHz for the seabed. Show that, for these values of κ, the inequality in Eq. (3.83) holds true at all frequencies for the typical sound velocities described in this chapter.

3.5 Expand the reflection coefficient for a homogeneous layer overlying an arbitrary horizontally stratified medium [cf. Eq. (3.77)] using the first three terms in the expansion in Eq. (3.40) (assume that there is a small amount of absorption in the layer, so that the condition for the validity of the expansion is satisfied). Provide physical interpretations for the terms in the reflection coefficient expansion.

Hint: The Rayleigh reflection coefficient associated with medium 0 and medium 1 has the following useful properties:

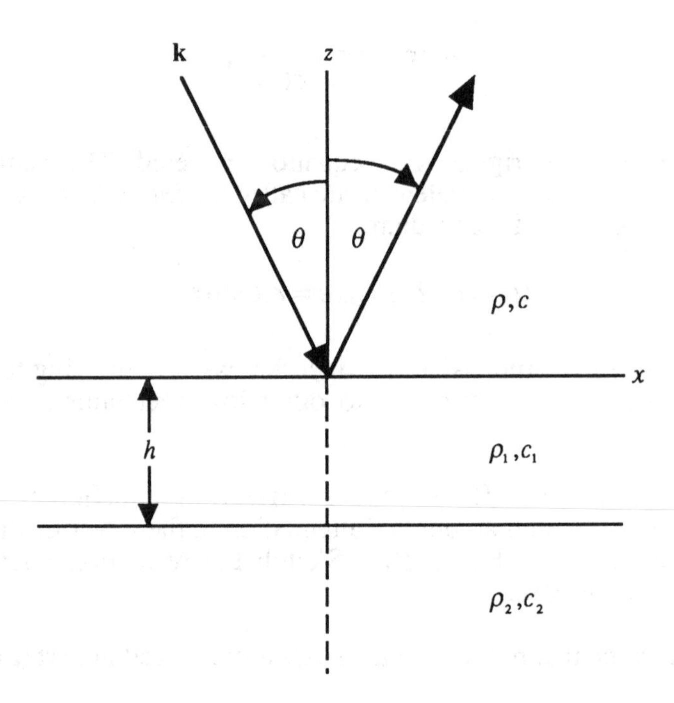

Figure 3.12 Plane wave incident upon a homogeneous fluid layer overlying a homogeneous fluid half-space.

$$\begin{aligned} R_{01} &= -R_{10} \\ T_{01} &= 1 + R_{01} \\ T_{10} &= 1 + R_{10} \ , \end{aligned} \tag{3.86}$$

3.6 Consider the reflection of a plane wave from a homogeneous fluid layer of thickness h overlying a homogeneous fluid half-space for which $c_1 < c < c_2$ and $\rho < \rho_1 < \rho_2$, as shown in Fig. 3.12.

(a) What is the critical angle in the upper medium?

(b) If $k_1 h << 1$, show that, to leading order, the plane wave reflection coefficient for the layer reduces to the reflection coefficient for the case without the layer present.

3.7 In Fig. 3.12, suppose that $\rho_2 = \rho < \rho_1$, $c_2 = c < c_1$, and that the plane wave is incident at an angle which equals or exceeds the critical angle associated with the upper medium and a half-space which has the acoustic properties of the layer, i.e., $\theta \geq \theta_c = \sin^{-1}(c/c_1)$.

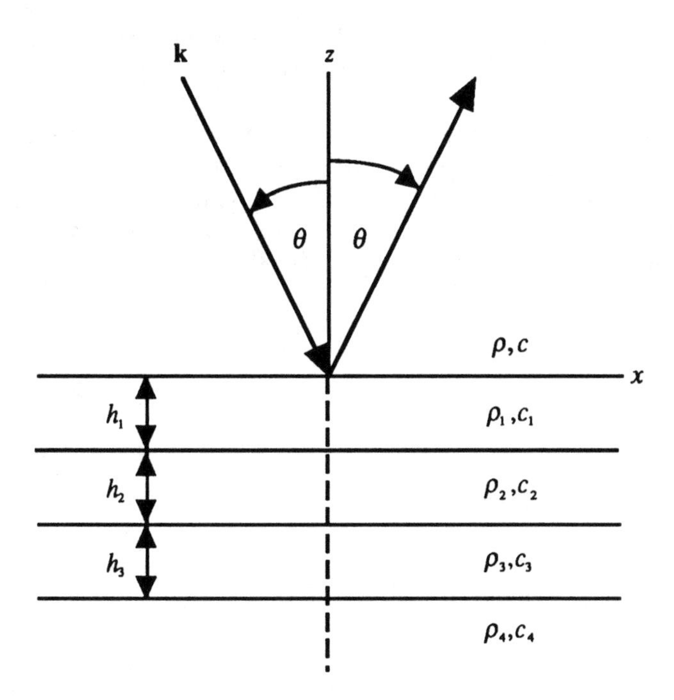

Figure 3.13 Plane wave incident upon a stack of homogeneous fluid layers.

(a) What is the angle of the transmitted wave in the lower half-space?

(b) What is the form of the solution in the layer for (i) $\theta = \theta_c$ and (ii) $\theta > \theta_c$?

(c) Calculate the reflection coefficient in the upper half-space and the transmission coefficient in the lower half-space for (i) $\theta = \theta_c$ and (ii) $\theta > \theta_c$.

(d) Show that energy is conserved in this problem.

(e) Derive the leading order behavior of the reflection and transmission coefficients for $\theta > \theta_c$ and $k_1 h >> 1$ and confirm that energy is conserved.

This problem is a fascinating one because it demonstrates the phenomenon of *wave tunneling*, that is, for incident angles which exceed the critical angle, a propagating wave in the upper medium creates a propagating wave in the lower half-space by tunneling through the layer via a non-propagating, inhomogeneous wave field.

This effect is diminished as the layer thickness increases and the resultant amplitude of the transmitted wave decreases.

3.8 Determine the plane wave reflection coefficient for the structure shown in Fig. 3.13. Express the result in terms of the Rayleigh reflection coefficients for the *i*-*j*th interface and the parameter $k_i h_i \cos\theta_i$, where k_i, h_i, and θ_i are the wavenumber, thickness, and angle of propagation in the *i*th layer ($i = 0,...3$, and $j = 1,...4$).

Chapter 4. Acoustic Sources and Green's Functions

Mathematics possesses not only truth, but supreme beauty - a beauty cold and austere, like that of sculpture, without appeal to any part of our weaker nature, yet sublimely pure, and capable of a stern perfection such as only the greatest art can show.
Bertrand Russell, 1903

4.1 Introduction

We have, until this point, considered solutions to the wave equation in the absence of sources. For example, we implicitly considered spherical wave solutions, as in Eq. (2.16), only for $r \neq 0$. However, we have now reached a stage in the theoretical development where it is crucial that we introduce acoustic sources and the mathematical formalism by which we can incorporate them into solutions of the wave equation, namely, the theory of Green's functions. This is a class of functions which is of great importance in solving a wide variety of ordinary and partial differential equations with boundary and initial conditions. The enormous power and generality of the theory of Green's functions make it one of the supreme achievements of mathematical physics. We begin by describing the manner in which sources are added to the wave equation, focusing on the point source and its representation via the Dirac delta function. Then we discuss the properties of the Green's function and formulate the general boundary value problem for the Helmholtz equation, concluding with a recipe for solving problems using Green's functions. A key element in this procedure is the construction of the Green's function, for which the method of images is described as an important technique. This method is applied to several problems, including the rudimentary example of a point source close to a pressure-release surface, a case which leads to the Lloyd mirror effect in ocean acoustics. The method of images is then used to solve the problem of a point source in a homogeneous fluid layer with impenetrable boundaries. A second technique for constructing Green's functions, which we call the endpoint method, is also described and, combined with a Fourier transform approach, is used to generate the solution for a point source in a homogeneous half-space overlying an arbitrary horizontally stratified medium. Finally, we show that the latter solution can also be expressed in terms of a Hankel transform, which we then evaluate asymptotically for the case of an underlying, homogeneous half-space. The results of this evaluation can be interpreted in terms of a specularly reflected wave and a lateral wave.

4.2 Wave Equations with Source Terms

Sources are introduced into the time-dependent acoustic wave equation by adding a source term $F(\mathbf{r},t)$ to the right-hand side of Eq. (1.2) [Kinsler et al., 1982; Boyles, 1984],

$$\nabla^2 P(\mathbf{r},t) - \frac{1}{c^2(\mathbf{r})}\frac{\partial^2 P(\mathbf{r},t)}{\partial t^2} = -4\pi F(\mathbf{r},t) \ . \tag{4.1}$$

Because of this additional term, Eq. (4.1) is called an *inhomogeneous wave equation.* If we assume harmonic time-dependence for both the field and the source,

$$P(\mathbf{r},t) = p(\mathbf{r},\omega)e^{-i\omega t},$$
$$F(\mathbf{r},t) = f(\mathbf{r},\omega)e^{-i\omega t} \quad , \tag{4.2}$$

then we obtain the *inhomogeneous Helmholtz equation*:

$$\left[\nabla^2 + k^2(\mathbf{r})\right]p(\mathbf{r},\omega) = -4\pi f(\mathbf{r},\omega) \ . \tag{4.3}$$

Of particular importance is a point source at $\mathbf{r} = \mathbf{r}_0$ which we represent by the *Dirac delta function* $\delta(\mathbf{r}-\mathbf{r}_0)$. In Cartesian coordinates, the delta function is given by

$$\delta(\mathbf{r}-\mathbf{r}_0) = \delta(x-x_0)\delta(y-y_0)\delta(z-z_0) \ . \tag{4.4}$$

The delta function has the following properties:

$$\delta(\mathbf{r}-\mathbf{r}_0) = 0 \quad \text{for } \mathbf{r} \neq \mathbf{r}_0 \ , \tag{4.5}$$

$$\int \delta(\mathbf{r}-\mathbf{r}_0)\,dV = 1 \ , \tag{4.6}$$

Sifting Property:

$$\int g(\mathbf{r})\delta(\mathbf{r}-\mathbf{r}_0)dV = g(\mathbf{r}_0) \ . \tag{4.7}$$

Here, dV is the volume element (e.g., $dV = dxdydz$ in Cartesian coordinates) and, in Eqs. (4.6) and (4.7), the volume of integration includes the point $\mathbf{r} = \mathbf{r}_0$. Although the delta function is not itself an analytic function, it can be obtained as a limiting case of various analytic functions, for example,

$$\delta(x) = \lim_{a\to\infty} \frac{\sin ax}{\pi x} = \lim_{a\to\infty} \frac{1}{2\pi} \int_{-a}^{a} e^{ixt} dt \ . \tag{4.8}$$

The solution of Eq. (4.3) for a point source is called a *Green's function* $G(\mathbf{r},\mathbf{r}_0)$:

$$\left[\nabla^2 + k^2(\mathbf{r})\right] G(\mathbf{r},\mathbf{r}_0) = -4\pi\delta(\mathbf{r}-\mathbf{r}_0) \ . \tag{4.9}$$

In the case of a homogeneous medium [i.e., $k(\mathbf{r}) = k$] without boundaries, the solution of Eq. (4.9) is the *free-space Green's function* $G_0(\mathbf{r},\mathbf{r}_0)$, which is given by

$$G_0(\mathbf{r},\mathbf{r}_0) = \frac{e^{ik|\mathbf{r}-\mathbf{r}_0|}}{|\mathbf{r}-\mathbf{r}_0|} \tag{4.10}$$

in three dimensions, and

$$G_0(x,x_0) = \frac{2i\pi}{k} e^{ik|x-x_0|} \tag{4.11}$$

in one dimension. It is clear that for a point source at $\mathbf{r}_0 = 0$, Eq. (4.10) reduces to the spherical wave solution discussed in Eq. (2.16).

4.3 Properties of the Green's Function

In addition to satisfying Eq. (4.9), the Green's function for the Helmholtz equation has the following properties:

(i) $G(\mathbf{r},\mathbf{r}_0)$ satisfies the Sommerfeld radiation condition in exterior problems. For boundary value problems in which we desire a

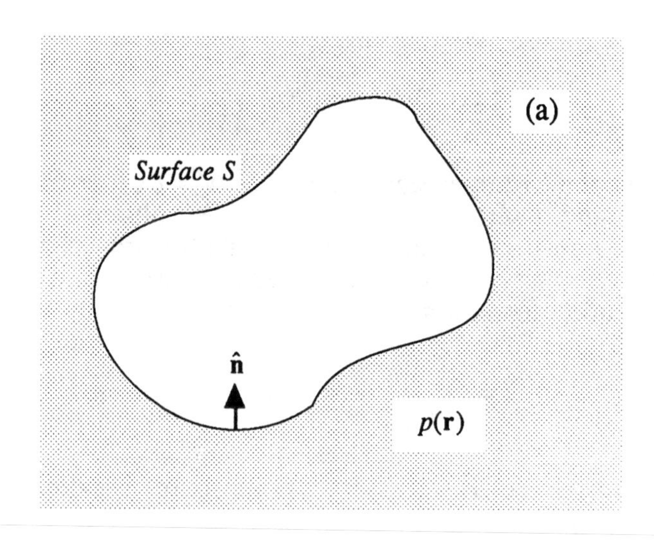

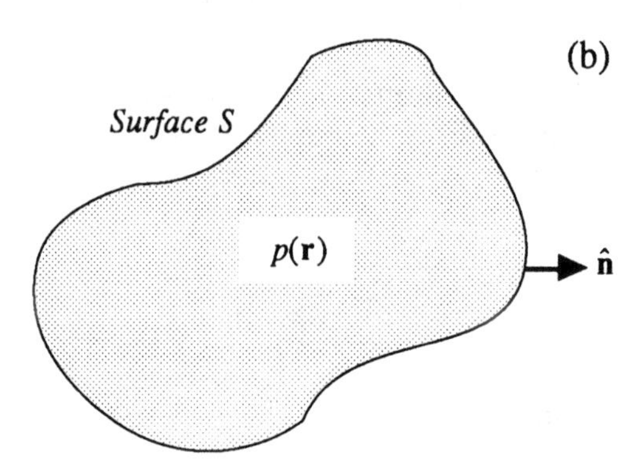

Figure 4.1 (a) Exterior and (b) interior problems for which solutions of the Helmholtz equation are desired in the stippled regions.

solution to the Helmholtz equation exterior to a given region [cf. Fig. 4.1], the Green's function satisfies Eq. (3.7).

(ii) $G(\mathbf{r},\mathbf{r}_0)$ *satisfies homogeneous boundary conditions on surfaces of the same type imposed on* $p(\mathbf{r})$. If $p(\mathbf{r})$ is specified to be a function $a(\mathbf{r})$ on the boundary S (a Dirichlet condition), then $G(\mathbf{r},\mathbf{r}_0)=0$ on S. If $\partial p(\mathbf{r})/\partial n$ is prescribed to be the function $b(\mathbf{r})$ on S (a Neumann condition), then $\partial G(\mathbf{r},\mathbf{r}_0)/\partial n = 0$ on S. In general, for a Cauchy condition, if

$$Ap(\mathbf{r})+B\frac{\partial p(\mathbf{r})}{\partial n}=d(\mathbf{r}) \text{ on } S, \tag{4.12}$$

then

$$AG(\mathbf{r},\mathbf{r}_0)+B\frac{\partial G(\mathbf{r},\mathbf{r}_0)}{\partial n}=0 \text{ on } S. \tag{4.13}$$

Here, $d(\mathbf{r})$ is an arbitrary function, and A and B are constants, either one of which can be zero. This condition is equivalent to the impedance boundary condition discussed in Eq. (3.6), since Eq. (4.13) can be rewritten as

$$\frac{G(\mathbf{r},\mathbf{r}_0)}{\partial G(\mathbf{r},\mathbf{r}_0)/\partial n}=-\frac{B}{A} \quad . \tag{4.14}$$

In Eqs. (4.12)-(4.14), the unit normal is taken to be outward from the region in which the solution is desired (cf. Fig. 4.1).

(iii) The derivative of $G(\mathbf{r},\mathbf{r}_0)$ *is discontinuous at the point* $\mathbf{r}=\mathbf{r}_0$. For the general linear, second-order, ordinary differential equation with coefficients $p_0(x)$, $p_1(x)$, and $p_2(x)$,

$$p_0(x)\frac{d^2\Gamma(x,x_0)}{dx^2}+p_1(x)\frac{d\Gamma(x,x_0)}{dx}+p_2(x)\Gamma(x,x_0)=-\delta(x-x_0) \ , \tag{4.15}$$

there is a *jump discontinuity* in the derivative of the Green's function $\Gamma(x,x_0)$ such that

$$\lim_{x\to x_0}\left[\frac{\partial\Gamma_<(x,x_0)}{\partial x}-\frac{\partial\Gamma_>(x,x_0)}{\partial x}\right]=\frac{1}{p_0(x_0)} \ , \tag{4.16}$$

where

$$\begin{aligned}\Gamma_<(x,x_0)&=\Gamma(x,x_0) \text{ for } x<x_0,\\ \Gamma_>(x,x_0)&=\Gamma(x,x_0) \text{ for } x>x_0 \quad .\end{aligned} \tag{4.17}$$

For the one-dimensional, free-space Green's function, we therefore have

$$\lim_{x\to x_0}\left[\frac{\partial G_{0_<}(x,x_0)}{\partial x}-\frac{\partial G_{0_>}(x,x_0)}{\partial x}\right]=4\pi \ . \tag{4.18}$$

(iv) $G(\mathbf{r},\mathbf{r}_0)$ *obeys the principle of reciprocity.* The Green's function is invariant under the interchange of the source and receiver positions, that is,

$$G(\mathbf{r},\mathbf{r}_0)=G(\mathbf{r}_0,\mathbf{r}) \ . \tag{4.19}$$

This property follows (cf. Prob. 4.2) from Eq. (4.9), property *(ii)*, and Green's theorem, which we will discuss in the next section.

4.4 General Solution of the Boundary Value Problem with Sources

We are now in a position to derive a formal solution of Eq. (4.3) with boundary conditions. We begin by exchanging the roles of $\mathbf{r}$ and $\mathbf{r}_0$ in Eq. (4.9), so that

$$\left[\nabla_0^2+k^2(\mathbf{r}_0)\right]G(\mathbf{r}_0,\mathbf{r})=-4\pi\delta(\mathbf{r}_0-\mathbf{r}) \ , \tag{4.20}$$

where ∇_0^2 operates on the variable $\mathbf{r}_0$. But reciprocity [cf. Eq. (4.19)] and the fact that the delta function is even,

$$\delta(\mathbf{r}-\mathbf{r}_0)=\delta(\mathbf{r}_0-\mathbf{r}) \ , \tag{4.21}$$

lead to the result

$$\left[\nabla_0^2+k^2(\mathbf{r}_0)\right]G(\mathbf{r},\mathbf{r}_0)=-4\pi\delta(\mathbf{r}-\mathbf{r}_0) \ . \tag{4.22}$$

We also rewrite Eq. (4.3) in terms of the variable $\mathbf{r}_0$ and obtain

$$\left[\nabla_0^2+k^2(\mathbf{r}_0)\right]p(\mathbf{r}_0)=-4\pi f(\mathbf{r}_0) \ . \tag{4.23}$$

Multiplying Eq. (4.23) by $G(\mathbf{r},\mathbf{r}_0)$ and Eq. (4.22) by $p(\mathbf{r}_0)$, subtracting the two results, and integrating over the volume V_0 where we seek a solution, we find that

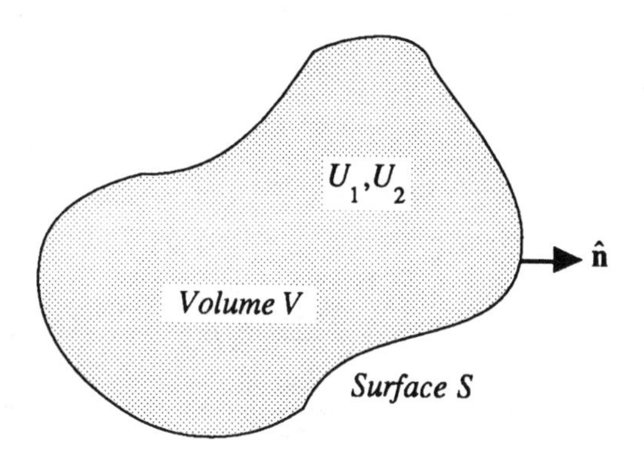

Figure 4.2 Geometry and functions associated with Green's Theorem.

$$\underbrace{\int_{V_0}\left[G(\mathbf{r},\mathbf{r}_0)\nabla_0^2 p(\mathbf{r}_0) - p(\mathbf{r}_0)\nabla_0^2 G(\mathbf{r},\mathbf{r}_0)\right]dV_0}_{Term\ I}$$

$$\underbrace{-4\pi\int_{V_0} p(\mathbf{r}_0)\delta(\mathbf{r}-\mathbf{r}_0)\,dV_0}_{Term\ II} + 4\pi\int_{V_0} f(\mathbf{r}_0)G(\mathbf{r},\mathbf{r}_0)\,dV_0 = 0 \quad .(4.24)$$

In order to evaluate *Term I*, we must use *Green's theorem* [Clay and Medwin, 1977; Arfken, 1985], which relates a volume integral of two scalar functions U_1 and U_2 to a surface integral of these functions over the surface S enclosing the volume V (cf. Fig. 4.2):

$$\int_V\left[U_1\nabla^2 U_2 - U_2\nabla^2 U_1\right]dV = \int_S\left[U_1\frac{\partial U_2}{\partial n} - U_2\frac{\partial U_1}{\partial n}\right]dS \quad . \quad (4.25)$$

Setting $U_1 = G$ and $U_2 = p$, we obtain the following result for *Term I*:

$$\int_{V_0}\left[G(\mathbf{r},\mathbf{r}_0)\nabla_0^2 p(\mathbf{r}_0) - p(\mathbf{r}_0)\nabla_0^2 G(\mathbf{r},\mathbf{r}_0)\right]dV_0$$

$$= \int_{S_0}\left[G(\mathbf{r},\mathbf{r}_0)\frac{\partial p(\mathbf{r}_0)}{\partial n_0} - p(\mathbf{r}_0)\frac{\partial G(\mathbf{r},\mathbf{r}_0)}{\partial n_0}\right]dS_0 \quad . \quad (4.26)$$

Applying the sifting property in Eq. (4.7) to *Term II* in Eq. (4.24), we determine that

$$-4\pi\int_{V_0} p(\mathbf{r}_0)\delta(\mathbf{r}-\mathbf{r}_0)\,dV_0 = -4\pi p(\mathbf{r}) \quad . \tag{4.27}$$

Equation (4.24) therefore becomes

$$p(\mathbf{r}) = \underbrace{\int_{V_0} f(\mathbf{r}_0)G(\mathbf{r},\mathbf{r}_0)\,dV_0}_{Source\ Term}$$
$$\underbrace{+\frac{1}{4\pi}\int_{S_0}\left[G(\mathbf{r},\mathbf{r}_0)\frac{\partial p(\mathbf{r}_0)}{\partial n_0} - p(\mathbf{r}_0)\frac{\partial G(\mathbf{r},\mathbf{r}_0)}{\partial n_0}\right]dS_0}_{Boundary\ Values} \ , \tag{4.28}$$

which expresses the total field as a volume integral over the source function and a surface integral over prescribed boundary values. The critical role which the Green's function plays in determining the solution demonstrates the importance of knowing the response of a physical system to an impulsive driving term. Although we introduced the concept of a Green's function in the context of acoustic sources, it is clear that the Green's function approach can be useful in solving problems even when no real sources are present, i.e., $f(\mathbf{r}_0) = 0$. Once we have determined the Green's function for the particular problem at hand, we can formally express the solution via Eq. (4.28). Although the detailed evaluation of the resulting volume and surface integrals may be nontrivial, frequently the most difficult part of the entire procedure is the determination of the Green's function itself. Finally, we point out that when we are dealing with a problem which involves an actual point source and homogeneous boundary conditions on $p(\mathbf{r})$, the Green's function *is* the solution to the problem. For example, suppose we have a point source at $\mathbf{r} = \mathbf{r}_i$ and no boundaries. Then

$$f(\mathbf{r}_0) = \delta(\mathbf{r}_0 - \mathbf{r}_i) \ , \tag{4.29}$$

and the Green's function is given by Eq. (4.10), so that

$$p(\mathbf{r}) = \int_{V_0} \delta(\mathbf{r}_0 - \mathbf{r}_i)\frac{e^{ik|\mathbf{r}-\mathbf{r}_0|}}{|\mathbf{r}-\mathbf{r}_0|}\,dV_0 = \frac{e^{ik|\mathbf{r}-\mathbf{r}_i|}}{|\mathbf{r}-\mathbf{r}_i|} \ , \tag{4.30}$$

as expected. This rudimentary example also illustrates the subtleties associated with identifying the source point, the observation point, and the integration variable.

For a Dirichlet condition, Eq. (4.28) becomes

$$p(\mathbf{r}) = \int_{V_0} f(\mathbf{r}_0)G(\mathbf{r},\mathbf{r}_0)\,dV_0 - \frac{1}{4\pi}\int_{S_0} p(\mathbf{r}_0)\frac{\partial G(\mathbf{r},\mathbf{r}_0)}{\partial n_0}dS_0, \quad (4.31)$$

while for a Neumann condition, it reduces to

$$p(\mathbf{r}) = \int_{V_0} f(\mathbf{r}_0)G(\mathbf{r},\mathbf{r}_0)\,dV_0 + \frac{1}{4\pi}\int_{S_0} G(\mathbf{r},\mathbf{r}_0)\frac{\partial p(\mathbf{r}_0)}{\partial n_0}dS_0. \quad (4.32)$$

For a Cauchy condition [cf. Eqs. (4.12) and (4.13)], we can write

$$G\frac{\partial p}{\partial n_0} - p\frac{\partial G}{\partial n_0} = \left[Ap + B\frac{\partial p}{\partial n_0}\right]\frac{G}{B} - \underbrace{\left[AG + B\frac{\partial G}{\partial n_0}\right]}_{=0}\frac{p}{B}, \quad (4.33)$$

and therefore,

$$p(\mathbf{r}) = \int_{V_0} f(\mathbf{r}_0)G(\mathbf{r},\mathbf{r}_0)\,dV_0 + \frac{1}{4\pi}\int_{S_0}\left[Ap(\mathbf{r}_0) + B\frac{\partial p(\mathbf{r}_0)}{\partial n_0}\right]\frac{G(\mathbf{r},\mathbf{r}_0)}{B}dS_0 \ . \quad (4.34)$$

4.5 A Recipe for Solving Problems Using Green's Functions

There are three major steps to follow in solving the inhomogeneous Helmholtz equation with boundary values:

(i) Identify the specific type of problem under consideration. We must determine if there are real sources in the problem and identify the type of boundary conditions being prescribed. The proper identification of the problem initially will help avoid confusion in the subsequent steps.

(ii) Construct the appropriate Green's function. As mentioned earlier, this may be the most difficult step, and, from one perspective, it may sometimes appear that we have replaced the original problem with an equally difficult problem of constructing the Green's function. However, once we have found the Green's function, we have not only

solved the problem posed originally, but have also solved, at least formally, a much broader class of problems.

(iii) Calculate the volume and surface integrals. When these integrals cannot be evaluated exactly, sometimes they are evaluated using approximate analytic methods, such as asymptotic analysis, or they may be computed numerically.

4.6 Construction of the Green's Function: The Method of Images

In this chapter, we will discuss three techniques for constructing Green's functions: (i) the method of images, (ii) the endpoint method, and (iii) a hybrid Fourier transform/endpoint method. A fourth approach, known as the method of eigenfunction expansion, will be the subject of the next chapter.

The method of images is well suited for dealing with homogeneous media, planar boundaries, and Dirichlet or Neumann boundary conditions. In this technique, the Green's function is decomposed into the free-space Green's function plus a sum of contributions from image sources which fall outside the region of interest:

$$G(\mathbf{r},\mathbf{r}_0)=\frac{e^{ik|\mathbf{r}-\mathbf{r}_0|}}{|\mathbf{r}-\mathbf{r}_0|}+\sum_{i=1}^{N}\mathrm{SGN}_i\frac{e^{ik|\mathbf{r}-\mathbf{r}_i|}}{|\mathbf{r}-\mathbf{r}_i|}\ , \tag{4.35}$$

where SGN_i is the sign of the ith image source. There is no simple recipe for determining the locations or signs of the image sources; the successful implementation of this technique is dependent on intuition, experience, and luck!

4.6.1 The Method of Images for a Plane with Dirichlet Conditions

Suppose we desire the solution $p(\mathbf{r})$ in a homogeneous medium for the region $x \geq 0$ satisfying the boundary condition $p(\mathbf{r}) = a(y,z)$ on the y-z plane, i.e., for $x = 0$ (cf. Fig. 4.3). This is a Dirichlet problem which therefore requires that $G(\mathbf{r},\mathbf{r}_0) = 0$ for $x = 0$. Our intuition suggests that we should try an image source symmetrically positioned about the y-z plane in relation to the source associated with the free-space Green's function and opposite in source strength, so that

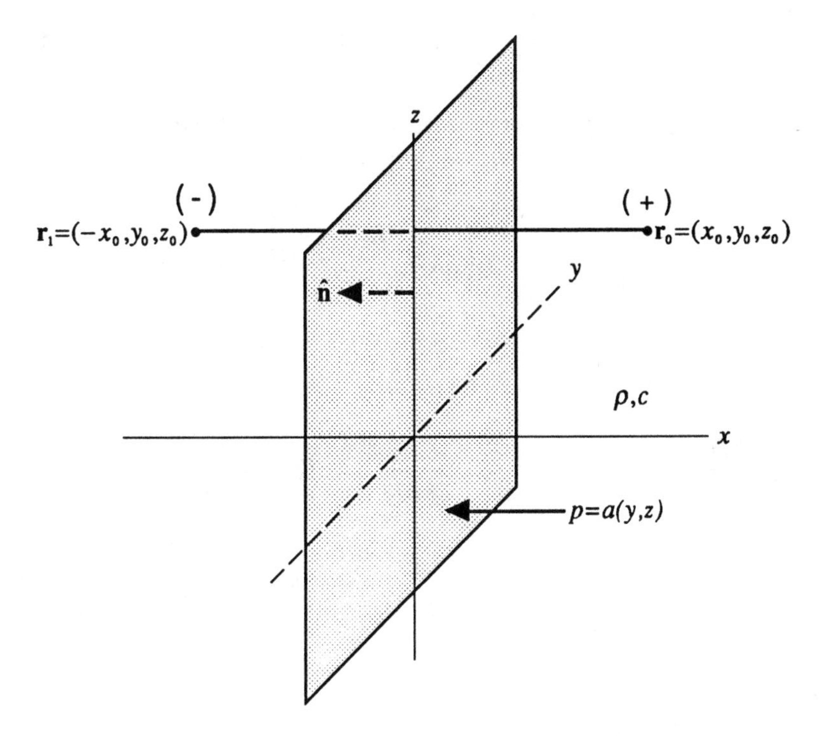

Figure 4.3 Geometry of the image method for a Dirichlet problem in a half-space.

the fields of the two sources cancel one another on the *y*-*z* plane. We therefore write

$$G(\mathbf{r},\mathbf{r}_0)=\frac{e^{ik|\mathbf{r}-\mathbf{r}_0|}}{|\mathbf{r}-\mathbf{r}_0|}-\frac{e^{ik|\mathbf{r}-\mathbf{r}_1|}}{|\mathbf{r}-\mathbf{r}_1|} \quad , \tag{4.36}$$

where

$$\mathbf{r}_0=(x_0,y_0,z_0),\ \ \mathbf{r}_1=(-x_0,y_0,z_0),$$
$$|\mathbf{r}-\mathbf{r}_0|=\sqrt{(x-x_0)^2+(y-y_0)^2+(z-z_0)^2},$$
$$|\mathbf{r}-\mathbf{r}_1|=\sqrt{(x+x_0)^2+(y-y_0)^2+(z-z_0)^2} \quad . \tag{4.37}$$

We can verify that Eq. (4.36) is in fact the correct Green's function since

$$\left|\mathbf{r}-\mathbf{r}_0\right|_{x=0}=\left|\mathbf{r}-\mathbf{r}_1\right|_{x=0} \ , \tag{4.38}$$

and therefore, $G(\mathbf{r},\mathbf{r}_0)=0$ at $x=0$, as required. Since there are no real sources in the problem and $\partial G/\partial n_0 = -\partial G/\partial x_0$, we obtain

$$p(\mathbf{r})=\frac{1}{4\pi}\int_{-\infty}^{\infty}\int_{-\infty}^{\infty} a(y_0,z_0)\left.\frac{\partial G(\mathbf{r},\mathbf{r}_0)}{\partial x_0}\right|_{x_0=0} dy_0 dz_0 \ , \tag{4.39}$$

with

$$\left.\frac{\partial G(\mathbf{r},\mathbf{r}_0)}{\partial x_0}\right|_{x_0=0}=\left\{(x-x_0)\frac{e^{ik|\mathbf{r}-\mathbf{r}_0|}}{\left|\mathbf{r}-\mathbf{r}_0\right|^2}\left[-ik+\frac{1}{\left|\mathbf{r}-\mathbf{r}_0\right|}\right]\right.$$
$$\left.+(x+x_0)\frac{e^{ik|\mathbf{r}-\mathbf{r}_1|}}{\left|\mathbf{r}-\mathbf{r}_1\right|^2}\left[-ik+\frac{1}{\left|\mathbf{r}-\mathbf{r}_1\right|}\right]\right\}_{x_0=0} \ , \tag{4.40}$$

which becomes

$$\left.\frac{\partial G(\mathbf{r},\mathbf{r}_0)}{\partial x_0}\right|_{x_0=0}=2x\frac{e^{ik\sqrt{x^2+(y-y_0)^2+(z-z_0)^2}}}{\left[x^2+(y-y_0)^2+(z-z_0)^2\right]}$$
$$\times\left\{-ik+\frac{1}{\sqrt{x^2+(y-y_0)^2+(z-z_0)^2}}\right\} . \tag{4.41}$$

4.6.2 The Method of Images for a Plane with Neumann Conditions

Let us consider the Neumann problem analogous to the Dirichlet problem discussed in Sec. 3.6.1. Here, we specify the condition $\partial p(\mathbf{r})/\partial n = b(y,z)$ on the y-z plane and wish to obtain the solution for $x \geq 0$ (cf. Fig. 4.4). This requires that $\partial G(\mathbf{r},\mathbf{r}_0)/\partial n = 0$ for $x=0$. We position the sources as in the Dirichlet case, but assign them equal source strengths, so that the Green's function is

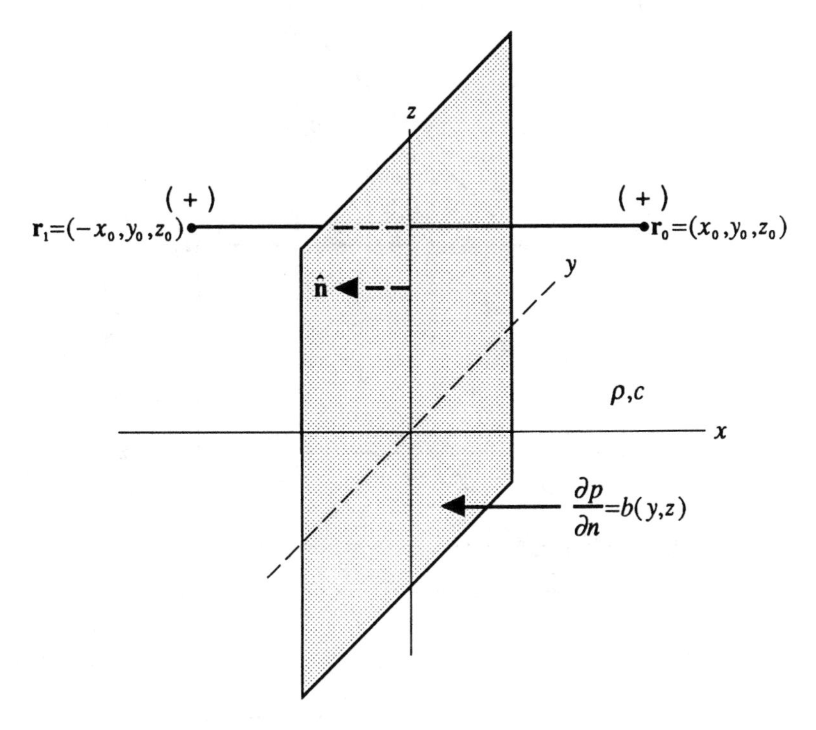

Figure 4.4 Geometry of the image method for a Neumann problem in a half-space.

$$G(\mathbf{r},\mathbf{r}_0)=\frac{e^{ik|\mathbf{r}-\mathbf{r}_0|}}{|\mathbf{r}-\mathbf{r}_0|}+\frac{e^{ik|\mathbf{r}-\mathbf{r}_1|}}{|\mathbf{r}-\mathbf{r}_1|} \quad . \tag{4.42}$$

We then have

$$\left.\frac{\partial G(\mathbf{r},\mathbf{r}_0)}{\partial x}\right|_{x=0}=\left\{(x-x_0)\frac{e^{ik|\mathbf{r}-\mathbf{r}_0|}}{|\mathbf{r}-\mathbf{r}_0|^2}\left[ik-\frac{1}{|\mathbf{r}-\mathbf{r}_0|}\right]\right.$$
$$\left.\left.+(x+x_0)\frac{e^{ik|\mathbf{r}-\mathbf{r}_1|}}{|\mathbf{r}-\mathbf{r}_1|^2}\left[ik-\frac{1}{|\mathbf{r}-\mathbf{r}_1|}\right]\right\}\right|_{x=0}=0, \tag{4.43}$$

as required. The solution of the problem is therefore

$$p(\mathbf{r})=\frac{1}{4\pi}\int_{-\infty}^{\infty}\int_{-\infty}^{\infty}b(y_0,z_0)G(\mathbf{r},\mathbf{r}_0)\Big|_{x_0=0}dy_0dz_0 \quad , \tag{4.44}$$

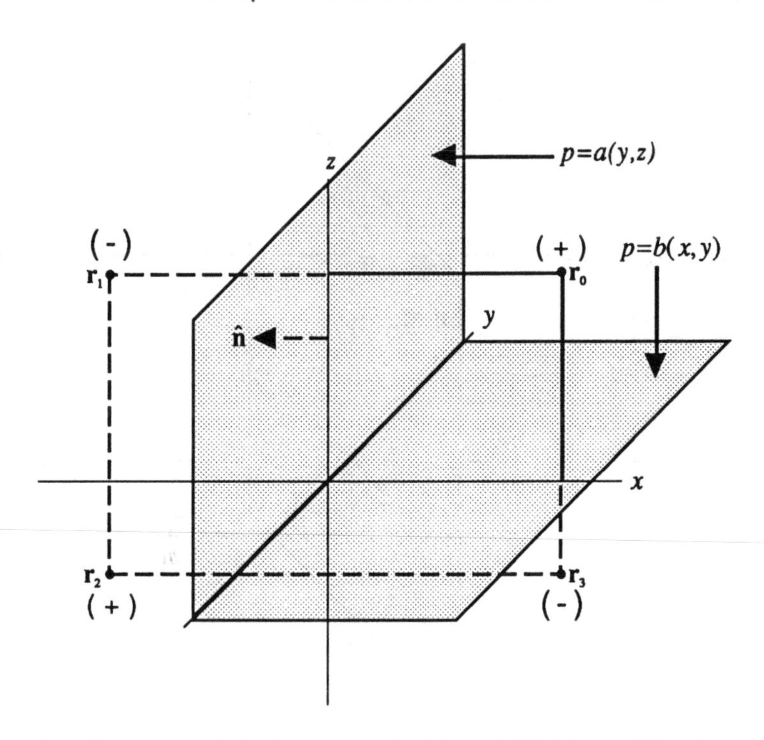

Figure 4.5 Geometry of the image method for a Dirichlet problem in a quadrant.

where

$$G(\mathbf{r},\mathbf{r}_0)\big|_{x_0=0} = 2\frac{e^{ik\sqrt{x^2+(y-y_0)^2+(z-z_0)^2}}}{\sqrt{x^2+(y-y_0)^2+(z-z_0)^2}} \quad . \tag{4.45}$$

4.6.3 The Method of Images for a Quadrant with Dirichlet Conditions

Suppose that we have a problem for which we specify $p(\mathbf{r}) = a(y,z)$ on the y-z plane for $z \geq 0$ and $p(\mathbf{r}) = b(x,y)$ on the x-y plane for $x \geq 0$, and we want to determine $p(\mathbf{r})$ in the quadrant $x \geq 0$, $z \geq 0$, as shown in Fig. 4.5. This is a Dirichlet problem for which $G(\mathbf{r},\mathbf{r}_0) = 0$ on the prescribed boundaries. In determining the Green's function for this problem, we can follow one of the few concrete rules which exist in the implementation of the image method, namely: *Sources symmetrically positioned about a Dirichlet surface have opposite signs, while sources symmetrically positioned about a*

Neumann surface have the same signs. We therefore propose the following form for the Green's function:

$$G(\mathbf{r},\mathbf{r}_0)=\frac{e^{ikR_0}}{R_0}-\frac{e^{ikR_1}}{R_1}+\frac{e^{ikR_2}}{R_2}-\frac{e^{ikR_3}}{R_3} \quad , \tag{4.46}$$

where

$$\begin{aligned} &R_i=|\mathbf{r}-\mathbf{r}_i|,\ \ i=0,1,2,3, \\ &\mathbf{r}_0=(x_0,y_0,z_0),\ \ \mathbf{r}_1=(-x_0,y_0,z_0), \\ &\mathbf{r}_2=(-x_0,y_0,-z_0),\ \ \mathbf{r}_3=(x_0,y_0,-z_0) \quad . \end{aligned} \tag{4.47}$$

The verification that Eqs (4.46) and (4.47) correspond to the correct Green's function as well as the details of the formal solution to this problem are left as exercises for the reader (cf. Prob. 4.3).

4.6.4 The Lloyd Mirror Effect

The *Lloyd mirror effect* [Officer, 1958] arises in the context of placing a point source in a homogeneous ocean close to the surface, which we model as a pressure-release boundary. This is simply a Dirichlet problem with homogeneous boundary conditions, and therefore the solution is the Green's function in Eq. (4.36). It is of interest to examine the behavior of this solution for source-receiver separations which are much greater than the depths of the source and receiver below the interface. First, we rewrite Eq. (4.36) in terms of cylindrical coordinates (cf. Fig. 4.6) for a source located at $\mathbf{r}_0=(0,z_0)$ and receiver located at $\mathbf{r}=(r,z)$:

$$p(\mathbf{r})=G(\mathbf{r},\mathbf{r}_0)=\frac{e^{ikR_0}}{R_0}-\frac{e^{ikR_1}}{R_1} \quad , \tag{4.48}$$

where

$$R_0=\sqrt{r^2+(z-z_0)^2},\ \ R_1=\sqrt{r^2+(z+z_0)^2} \quad , \tag{4.49}$$

and $r=\sqrt{x^2+y^2}$. It is clear from Fig. 4.6 that the total field at the receiver can be interpreted as the sum of a direct arrival with path

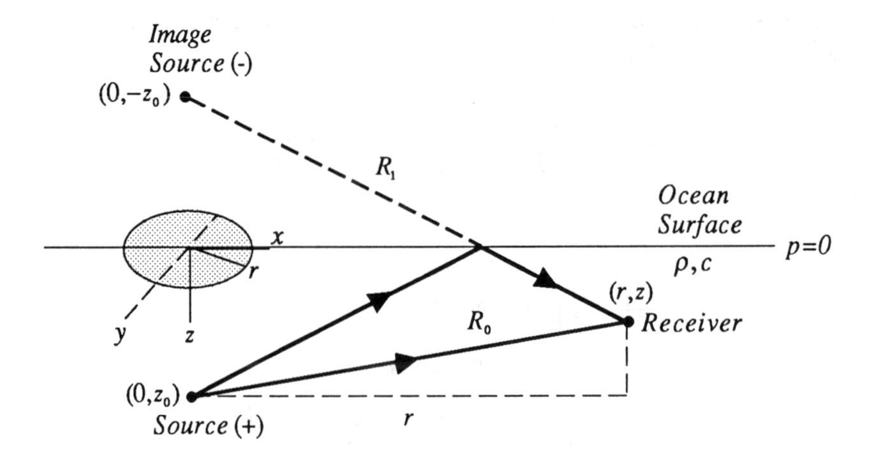

Figure 4.6 Geometry of the Lloyd mirror effect.

length R_0 and a surface-reflected arrival with path length R_1. Using the first two terms in the binomial expansion [Arfken, 1985],

$$(1 \pm z)^{1/2} = 1 \pm \frac{1}{2}z - \frac{1}{8}z^2 \pm \frac{1}{16}z^3 - \ldots \quad \text{for } |z| < 1 \ , \tag{4.50}$$

we find that for $r >> z, z_0$, Eq. (4.49) becomes

$$R_0 = r\sqrt{1 + \frac{(z - z_0)^2}{r^2}} \approx r\left[1 + \frac{1}{2}\frac{(z - z_0)^2}{r^2}\right],$$

$$R_1 = r\sqrt{1 + \frac{(z + z_0)^2}{r^2}} \approx r\left[1 + \frac{1}{2}\frac{(z + z_0)^2}{r^2}\right] \ , \tag{4.51}$$

and therefore

$$R_0 \approx r\left[1 + \frac{(z^2 + z_0^2)}{2r^2} - \frac{zz_0}{r^2}\right],$$

$$R_1 \approx r\left[1 + \frac{(z^2 + z_0^2)}{2r^2} + \frac{zz_0}{r^2}\right] \ , \tag{4.52}$$

which implies that

$$R_1 - R_0 \approx \frac{2zz_0}{r}, \quad R_1 + R_0 \approx 2r \ . \tag{4.53}$$

Keeping only the leading order terms of Eq. (4.52) in the amplitudes,

$$\frac{1}{R_0} \approx \frac{1}{R_1} \approx \frac{1}{r} \ , \tag{4.54}$$

we see that Eq. (4.48) becomes

$$p(\mathbf{r}) \approx \frac{e^{ik\left(\frac{R_0+R_1}{2}\right)}}{r}\left[e^{-ik\left(\frac{R_1-R_0}{2}\right)} - e^{ik\left(\frac{R_1-R_0}{2}\right)}\right] , \tag{4.55}$$

which, using Eq. (4.53), reduces to

$$p(\mathbf{r}) \approx -2i \sin\frac{kzz_0}{r}\frac{e^{ikr}}{r} \ . \tag{4.56}$$

Thus, the field associated with a point source in free space has been significantly modified due to the proximity of the pressure-release boundary, which has introduced a sine envelope dependent on wavelength and source/receiver geometry. A previously spherically spreading field now falls off as $1/r^2$ when $r >> kzz_0$:

$$p(\mathbf{r}) \approx -2ikzz_0 \frac{e^{ikr}}{r^2} \ , \tag{4.57}$$

where we have used the approximation $\sin\vartheta \approx \vartheta$ for $|\vartheta| << 1$. Furthermore, the intensity, instead of falling off as $1/r^2$, falls off as $1/r^4$ under these conditions:

$$I \approx \frac{2(kzz_0)^2}{\rho c r^4} \ . \tag{4.58}$$

In general, for fixed source/receiver depths, the field has nulls at the positions $r = r_n$ such that

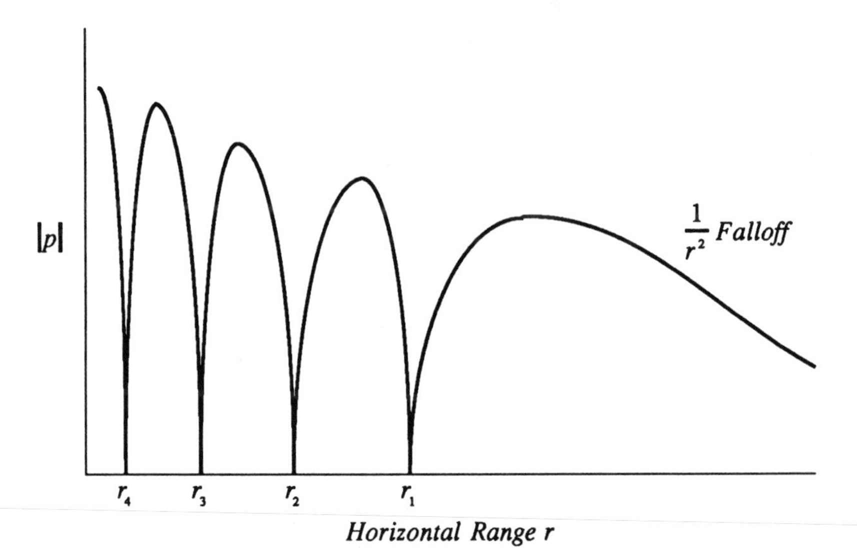

Figure 4.7 The Lloyd mirror interference pattern.

$$\sin\frac{kzz_0}{r_n} = 0 \ , \tag{4.59}$$

and therefore

$$\frac{kzz_0}{r_n} = n\pi, \ \ n = 0,1,2,\ldots \ , \tag{4.60}$$

so that

$$r_n = \frac{2zz_0}{n\lambda}, \ \ n = 0,1,2,\ldots \ . \tag{4.61}$$

The null spacing,

$$r_n - r_{n+1} = \frac{2zz_0}{n(n+1)\lambda} \ , \tag{4.62}$$

clearly decreases with increasing index n, as shown in Fig. 4.7. For example, if f = 100 Hz, c = 1500 m/s, and $z = z_0 = 100$ m, then

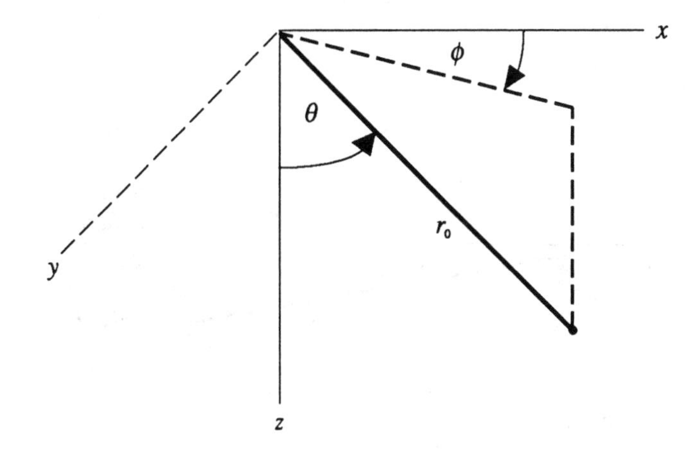

Figure 4.8 Spherical coordinate system.

$$r_0 = \infty, \quad r_3 = 444 \text{ m},$$
$$r_1 = 1333 \text{ m}, \quad r_4 = 333 \text{ m},$$
$$r_2 = 667 \text{ m}, \quad \vdots \quad . \tag{4.63}$$

In practice, this interference pattern is observed only at ranges less than several kilometers, source/receiver depths less than several hundred meters, and frequencies less than several kilohertz in moderately calm seas [Urick, 1975]. For longer ranges and greater source/receiver depths, the continuously refracting nature of the real ocean must be taken into account, and other propagation effects prevail. For higher frequencies and rough seas, scattering from the rough ocean surface dominates.

It is informative to compare the total power radiated from the Lloyd mirror *dipole source*, consisting of the real source and its negative image, with the power radiated from a point source in free space. For a unit point source at the origin of the spherical coordinate system (r_0, θ, ϕ) shown in Fig. 4.8, the field is given by

$$p(\mathbf{r}) = \frac{e^{ikr_0}}{r_0} \ , \tag{4.64}$$

and the total radiated power P_0 is

$$P_0 = \int_S I_0 \, dS \ , \tag{4.65}$$

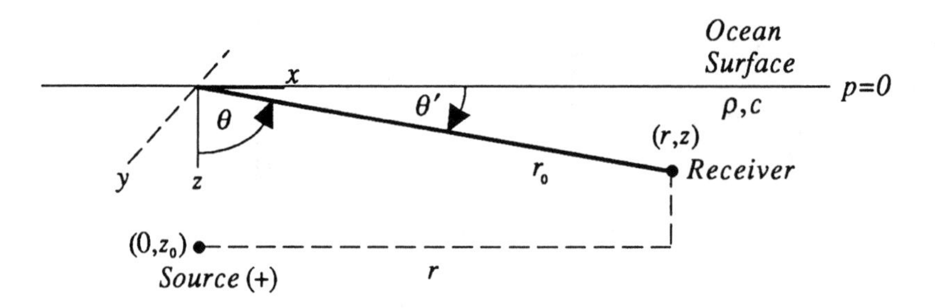

Figure 4.9 Geometry for calculation of the power radiated by the Lloyd mirror dipole source.

where the intensity I_0 is

$$I_0 = \frac{1}{2\rho c r_0^2} \ , \tag{4.66}$$

and S is the surface of a sphere at a distance r_0 from the origin. Equation (4.64) therefore becomes

$$P_0 = \int_0^{2\pi} d\phi \int_0^{\pi} d\theta \, \frac{r_0^2 \sin\theta}{2\rho c r_0^2} = \frac{2\pi}{\rho c} \ . \tag{4.67}$$

For the Lloyd mirror dipole source centered at the origin of the coordinate system, the intensity I is

$$I = \frac{2}{\rho c r^2} \sin^2 \frac{kzz_0}{r} = \frac{1}{\rho c r^2}\left[1 - \cos\frac{2kzz_0}{r}\right] \ , \tag{4.68}$$

where we have used the identity

$$2\sin^2\frac{\vartheta}{2} = 1 - \cos\vartheta \ . \tag{4.69}$$

For $r >> z, z_0$, we can make the approximation (cf. Fig. 4.9)

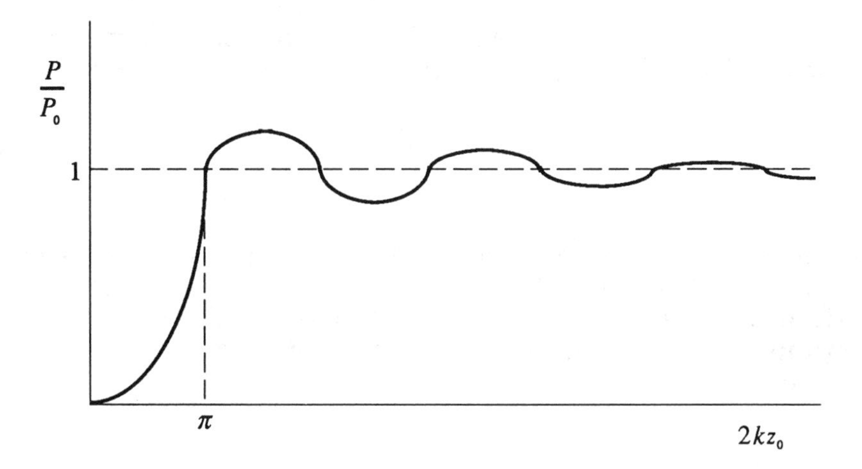

Figure 4.10 Power radiated by the Lloyd mirror dipole source normalized to the power radiated by a point source in free space.

$$\sin\theta' = \frac{z}{r_0} \approx \frac{z}{r} \ , \tag{4.70}$$

so that

$$I \approx \frac{1}{\rho c r^2}\left[1-\cos(2kz_0 \sin\theta')\right] \ , \tag{4.71}$$

where $\theta' = \pi/2 - \theta$. The total power P radiated by the dipole source is therefore

$$P \approx \int_0^{2\pi} d\phi \int_0^{\pi/2} d\theta' \, \frac{r_0^2 \cos\theta'}{\rho c r^2}\left[1-\cos(2kz_0 \sin\theta')\right] \ , \tag{4.72}$$

where we integrate over a hemisphere of radius r_0 in the water column. Using the fact that $r_0 \approx r$ and changing variables, we obtain

$$P \approx \frac{2\pi}{\rho c}\int_0^1 d(\sin\theta') \left[1-\cos(2kz_0 \sin\theta')\right] = \frac{2\pi}{\rho c}\left[1 - \frac{\sin 2kz_0}{2kz_0}\right]. \tag{4.73}$$

We note that although $r_0 \approx r$ is a poor approximation as θ' approaches $\pi/2$, it does not introduce a significant error in the calculation because the integrand approaches 0 in this angular region due to the $\cos\theta'$

factor. The ratio of the power radiated by the Lloyd mirror dipole source to that radiated by the point source in free space is therefore

$$\frac{P}{P_0} = 1 - \frac{\sin 2kz_0}{2kz_0} \ , \tag{4.74}$$

a function which is sketched in Fig. 4.10. The oscillatory behavior of P/P_0 about the value of 1 begins when $2kz_0 = \pi$, or $z_0 = \lambda/4$. The effective power of a point source is therefore diminished when it is positioned within one quarter-wavelength of a pressure-release surface.

4.6.5 A Point Source in a Homogeneous Fluid Layer with Impenetrable Boundaries

The method of images can also be used to solve the problem of a point source in a homogeneous fluid layer with impenetrable boundaries [Brekhovskikh, 1980; Brekhovskikh and Lysanov, 1991]. This is the first *waveguide* case we will examine which incorporates both the presence of a source and upper and lower reflecting boundaries. Although it is only a crude approximation to the real ocean, it nevertheless is a very instructive example.

First, we consider the case, shown in Fig. 4.11, where both the upper and lower interfaces are characterized by homogeneous Neumann conditions, i.e., $\partial p/\partial z = 0$ at $z = 0$ and $z = h$. We begin by introducing an image source S_{02} at a position symmetric to the real source S_{01} about the plane $z = h$ and with the same source strength:

$$p(\mathbf{r}) = \frac{e^{ikR_{01}}}{R_{01}} + \frac{e^{ikR_{02}}}{R_{02}} + \ldots \ , \tag{4.75}$$

where

$$R_{01} = \sqrt{r^2 + (z - z_0)^2}, \ R_{02} = \sqrt{r^2 + (z + z_0 - 2h)^2} \ . \tag{4.76}$$

The total field satisfies the inhomogeneous Helmholtz equation and the boundary condition at $z = h$, but does not satisfy the condition at $z = 0$. This situation is rectified by adding a pair of image sources, S_{03} and S_{04}, symmetrically positioned in relation to S_{01} and S_{02} about the plane $z = 0$ and with the same strength:

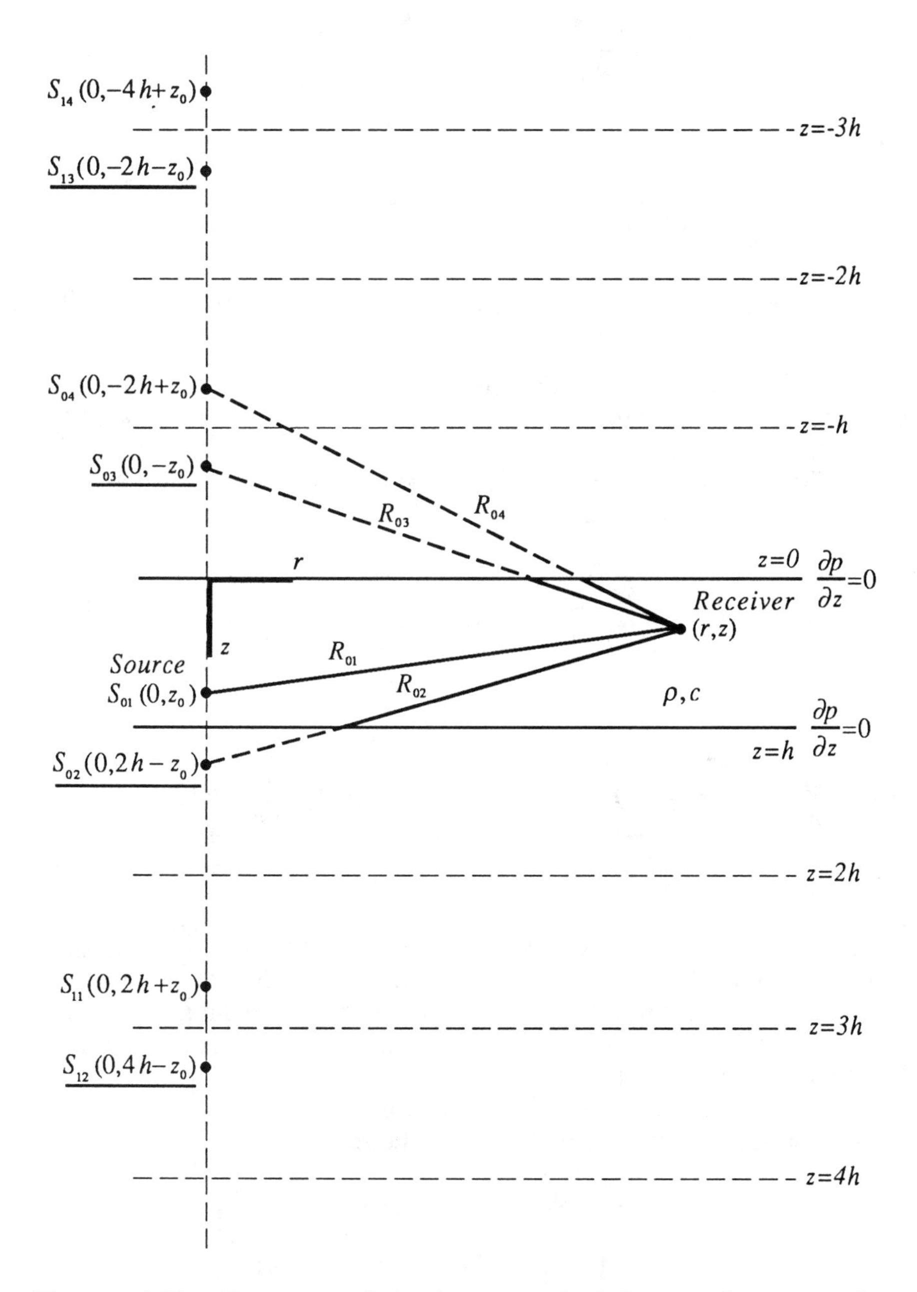

Figure 4.11 Geometry of the image method for a point source in a homogeneous fluid layer with Neumann boundaries. Underlined image sources change sign for the corresponding Dirichlet problem.

$$p(\mathbf{r}) = \frac{e^{ikR_{01}}}{R_{01}} + \frac{e^{ikR_{02}}}{R_{02}} + \frac{e^{ikR_{03}}}{R_{03}} + \frac{e^{ikR_{04}}}{R_{04}} + \dots \ , \tag{4.77}$$

where

$$R_{01} = \sqrt{r^2 + (z - z_0)^2}, \ R_{02} = \sqrt{r^2 + (z + z_0 - 2h)^2},$$
$$R_{03} = \sqrt{r^2 + (z + z_0)^2}, \ R_{04} = \sqrt{r^2 + (z - z_0 + 2h)^2} \ . \tag{4.78}$$

This combination then satisfies the wave equation and the boundary condition at the upper interface but does not satisfy the condition at the lower interface. This procedure continues until we have an infinite sum of source contributions which satisfy the wave equation and the conditions at both boundaries:

$$p(\mathbf{r}) = \sum_{n=0}^{\infty} \left[\frac{e^{ikR_{n1}}}{R_{n1}} + \frac{e^{ikR_{n2}}}{R_{n2}} + \frac{e^{ikR_{n3}}}{R_{n3}} + \frac{e^{ikR_{n4}}}{R_{n4}} \right] , \tag{4.79}$$

where

$$R_{n1} = \sqrt{r^2 + (z - z_0 - 2nh)^2}, \ R_{n2} = \sqrt{r^2 + [z + z_0 - 2(n+1)h]^2},$$
$$R_{n3} = \sqrt{r^2 + (z + z_0 + 2nh)^2}, \ R_{n4} = \sqrt{r^2 + [z - z_0 + 2(n+1)h]^2} \ . \tag{4.80}$$

Furthermore, the sum is convergent because each additional source introduced into the sum has a decreasing amplitude due to spherical spreading. Each term in the sum can be interpreted as an arrival at the receiver which has undergone a specific number of reflections from the boundaries (cf. Prob. 4.7).

For the homogeneous Dirichlet conditions, $p = 0$ at $z = 0$ and $z = h$, it is straightforward to show that the solution is

$$p(\mathbf{r}) = \sum_{n=0}^{\infty} \left[\frac{e^{ikR_{n1}}}{R_{n1}} - \frac{e^{ikR_{n2}}}{R_{n2}} - \frac{e^{ikR_{n3}}}{R_{n3}} + \frac{e^{ikR_{n4}}}{R_{n4}} \right] . \tag{4.81}$$

Thus, the image sources which are underlined in Fig. 4.11 change sign when compared with the Neumann problem.

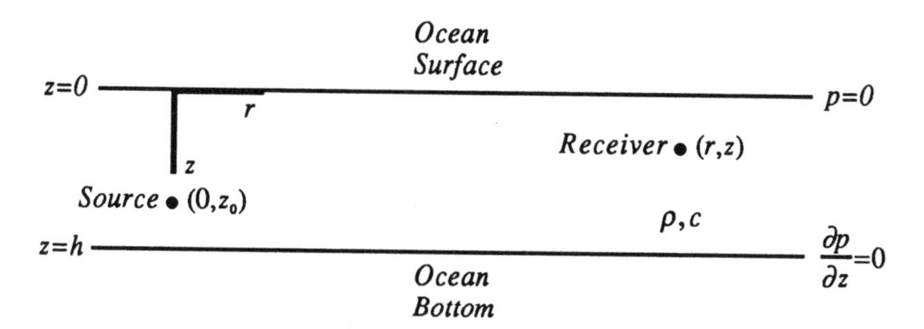

Figure 4.12 A zeroth order model for the ocean acoustic waveguide.

It is left as an exercise for the reader (cf. Prob. 4.8) to prove that

$$p(\mathbf{r}) = \sum_{n=0}^{\infty}(-1)^n\left[\frac{e^{ikR_{n1}}}{R_{n1}}+\frac{e^{ikR_{n2}}}{R_{n2}}-\frac{e^{ikR_{n3}}}{R_{n3}}-\frac{e^{ikR_{n4}}}{R_{n4}}\right] \tag{4.82}$$

is the solution to the problem shown in Fig. 4.12, where the upper interface is soft and the lower interface is hard. This case is extremely important and will be referred to frequently because it is a zeroth order model for the realistic ocean acoustic waveguide.

4.7 Construction of the Green's Function: The Endpoint Method

The endpoint method is more powerful than the method of images because it can be applied to problems in inhomogeneous media with Cauchy boundary conditions. It is used to generate the one-dimensional Green's function $\Gamma(x,x_0)$ which satisfies Eq. (4.15),

$$p_0(x)\frac{d^2\Gamma(x,x_0)}{dx^2}+p_1(x)\frac{d\Gamma(x,x_0)}{dx}+p_2(x)\Gamma(x,x_0)=-\delta(x-x_0),\tag{4.15}$$

on the interval $a \le x \le b$ and homogeneous boundary conditions or the Sommerfeld radiation condition at $x=a$ and $x=b$. The Green's function is constructed from linearly independent solutions $u_a(x)$ and $u_b(x)$ of the homogeneous version (zero on the right-hand side) of Eq. (4.15) which satisfy the boundary conditions at $x=a$ and $x=b$, respectively:

$$\Gamma(x,x_0)=\begin{cases}\Gamma_<(x,x_0)=-\dfrac{1}{C}u_a(x)u_b(x_0), & a\le x\le x_0,\\[2ex] \Gamma_>(x,x_0)=-\dfrac{1}{C}u_a(x_0)u_b(x), & x_0\le x\le b\end{cases} \quad , \tag{4.83}$$

where

$$C=p_0(x_0)W\left[u_a(x_0),\ u_b(x_0)\right] \ , \tag{4.84}$$

and the *Wronskian* W is defined as the determinant

$$W=\begin{vmatrix}u_a(x_0) & u_b(x_0)\\ u_a'(x_0) & u_b'(x_0)\end{vmatrix}=u_a(x_0)u_b'(x_0)-u_a'(x_0)u_b(x_0) \ , \tag{4.85}$$

with the primes denoting differentiation with respect to x_0.

As an example, we consider the solution of the one-dimensional, inhomogeneous Helmholtz equation,

$$\frac{d^2\Gamma(x,x_0)}{dx^2}+k^2\Gamma(x,x_0)=-\delta(x-x_0) \ , \tag{4.86}$$

on the interval $0\le x\le b$ and satisfying homogeneous Dirichlet conditions at the endpoints. Then $\Gamma=0$ at $x=0$ and $x=b$, and the linearly independent solutions of the homogeneous version of Eq. (4.86) are (A and B are arbitrary constants)

$$u_a(x)=A\sin kx,\ \ u_b(x)=B\sin k(x-b) \ , \tag{4.87}$$

and therefore

$$u_a'(x)=Ak\cos kx,\ \ u_b'(x)=Bk\cos k(x-b) \ , \tag{4.88}$$

with $C=ABk\sin kb$. The Green's function is then given by

$$\Gamma(x,x_0)=-\frac{\sin kx_<\sin k\left(x_>-b\right)}{k\sin kb} \ , \tag{4.89}$$

where $x_<$ $\left(x_>\right)$ is the lesser (greater) of x and x_0.

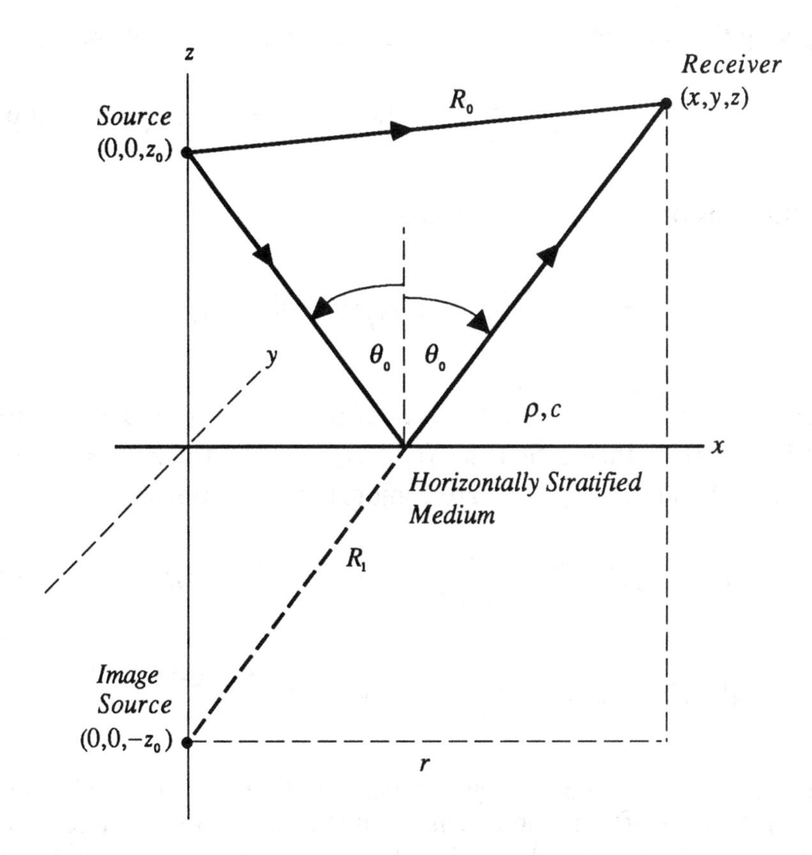

Figure 4.13 Geometry for the reflection of a spherical wave from a horizontally stratified medium.

4.8 A Point Source in a Homogeneous Fluid Half-Space Overlying an Arbitrary Horizontally Stratified Medium

The problem of determining the field due to a point source in a homogeneous fluid half-space overlying an arbitrary horizontally stratified medium provides an ideal practical example for demonstrating a hybrid Fourier transform/endpoint technique for constructing the Green's function. We assume that the source has Cartesian coordinates $(0,0,z_0)$ and the receiver is located at the position (x,y,z), as shown in Fig. 4.13. The inhomogeneous Helmholtz equation [cf. Eqs. (4.3) and (4.9)] then becomes

$$\left[\frac{\partial^2}{\partial x^2}+\frac{\partial^2}{\partial y^2}+\frac{\partial^2}{\partial z^2}+k^2\right]p(\mathbf{r})=-4\pi\delta(x)\delta(y)\delta(z-z_0) \quad . \quad (4.90)$$

Applying the two-dimensional inverse Fourier transform operator,

$$I.F.T.\{\bullet\} = \frac{1}{2\pi}\int_{-\infty}^{\infty}\int_{-\infty}^{\infty}\{\bullet\}e^{-i(k_x x + k_y y)}\,dx\,dy \ , \tag{4.91}$$

to both sides of Eq. (4.90), we obtain

$$\left[\frac{d^2}{dz^2} + k_z^2\right] g(k_x,k_y;z,z_0) = -2\delta(z-z_0) \ , \tag{4.92}$$

where $k_z = \sqrt{k^2 - k_x^2 - k_y^2}$ is the *vertical wavenumber*, and we require that $\mathrm{Im}\,k_z \geq 0$ to insure that the field is bounded for $z \to \infty$. Here $p(x,y;z,z_0)$ and $g(k_x,k_y;z,z_0)$ are conjugate Fourier transform pairs

$$p(x,y;z,z_0) = \frac{1}{2\pi}\int_{-\infty}^{\infty}\int_{-\infty}^{\infty} g(k_x,k_y;z,z_0)e^{i(k_x x + k_y y)}\,dk_x\,dk_y \ , \tag{4.93}$$

$$g(k_x,k_y;z,z_0) = \frac{1}{2\pi}\int_{-\infty}^{\infty}\int_{-\infty}^{\infty} p(x,y;z,z_0)e^{-i(k_x x + k_y y)}\,dx\,dy \ . \tag{4.94}$$

The quantity g is called the *depth-dependent Green's function* because after Fourier transformation in x and y, it still retains its dependence on the depth variables z and z_0. We now solve Eq. (4.92) using the endpoint method. The solution $u_a(z)$ and its derivative $u_a'(z)$ are

$$u_a(z) = A\left[e^{-ik_z z} + Re^{ik_z z}\right] \ , \tag{4.95}$$

$$u_a'(z) = -ik_z A\left[e^{-ik_z z} - Re^{ik_z z}\right] \ , \tag{4.96}$$

where A is an arbitrary constant, and we have incorporated the boundary condition at $z = 0$ through the plane-wave reflection coefficient R for the arbitrary horizontally stratified medium. The solution $u_b(z)$ must satisfy the radiation condition, so that

$$u_b(z) = Be^{ik_z z}, \quad u_b'(z) = ik_z Be^{ik_z z} \ , \tag{4.97}$$

where B is an arbitrary constant. It is straightforward to determine that $C = ik_z AB$, and therefore the depth-dependent Green's function is

$$g(k_x,k_y;z,z_0)=\begin{cases}\frac{i}{k_z}\left[e^{-ik_z z}+Re^{ik_z z}\right]e^{ik_z z_0}, & 0\le z\le z_0,\\ \frac{i}{k_z}\left[e^{-ik_z z_0}+Re^{ik_z z_0}\right]e^{ik_z z}, & z_0\le z<\infty\end{cases}, \quad (4.98)$$

which can be rewritten as

$$g(k_x,k_y;z,z_0)=\frac{i}{k_z}\left[e^{ik_z|z-z_0|}+Re^{ik_z(z+z_0)}\right],\ 0\le z<\infty\ . \quad (4.99)$$

We can then decompose the total field in Eq. (4.93) into the sum of an incident field and a reflected field,

$$p(\mathbf{r})=p_i(\mathbf{r})+p_r(\mathbf{r})\ , \quad (4.100)$$

where the incident field $p_i(\mathbf{r})$ corresponds to that portion of the Green's function that would exist if there were no boundary present (i.e., $R=0$) and therefore must equal the free-space Green's function [cf. Eq. (4.10)]

$$p_i(\mathbf{r})=\frac{1}{2\pi}\int_{-\infty}^{\infty}\int_{-\infty}^{\infty}\frac{i}{k_z}e^{ik_z|z-z_0|}e^{i(k_x x+k_y y)}\,dk_x dk_y=\frac{e^{ikR_0}}{R_0}\ ,(4.101)$$

where the path length $R_0=\sqrt{x^2+y^2+(z-z_0)^2}$ is shown in Fig. 4.13. The reflected field $p_r(\mathbf{r})$ is then given by

$$p_r(\mathbf{r})=\frac{1}{2\pi}\int_{-\infty}^{\infty}\int_{-\infty}^{\infty}\frac{i}{k_z}R(k_x,k_y)e^{ik_z(z+z_0)}e^{i(k_x x+k_y y)}\,dk_x dk_y.(4.102)$$

The Fourier integrals in Eqs. (4.101) and (4.102) have useful physical interpretations as spectral decompositions into plane waves. Specifically, the spherical wave in Eq. (4.101) consists of a spectrum of plane waves emanating from the source plane $z=z_0$ with a weighting factor i/k_z and incident upon the stratified medium at an angle θ such that $k_z=k\cos\theta$ (cf. Fig. 3.1). The reflected field in Eq. (4.102) is composed of a spectrum of plane waves emanating from the image source plane $z=-z_0$ with the same weighting factor (associated with the point source) and multiplied by the reflection coefficient at the angle θ, which incorporates the effect of the boundary. It is important to note that since $-\infty < k_x, k_y < \infty$, these spectral decompositions include both homogeneous (k_z real) and inhomogeneous (k_z imaginary) plane waves. Thus, the full spherical wave interaction with the boundary must be synthesized from the interactions of both propagating and exponentially decaying plane waves.

For hard or soft boundaries, where $R = \pm 1$, Eq. (4.102) becomes

$$p_r(\mathbf{r}) = \pm \frac{1}{2\pi} \int_{-\infty}^{\infty} \int_{-\infty}^{\infty} \frac{i}{k_z} e^{ik_z(z+z_0)} e^{i(k_x x + k_y y)} \, dk_x dk_y \ , \tag{4.103}$$

which, by comparison with Eq. (4.101), we can evaluate as

$$p_r(\mathbf{r}) = \pm \frac{e^{ikR_1}}{R_1}, \quad R_1 = \sqrt{x^2 + y^2 + (z + z_0)^2} \ , \tag{4.104}$$

where the positive sign corresponds to the hard interface. The path length R_1 is associated with the image source shown in Fig. 4.13 and can be interpreted as an arrival which is specularly reflected from the interface at the specular angle

$$\theta_0 = \tan^{-1}\left[\sqrt{x^2 + y^2} / (z + z_0)\right] \ . \tag{4.105}$$

Thus, in the case of impenetrable boundaries, the Green's functions obtained by the hybrid method reduce to the image method results. In general, however, for an arbitrary horizontally stratified medium, the reflection coefficient is not constant, and Eq. (4.102) is evaluated by approximate analytic methods, numerical techniques, or some combination of these two approaches. In the next section, for example, we will use asymptotic analysis to determine the reflected field from a homogeneous fluid half-space.

Finally, we point out that the cylindrical symmetry associated with this problem enables us to express the two-dimensional Fourier transforms in Eqs. (4.93) and (4.94) as one-dimensional Hankel transforms. Rewriting Eq. (4.93) in terms of cylindrical coordinates

$$\begin{aligned} k_x &= k_r \cos\alpha, \quad x = r\cos\beta, \\ k_y &= k_r \sin\alpha, \quad y = r\sin\beta, \\ k_r &= \sqrt{k_x^2 + k_y^2}, \quad r = \sqrt{x^2 + y^2} \ , \end{aligned} \tag{4.106}$$

we obtain

$$p(r) = \frac{1}{2\pi}\int_0^\infty \int_0^{2\pi} g(k_r) e^{ik_r r\cos(\alpha-\beta)} k_r \, dk_r d\alpha \quad , \tag{4.107}$$

which becomes

$$p(r) = \int_0^\infty g(k_r) J_0(k_r r) \, k_r \, dk_r \quad , \tag{4.108}$$

where we have used the following integral representation for the zero order Bessel function J_0 [Abramowitz and Stegun, 1964; Brekhovskikh, 1980]:

$$J_0(z) = \frac{1}{2\pi}\int_0^{2\pi} e^{iz\cos(\alpha-\beta)} \, d\alpha \quad . \tag{4.109}$$

Equation (4.108) is a *zero order Hankel transform* [Arfken, 1985] which has the inverse

$$g(k_r) = \int_0^\infty p(r) J_0(k_r r) r \, dr \quad , \tag{4.110}$$

corresponding to Eq. (4.94). The horizontal wavenumber k_r is defined as

$$k_r = \sqrt{k^2 - k_z^2} = k \sin\theta \quad . \tag{4.111}$$

We shall see that these Hankel transform representations are extremely useful in describing the wave field for both homogeneous and inhomogeneous media.

4.8.1 Asymptotic Analysis for Reflection of a Spherical Wave from a Homogeneous Fluid Half-Space

The asymptotic evaluation of Eq. (4.108) at high frequencies provides a physical interpretation for the interaction of a spherical wave with a horizontally stratified medium and illustrates the intimate connection between the characteristics of the reflected field and the analytic properties of the depth-dependent Green's function. The specific example of reflection from a homogeneous fluid half-space is particularly illuminating because it opens avenues of solution and physical understanding which can then be applied to more complicated cases. Our development is related to the work of Brekhovskikh [1980, 1982], Aki and Richards [1980], and Felsen and Marcuvitz [1973], and its mathematical underpinnings lie in the work of Erdelyi [1956] and Bleistein and Handelsman [1986].

We begin by putting Eq. (4.102) for the reflected field into a form suitable for the application of asymptotic methods. First, we rewrite Eq. (4.102) as a Hankel transform,

$$p_r(r)=\int_0^{\infty}\frac{i}{k_z}R(k_r)e^{ik_z(z+z_0)}J_0(k_r r)k_r\,dk_r \ , \tag{4.112}$$

and use the identity

$$J_0(k_r r)=\frac{1}{2}\left[H_0^{(1)}(k_r r)+H_0^{(2)}(k_r r)\right] \tag{4.113}$$

to obtain

$$p_r(r)=I_1(r)+I_2(r) \ , \tag{4.114}$$

where

$$I_1(r)=\frac{1}{2}\int_0^{\infty}\frac{i}{k_z}R(k_r)e^{ik_z(z+z_0)}H_0^{(1)}(k_r r)k_r\,dk_r,$$

$$I_2(r)=\frac{1}{2}\int_0^{\infty}\frac{i}{k_z}R(k_r)e^{ik_z(z+z_0)}H_0^{(2)}(k_r r)k_r\,dk_r \ . \tag{4.115}$$

In the expression for $I_2(r)$, we let $\xi=-k_r$, so that

$$I_2(r)=\frac{1}{2}\int_0^{-\infty}\frac{i}{k_z}R(-\xi)e^{ik_z(z+z_0)}H_0^{(2)}(-\xi r)\xi\,d\xi \ . \tag{4.116}$$

But $H_0^{(2)}(e^{-i\pi}z)=-H_0^{(1)}(z)$ [Abramowitz and Stegun, 1964] and $R(-\xi)=R(\xi)$ for the Rayleigh reflection coefficient in Eq. (3.33), and therefore we find that

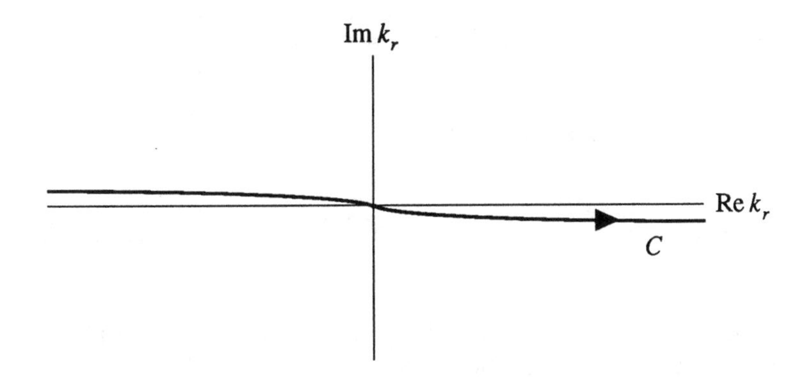

Figure 4.14 Path of integration C in the complex k_r-plane.

$$I_2(r) = \frac{1}{2}\int_{-\infty}^{0} \frac{i}{k_z} R(\xi) e^{ik_z(z+z_0)} H_0^{(1)}(\xi r)\xi \, d\xi \ , \tag{4.117}$$

and

$$p_r(r) = \frac{1}{2}\int_{-\infty}^{\infty} \frac{i}{k_z} R(k_r) e^{ik_z(z+z_0)} H_0^{(1)}(k_r r) k_r \, dk_r \ . \tag{4.118}$$

We note that the path of integration C is slightly displaced into the second and fourth quadrants [Sommerfeld, 1949], as shown in Fig. 4.14, in order that the radiation condition, corresponding to $\mathrm{Re}\, k_z > 0$, be satisfied in Eq. (4.118) (cf. Prob. 4.12). We now substitute the asymptotic form for the Hankel function [cf. Eq. (2.28)] for $k_r r >> 1$ (we shall see that this condition is satisfied for the region of the k_r-plane which dominates the integral),

$$H_0^{(1)}(k_r r) \sim \sqrt{\frac{2}{\pi k_r r}} e^{i(k_r r - \pi/4)} \ , \tag{4.119}$$

and obtain

$$p_r(r) \sim \frac{e^{i\pi/4}}{\sqrt{2\pi r}} \int_{-\infty}^{\infty} \frac{\sqrt{k_r}}{k_z} R(k_r) e^{ik_z(z+z_0)} e^{ik_r r} \, dk_r \ . \tag{4.120}$$

Finally, we change variables

$$K_r = \frac{k_r}{k} = \sin\theta, \; K_z = \frac{k_z}{k} = \cos\theta = \sqrt{1-K_r^2} \;, \qquad (4.121)$$

so that Eq. (4.120) becomes

$$p_r(r) \sim \frac{\sqrt{k}e^{i\pi/4}}{\sqrt{2\pi r}} \int_{-\infty}^{\infty} \frac{\sqrt{K_r}}{K_z} R(K_r) e^{ik[K_z(z+z_0)+K_r r]} \, dK_r \;. \qquad (4.122)$$

The integral $I(r)$ in Eq. (4.122) is now in a canonical form for the application of asymptotic methods for large values of the parameter k:

$$I(r) = \int_a^b f(K_r) e^{ikh(K_r)} dK_r \;, \qquad (4.123)$$

where, in this case, $a = -\infty$, $b = \infty$, and

$$f(K_r) = \frac{\sqrt{K_r}}{\sqrt{1-K_r^2}} R(K_r) = \frac{\sqrt{K_r}}{\sqrt{1-K_r^2}} \left[\frac{m\sqrt{1-K_r^2} - \sqrt{n^2 - K_r^2}}{m\sqrt{1-K_r^2} + \sqrt{n^2 - K_r^2}} \right],$$

$$h(K_r) = \sqrt{1-K_r^2}\,(z+z_0) + K_r r \;. \qquad (4.124)$$

Here, we recall that $m = \rho_1/\rho$, $n = k_1/k$, and ρ_1 and k_1 are the density and wavenumber, respectively, in the lower half-space.

The asymptotic behavior of integrals of the type shown in Eq. (4.123) is frequently determined by specific *critical points* which dominate the integral. In some instances, when these points are the endpoints of the integration interval, an integration by parts procedure can be used to evaluate the integral. In other cases, when the integral satisfies certain specific conditions, the critical points occur at *saddle points* or *stationary points* K_s such that $h'(K_s) = 0$, and saddle point methods are used in the evaluation. In general, the critical points of an integral cannot be identified a priori, and possible critical points are explored to see if a self-consistent result is obtained. From the theory of complex variables [Churchill and Brown, 1984], we know that the original path of integration can be deformed into a new path which passes through the critical points, as long as the contributions of any singularities (e.g., poles and branch points) encountered in the deformation process are taken into account. Therefore, the analytic properties of the integrand, specifically the reflection coefficient, play a critical role in both the mathematical evaluation of the integral and its subsequent physical interpretation.

Singularities of the Integrand

The singularities of the integrand in Eqs. (4.123) and (4.124) consist of branch points at the points $K_r = 0,\ \pm 1,\ \pm n$, and poles corresponding to zeros of the denominator of the reflection coefficient. The branch point at $K_r = 0$ can be avoided by a slight deformation of the original path of integration and is therefore of no consequence. For convenience, we select branch cuts emanating from the branch points $K_r = \pm 1$ and $K_r = \pm n$ which extend to infinity along the lines

$$\mathrm{Im}\sqrt{1-K_r^2} = 0, \quad \mathrm{Im}\sqrt{n^2-K_r^2} = 0 \ , \tag{4.125}$$

respectively (cf. Fig. 4.15). These are sometimes referred to as the *EJP cuts*, after Ewing, Jardetzky, and Press [1957]. For each of the square root functions in Eq. (4.125), the branch lines delineate two *Riemann sheets* which eliminate the multivalued nature of these functions and make them analytic. We define *physical sheets* corresponding to

$$\mathrm{Im}\sqrt{1-K_r^2} > 0, \quad \mathrm{Im}\sqrt{n^2-K_r^2} > 0 \ , \tag{4.126}$$

which ensure that the reflected field is bounded for $z \to \infty$ and the transmitted field is bounded for $z \to -\infty$. With these choices, all regions of the K_r-plane which we subsequently display correspond to physical sheets. By also requiring that at least the endpoints of the integration path lie on physical sheets in any contour deformation process, we guarantee that only physically meaningful solutions are obtained. It is left as an exercise for the reader (cf. Prob. 4.13) to show that the poles of the integrand do not contribute for the cases of interest to us because they either lie on unphysical sheets or do not interfere with the deformation of the integration path.

The asymptotic evaluation of the integral in Eq. (4.123) is extremely sensitive to the relative positions of the branch points and the critical point. We will therefore consider separately the two distinct cases $n > 1$ and $n < 1$.

Reflection from a Lower Velocity Half-Space ($n > 1$)

Equations (4.123) and (4.124) appear to be in a canonical form for the application of saddle point methods, and we therefore explore the ramifications of choosing the saddle point as the critical point. The saddle point K_s satisfies

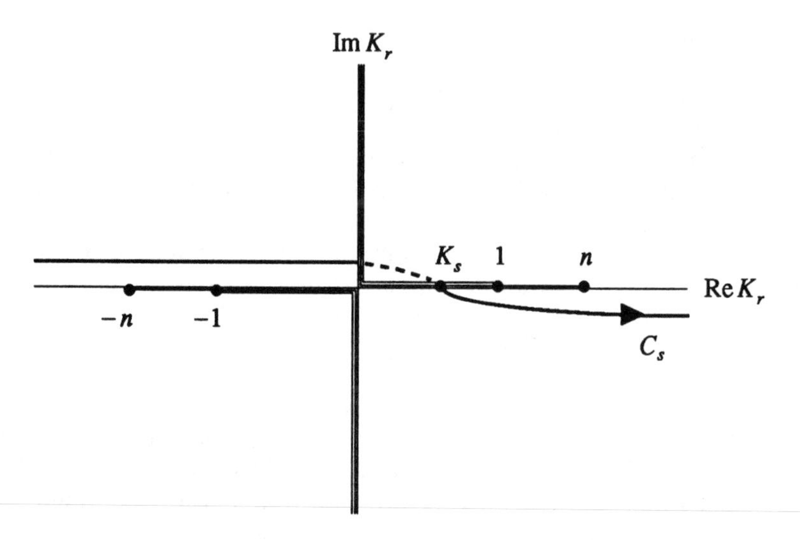

Figure 4.15 Saddle point contour C_s, branch points, and branch cuts in the complex K_r-plane for spherical wave reflection from a lower velocity half-space ($n>1$). Dashed line indicates portion of integration path which lies on unphysical Riemann sheets.

$$h'(K_s) = -\frac{K_s}{\sqrt{1-K_s^2}}(z+z_0)+r = -(z+z_0)\tan\theta_s + r = 0 \quad , \quad (4.127)$$

where we have defined $K_s = \sin\theta_s$. It is clear from Eqs. (4.105) and (4.127) that the saddle point occurs at the point $K_s = \sin\theta_0$, where θ_0 is the specular angle. The original path of integration C may be deformed into a path C_s which passes through the saddle point, as shown in Fig. 4.15. Since this new path intersects both branch cuts twice with its endpoints lying on physical sheets, we are assured of a physically meaningful solution. This path begins on physical sheets in the second quadrant, intersects the branch cuts at the ImK_r axis, drops down to unphysical sheets, intersects the cuts again at the ReK_r axis, and reemerges on physical sheets, passing through the saddle point. Along this path, the functions $f(K_r)$ and $h(K_r)$ are well-behaved, analytic functions as long as $K_s << 1$ (thereby avoiding the branch point at $K_r = 1$), and saddle point methods may in fact be used for the evaluation.

The subsequent calculation corresponds to a modified version of the *method of stationary phase*. First, we note that for large k and $|K_r| < 1$, the rapidly oscillating nature of the phase factor $\exp[ikh(K_r)]$ in the integrand of Eqs. (4.123) and (4.124) will cause the positive and negative contributions to the integral to cancel one another everywhere except in the vicinity of the saddle point, where the phase is stationary. These contributions are further diminished for $|K_r| > 1$ because of the exponentially decaying factor introduced by the square root term in $h(K_r)$:

$$h(K_r) = i\sqrt{K_r^2 - 1}\left(z + z_0\right) + K_r r \quad \text{for } |K_r| > 1 \ . \tag{4.128}$$

Since the dominant contributions to the integral occur in the neighborhood of the saddle point, we expand the integrand in a Taylor series about this point, keeping one higher order term in the phase $h(K_r)$ than in the amplitude $f(K_r)$,

$$\begin{aligned} f(K_r) &\approx f(K_s), \\ h(K_r) &\approx h(K_s) + \frac{1}{2} h''(K_s)(K_r - K_s)^2 \ , \end{aligned} \tag{4.129}$$

so that Eq. (4.123) becomes

$$I(r) \approx f(K_s) e^{ikh(K_s)} \int_{-\infty}^{\infty} e^{i\frac{k}{2}h''(K_s)(K_r - K_s)^2} dK_r \ . \tag{4.130}$$

Changing variables,

$$\tau = K_r - K_s \Rightarrow d\tau = dK_r \ , \tag{4.131}$$

we find that

$$I(r) \approx f(K_s) e^{ikh(K_s)} \int_{-\infty}^{\infty} e^{i\frac{k}{2}h''(K_s)\tau^2} d\tau \ , \tag{4.132}$$

and finally, using Eq. (2.67), obtain

$$I(r) \approx f(K_s) \sqrt{\frac{2\pi}{kh''(K_s)}} e^{i\pi/4} e^{ikh(K_s)} \ . \tag{4.133}$$

For the specific case under consideration, we have

$$h(K_s) = (z + z_0)\cos\theta_0 + r\sin\theta_0 = R_1 \ , \tag{4.134}$$

since, from Fig. 4.13, it is clear that

$$\sin\theta_0 = r/R_1, \;\; \cos\theta_0 = (z+z_0)/R_1 \;\; . \tag{4.135}$$

It is also straightforward to show that

$$f(K_s) = \frac{\sqrt{\sin\theta_0}}{\cos\theta_0} R(\theta_0), \;\; h''(K_s) = \frac{-R_1}{\cos^2\theta_0} \;\; , \tag{4.136}$$

so that the reflected field in Eq. (4.122) becomes

$$p_r(r) \sim R(\theta_0)\frac{e^{ikR_1}}{R_1} \;\; . \tag{4.137}$$

Physically, Eq. (4.137) states that the reflected field consists of a spherical wave emanating from the image source and multiplied by the plane wave reflection coefficient evaluated at the specular angle. This result is called the *geometrical acoustics approximation* to the total reflected field. It shows that the energy in the reflected field is concentrated about the specular angle, even though the incident spherical wave contains a spectrum of plane waves distributed over a broad range of angles. Mathematically, this specularly reflected arrival corresponds to the first term in an *asymptotic expansion* of the field in inverse powers of k. Additional terms in the expansion can be obtained by keeping higher order terms in the Taylor expansion of the integrand in Eq. (4.123). Finally, the case of grazing incidence ($K_s \to 1$) can be treated by using *uniform asymptotic expansions* [Bleistein and Handelsman, 1986], which are uniformly valid for arbitrary values of the saddle point, including regions containing singularities.

Reflection from a Higher Velocity Half-Space ($n < 1$)

In this case, for plane wave incidence, we know that total internal reflection can occur, and this effect significantly complicates the spherical wave analysis. The relative positions of the branch points at $K_r = \pm 1$ and $K_r = \pm n$ are reversed when compared with those in the previous example (cf. Fig. 4.16), and the location of the saddle point (which still occurs at the specular angle) in relation to the corresponding branch cut structure plays a critical role in determining the nature of the solution.

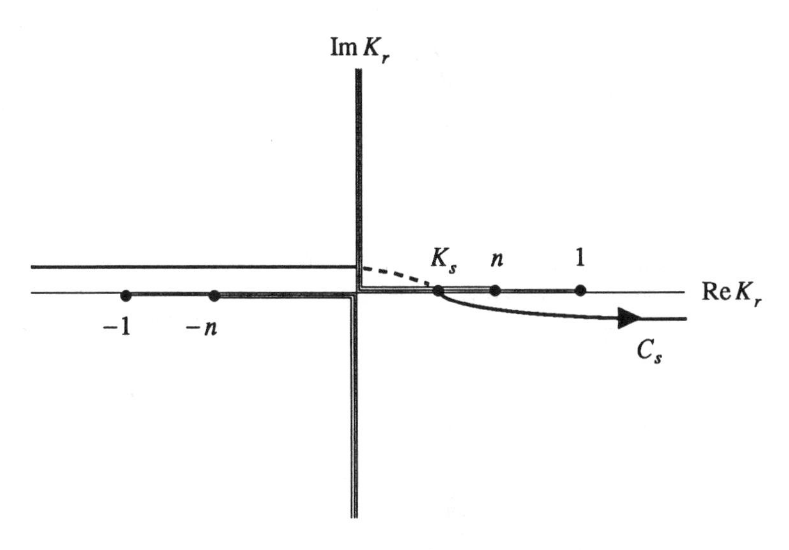

Figure 4.16 Saddle point contour C_s, branch points, and branch cuts in the complex K_r-plane for spherical wave reflection from a higher velocity half-space (n<1) at pre-critical angles (K_s<<n). Dashed line indicates portion of integration path which lies on unphysical Riemann sheets.

Reflection at Pre-Critical Angles: When the specular angle is much less than the critical angle ($K_s << n$), as shown in Fig. 4.16, the original integration path can be deformed into the saddle point contour in a manner analogous to that described in the previous section for $n > 1$. The deformed path intersects both branch cuts twice, and only the saddle point contributes to the final result, given by Eq. (4.137).

Reflection at Post-Critical Angles and the Lateral Wave: When the specular angle is much greater than the critical angle ($n << K_s << 1$), and we attempt a simple contour deformation of the type described in the previous cases (cf. Fig. 4.17), a complication arises. We see that while the branch cut originating at $K_r = 1$ is crossed twice by the deformed contour, the branch cut associated with $K_r = n$ is crossed only once. The saddle point contour in Fig. 4.17 is therefore unacceptable because it violates the condition that the endpoints of the integration path must lie on physical sheets. Consequently, we search for an alternative contour and obtain the result shown in Fig. 4.18. Here, for clarity, we display the intersection of the deformed path with the physical and unphysical sheets for each branch point separately. The total reflected field is therefore composed of two terms,

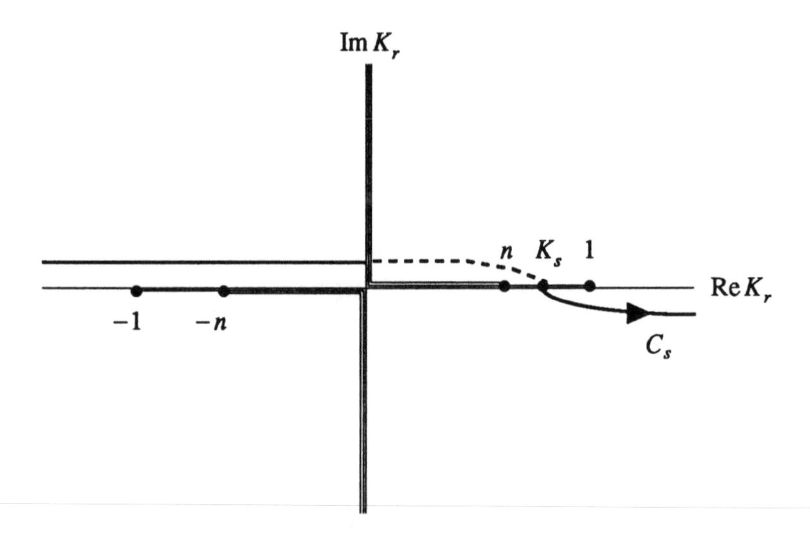

Figure 4.17 Unacceptable saddle point contour C_s, branch points, and branch cuts in the complex K_r-plane for spherical wave reflection from a higher velocity half-space ($n<1$) at post-critical angles ($n<<K_s<<1$). Dashed line indicates portion of integration path which lies on the unphysical Riemann sheet for the branch point $K_r=1$.

$$p_r(r) \sim \frac{\sqrt{k}e^{i\pi/4}}{\sqrt{2\pi r}} \int_{-\infty}^{\infty} \frac{\sqrt{K_r}}{K_z} R(K_r) e^{ik[K_z(z+z_0)+K_r r]} \, dK_r$$
$$+\frac{\sqrt{k}e^{i\pi/4}}{\sqrt{2\pi r}} \int_{C_B} \frac{\sqrt{K_r}}{K_z} R(K_r) e^{ik[K_z(z+z_0)+K_r r]} \, dK_r \quad , \quad (4.138)$$

with the first term yielding the specularly reflected result as before, and the second term, corresponding to integration along the path C_B around the branch line originating at $K_r = n$, giving rise to the lateral wave $p_L(r)$:

$$p_r(r) \sim R(\theta_0) \frac{e^{ikR_1}}{R_1} + p_L(r) \quad . \quad (4.139)$$

The up- and down-going portions of the branch line integral along the $\text{Im}\,K_r$ axis cancel out one another (cf. Fig. 4.18), since the integrand is continuous in passing from the physical to the unphysical sheets. The part of the integral which remains to be computed is due to the discontinuous behavior of the integrand across the branch line emanating from $K_r = n$ in the interval $0 \le K_r \le n$:

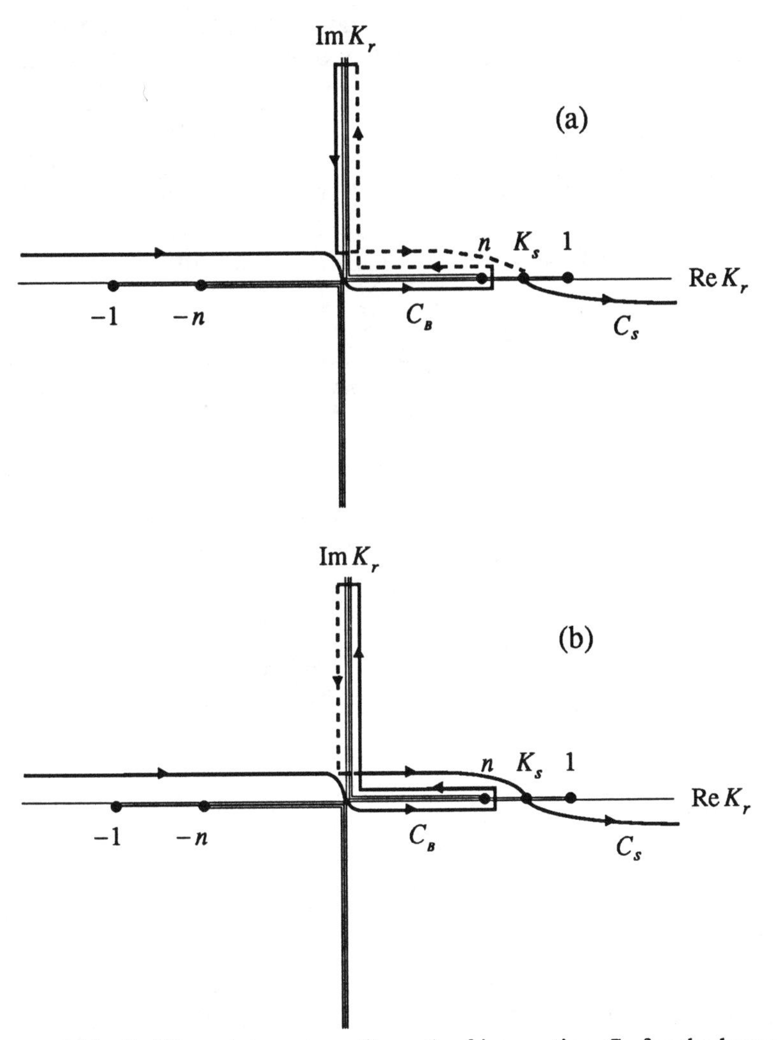

Figure 4.18 Saddle point contour C_s, path of integration C_B for the branch line integral, branch points, and branch cuts in the complex K_r-plane for spherical wave reflection from a higher velocity half-space (n<1) at post-critical angles (n<<K_s<<1). Dashed lines indicate portions of integration path which lie on unphysical Riemann sheets for the branch points (a) K_r=1 and (b) K_r=n.

$$p_L(r) \sim \frac{\sqrt{k}e^{i\pi/4}}{\sqrt{2\pi r}} \int_0^n \frac{\sqrt{K_r}}{K_z} R_+(K_r) e^{ik[K_z(z+z_0)+K_r r]}\, dK_r$$

$$+ \frac{\sqrt{k}e^{i\pi/4}}{\sqrt{2\pi r}} \int_n^0 \frac{\sqrt{K_r}}{K_z} R_-(K_r) e^{ik[K_z(z+z_0)+K_r r]}\, dK_r , \qquad (4.140)$$

where $R_+(K_r)$ and $R_-(K_r)$ are the values of the reflection coefficient on the lower and upper sides of the cut, respectively (cf. Prob. 4.12):

$$R_+ = \frac{m\sqrt{1-K_r^2} - \sqrt{n^2 - K_r^2}}{m\sqrt{1-K_r^2} + \sqrt{n^2 - K_r^2}}, \quad R_- = \frac{m\sqrt{1-K_r^2} + \sqrt{n^2 - K_r^2}}{m\sqrt{1-K_r^2} - \sqrt{n^2 - K_r^2}} \quad . \tag{4.141}$$

Substituting Eq. (4.141) into Eq. (4.140), we obtain

$$p_L(r) \sim -\frac{\sqrt{k}e^{i\pi/4}}{\sqrt{2\pi r}} \int_0^n \frac{\sqrt{K_r}}{K_z} \left[\frac{4mK_z K_{1z}}{m^2 K_z^2 - K_{1z}^2} \right] e^{ik[K_z(z+z_0)+K_r r]} \, dK_r \quad , \tag{4.142}$$

where

$$K_{1z} = \sqrt{n^2 - K_r^2} = \sqrt{n^2 - 1 + K_z^2} \quad . \tag{4.143}$$

Changing variables,

$$K_z = \sqrt{1-K_r^2} \Rightarrow dK_z = -\frac{K_r}{K_z} dK_r \quad , \tag{4.144}$$

we find that Eq. (4.142) becomes

$$p_L(r) \sim -\frac{\sqrt{k}e^{i\pi/4}}{\sqrt{2\pi r}} \int_{\sqrt{1-n^2}}^1 \frac{1}{\sqrt{K_r}} \left[\frac{4mK_z K_{1z}}{m^2 K_z^2 - K_{1z}^2} \right] e^{ik[K_z(z+z_0)+K_r r]} \, dK_z \, . \tag{4.145}$$

Extending the upper limit of integration to ∞ in Eq. (4.145),

$$p_L(r) \sim -\frac{\sqrt{k}e^{i\pi/4}}{\sqrt{2\pi r}} \int_{\sqrt{1-n^2}}^\infty \frac{1}{\sqrt{K_r}} \left[\frac{4mK_z K_{1z}}{m^2 K_z^2 - K_{1z}^2} \right] e^{ik[K_z(z+z_0)+K_r r]} \, dK_z \, , \tag{4.146}$$

adds only a negligible contribution to the integral because, for $K_z > 1$, the phase term oscillates rapidly and contains an exponentially decaying factor :

$$h(K_z) = K_z(z + z_0) + i\sqrt{K_z^2 - 1}\, r \quad \text{for } K_z > 1 \quad . \tag{4.147}$$

The integral $I(r)$ in Eq. (4.146) is in a canonical form [cf. Eq. (4.123)] for the application of asymptotic methods for large values of the parameter k:

$$I(r)=\int_{\sqrt{1-n^2}}^{\infty} f(K_z)e^{ikh(K_z)}dK_z \ , \tag{4.148}$$

where

$$f(K_z)=\frac{1}{\left(1-K_z^2\right)^{1/4}}\left[\frac{4mK_z\sqrt{n^2-1+K_z^2}}{m^2K_z^2-\left(n^2-1+K_z^2\right)}\right],$$

$$h(K_z)=K_z(z+z_0)+\sqrt{1-K_z^2}\,r \ . \tag{4.149}$$

However, the saddle point, which is again given by Eq. (4.127), is not the critical point in this case. Instead, we consider the possibility that the critical point K_c is the branch point corresponding to the endpoint of the path of integration:

$$K_c=\sqrt{1-n^2} \ . \tag{4.150}$$

We therefore expand $h(K_z)$ in a Taylor series about the critical point,

$$h(K_z)\approx h(K_c)+h'(K_c)(K_z-K_c) \ , \tag{4.151}$$

and, using Eq. (4.147), obtain

$$h(K_z)\approx K_c(z+z_0)+\sqrt{1-K_c^2}\,r+\left[z+z_0-\frac{K_c}{\sqrt{1-K_c^2}}r\right](K_z-K_c). \tag{4.152}$$

We approximate $f(K_z)$ close to the critical point by noting first that

$$K_{1z}=\sqrt{K_z^2-K_c^2}=\sqrt{K_z-K_c}\sqrt{K_z+K_c} \ , \tag{4.153}$$

and therefore

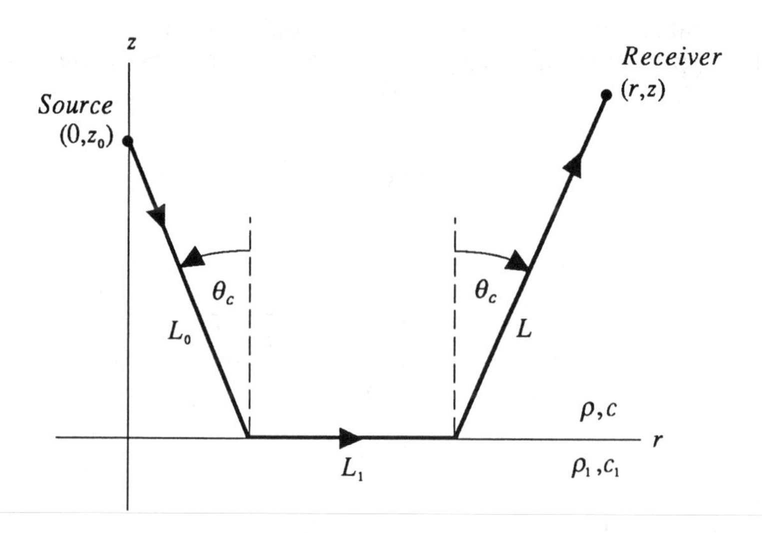

Figure 4.19 Lateral wave excitation at the critical angle θ_c for the case of spherical wave reflection from a higher velocity half-space ($n<1$) at post-critical angles ($n<<K_s<<1$).

$$K_{1z}\big|_{K_z\approx K_c} \approx \sqrt{2K_c}\sqrt{K_z-K_c} \ , \tag{4.154}$$

so that

$$f(K_z)\big|_{K_z\approx K_c} \approx \frac{1}{\left(1-K_c^2\right)^{1/4}}\frac{4\sqrt{2}\sqrt{K_z-K_c}}{m\sqrt{K_c}} \ . \tag{4.155}$$

The branch line integral in Eq. (4.146) then becomes

$$p_L(r) \sim -\frac{\sqrt{k}e^{i\pi/4}}{\sqrt{\pi r}}\frac{4}{\left(1-K_c^2\right)^{1/4}m\sqrt{K_c}}e^{ik\overbrace{\left[K_c(z+z_0)+\sqrt{1-K_c^2}\,r\right]}^{Term\ I}}$$

$$\times\int_{\sqrt{1-n^2}}^{\infty}\sqrt{K_z-K_c}\,e^{ik\overbrace{\left[z+z_0-K_cr/\sqrt{1-K_c^2}\right]}^{Term\ II}(K_z-K_c)}\,dK_z , \tag{4.156}$$

where the phase terms can be simplified by recognizing that the critical angle θ_c satisfies

$$K_c = \cos\theta_c = \sqrt{1-n^2}, \quad \sqrt{1-K_c^2} = \sin\theta_c = n, \qquad (4.157)$$

and by using the geometry shown in Fig. 4.19:

$$\begin{aligned} Term\, I &= k(z+z_0)\cos\theta_c + k_1\left[L_1 + (z+z_0)\tan\theta_c\right] \\ &= k(L+L_0)\cos^2\theta_c + k_1\left[L_1 + (L+L_0)\sin\theta_c\right] \\ &= k(L+L_0)\cos^2\theta_c + k_1 L_1 + k(L+L_0)\sin^2\theta_c \\ &= k(L+L_0) + k_1 L_1, \end{aligned} \qquad (4.158)$$

$$\begin{aligned} Term\, II &= -k\frac{K_c}{\sqrt{1-K_c^2}}\left[r - \frac{\sqrt{1-K_c^2}}{K_c}(z+z_0)\right] \\ &= -k\frac{K_c}{\sqrt{1-K_c^2}}\left[r-(z+z_0)\tan\theta_c\right] \\ &= -k\frac{K_c}{\sqrt{1-K_c^2}}L_1 \ . \end{aligned} \qquad (4.159)$$

Substituting Eqs. (4.158) and (4.159) into Eq. (4.156), we obtain

$$\begin{aligned} p_L(r) \sim -\frac{\sqrt{k}e^{i\pi/4}}{\sqrt{\pi r}}\frac{4}{\left(1-K_c^2\right)^{1/4} m\sqrt{K_c}} e^{i\left[k(L+L_0)+k_1L_1\right]} \\ \times \int_{\sqrt{1-n^2}}^{\infty} \sqrt{K_z - K_c}\, e^{-ik\left[K_c/\sqrt{1-K_c^2}\right]L_1(K_z-K_c)}\, dK_z \ , \end{aligned} \qquad (4.160)$$

and changing variables,

$$\tau = \sqrt{K_z - K_c} \Rightarrow d\tau = \frac{1}{2}\frac{dK_z}{\sqrt{K_z-K_c}} = \frac{1}{2\tau}dK_z \ , \qquad (4.161)$$

we find that Eq. (4.160) becomes

$$p_L(r) \sim -\frac{\sqrt{k}e^{i\pi/4}}{\sqrt{\pi r}} \frac{8}{\left(1-K_c^2\right)^{1/4} m\sqrt{K_c}} e^{i[k(L+L_0)+k_1 L_1]}$$

$$\times \int_0^\infty \tau^2 e^{-ik\left[K_c/\sqrt{1-K_c^2}\right]L_1\tau^2} d\tau \ . \qquad (4.162)$$

Using Eq. (2.67), we determine that

$$\int_0^\infty x^2 e^{-\alpha x^2} dx = -\frac{d}{d\alpha}\int_0^\infty e^{-\alpha x^2} dx = -\frac{d}{d\alpha}\left[\frac{1}{2}\sqrt{\frac{\pi}{\alpha}}\right] = \frac{1}{4}\frac{\sqrt{\pi}}{\alpha^{3/2}} \ , \qquad (4.163)$$

and therefore the result for the branch line integral in Eq. (4.162) is

$$p_L(r) \sim \frac{2in}{km(1-n^2)\sqrt{r}L_1^{3/2}} e^{i[k(L+L_0)+k_1 L_1]} \ , \qquad (4.164)$$

which agrees with the results of Brekhovskikh and Lysanov [1991] and Stickler [1976].

From an examination of the phases in Eq. (4.164) and the geometry in Fig. 4.19, it is clear that the branch line contribution corresponds to a wave which is excited at the critical angle, propagates along the interface at the speed in the lower medium, and radiates back into the upper medium, via a *head wave* [Grant and West, 1965; Überall, 1973], at the same angle. Thus, the particular spectral component of the incoming spherical wave which is incident at the critical angle is singled out and gives rise to the lateral wave. It constitutes a *diffracted component* of the total reflected field in that its amplitude cannot be obtained from simple geometrical acoustic considerations. At long ranges ($r >> z + z_0$), the amplitude of the lateral wave falls off as $1/r^2$, in contrast to the direct and specularly reflected arrivals, which decrease as $1/r$. On the other hand, under certain conditions for an impulsive source, the lateral wave may arrive at the receiver before the direct or specularly reflected waves (cf. Prob. 4.15).

Finally, the cases of grazing incidence ($K_s \to 1$) and incidence at the critical angle ($K_s \to n$) can both be treated by using uniform asymptotic expansions. In the latter case, it is important to note that no terms in the reflected field explicitly contain the plane wave reflection coefficient [Stickler, 1976, 1977].

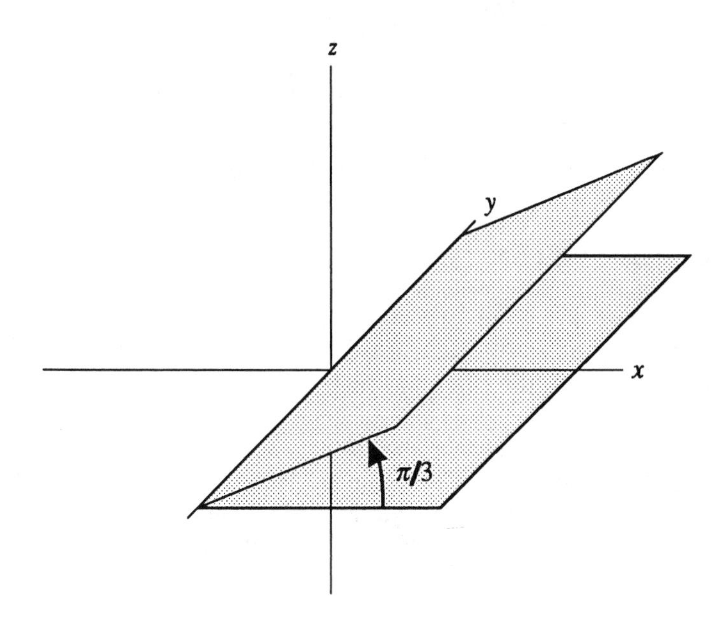

Figure 4.20 Wedge geometry for Prob. 4.6.

PROBLEMS

4.1 Verify the following properties of the one-dimensional free-space Green's function $G_0(x,x_0)$ in Eq. (4.11):

(a) The discontinuity in its first derivative at $x = x_0$ is equal to 4π.

(b) Its second derivative is continuous at $x = x_0$.

4.2 Given the inhomogeneous Helmholtz equation in Eq. (4.9), Green's theorem in Eq. (4.25), and the homogeneous boundary conditions in Eq. (4.13), prove the reciprocity relation in Eq. (4.19).

4.3 Verify that Eqs. (4.46) and (4.47) correspond to the correct Green's function for the problem of a quadrant with Dirichlet conditions described in Sec. 4.6.3. Write down the formal solution for the pressure in the quadrant, expressing any surface integrals as integrals over the boundary values and Green's function evaluated on the boundaries.

4.4 What is the Green's function for a quadrant with Neumann boundary conditions?

4.5 Consider a point source of positive unit strength at the position $\mathbf{r}_a = (x_a, y_a, z_a)$ and a point source of negative unit strength at the position $\mathbf{r}_b = (x_b, y_b, z_b)$, both of which are located in the first quadrant, $x \geq 0$, $z \geq 0$ (cf. Fig. 4.5). We are given the boundary values $p(\mathbf{r}) = 0$ on the y-z plane and $\partial p(\mathbf{r})/\partial n = b(x,y)$ on the x-y plane.

(a) Calculate the Green's function for this problem.

(b) Calculate the total pressure in the quadrant, expressing any surface integrals as integrals over the boundary values and Green's function evaluated on the boundaries.

4.6 Determine the Green's function for the interior of a 60° wedge with pressure-release boundaries, as shown in Fig. 4.20. Note that the problem of a wedge with impenetrable boundaries has a solution with a finite number of images as long as the wedge angle is π times a rational number.

4.7 Provide a pictorial, physical interpretation for the $n = 0$ and $n = 1$ terms in Eqs. (4.79) and (4.80) which shows that they correspond to arrivals at the receiver which have undergone a specific number of reflections from the boundaries.

4.8 Derive the image method solution in Eq. (4.82) for the zeroth order model of the ocean acoustic waveguide in Fig. 4.12.

4.9 Derive the one-dimensional, free-space Green's function in Eq. (4.11) using the endpoint method.

4.10 The two-dimensional, free-space Green's function is given by [Sommerfeld, 1949]

$$G_0(\mathbf{r},\mathbf{r}_0) = i\pi H_0^{(1)}\left(k|\mathbf{r} - \mathbf{r}_0|\right) \ , \tag{4.165}$$

where $\mathbf{r} = (x,z)$ and $\mathbf{r}_0 = (x_0, z_0)$, and corresponds to the radiation from a line source (cf. Sec. 2.4) of infinite extent in the y direction in Fig. 4.13. There are several useful integral representations for this Green's function. Probably the most physical one arises from the integral over an infinite line array of point sources:

$$G_0(\mathbf{r},\mathbf{r}_0) = i\pi H_0^{(1)}(kr) = \int_{-\infty}^{\infty} \frac{e^{ikR}}{R} dy_0 = \int_{-\infty}^{\infty} \frac{e^{ik\sqrt{r^2+y_0^2}}}{\sqrt{r^2+y_0^2}} dy_0, \quad (4.166)$$

where $r = \sqrt{(x-x_0)^2 + (z-z_0)^2}$.

Using Fourier transforms and the endpoint method, derive the following plane wave expansion for the two-dimensional, free-space Green's function for a line source located at $\mathbf{r}_0 = (0, z_0)$:

$$G_0(\mathbf{r},\mathbf{r}_0) = \int_{-\infty}^{\infty} \frac{i}{k_z} e^{ik_z|z-z_0|} e^{ik_x x}\, dk_x \ . \quad (4.167)$$

4.11 The nature of pulse propagation is dependent on the dimensionality of the space, even for a homogeneous medium [Morse and Feshbach, 1953]. In order to illustrate this effect, we will examine the propagation of a delta function impulse in both space and time originating at $t = 0$. In doing so, we will determine the free-space Green's functions $G_0(\mathbf{r},\mathbf{r}_0,t)$ for the time-dependent acoustic wave equation in Eq. (4.1).

Hint: Use the Fourier transform to determine the time-dependent behavior.

(a) Show that in three dimensions, the Green's function is given by

$$G_0(\mathbf{r},\mathbf{r}_0,t) = \frac{\delta\left(|\mathbf{r}-\mathbf{r}_0|/c - t\right)}{|\mathbf{r}-\mathbf{r}_0|} , \quad (4.168)$$

and therefore an impulse propagates without distortion.

(b) Using Eq. (4.166), show that in two dimensions, we obtain

$$G_0(\mathbf{r},\mathbf{r}_0,t) = \begin{cases} \dfrac{2c}{\sqrt{c^2t^2 - r^2}}, & r < ct, \\ 0, & r > ct \end{cases} , \quad (4.169)$$

and therefore the shape of the pulse changes with time. Thus, the propagation in this case is *dispersive* in nature, as if the medium had frequency-dependent properties.

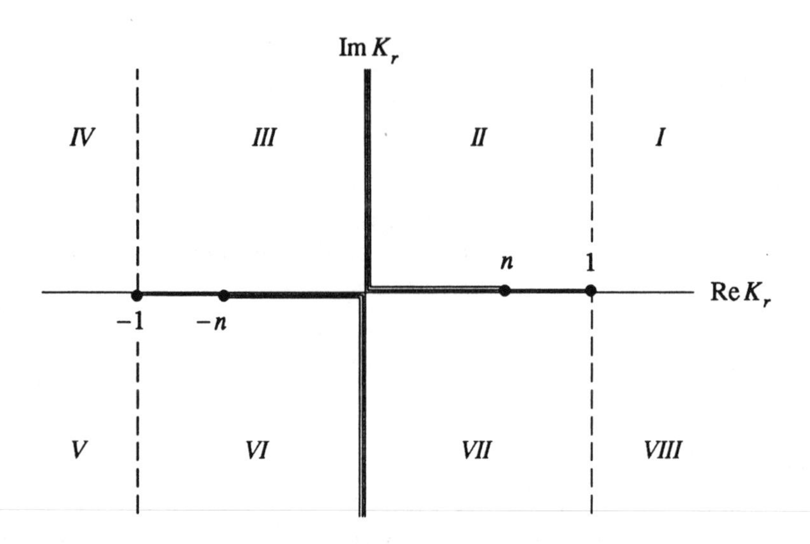

Figure 4.21 Regions in the complex K_r-plane for Prob. 4.12.

(c) Show that in one dimension, the result is

$$G_0(x,x_0,t)=\begin{cases} 2\pi c, & |x-x_0|/c<t, \\ 0, & |x-x_0|/c>t \end{cases} , \tag{4.170}$$

so that the pulse is again distorted, being uniformly distributed for times $|x-x_0|/c<t$.

Hint: Introduce a small amount of attenuation (cf. Prob. 3.4) and evaluate the integral using Cauchy's residue theorem [Churchill and Brown, 1984]. Let the attenuation go to zero at the end of the calculation.

(d) Provide physical interpretations for the behavior of the Green's functions in the above three cases.

Note that all three solutions incorporate the principle of *causality*, that is, the wave field arrives at the receiver only for positive times which are consistent with the finite speed of propagation in the medium.

4.12 Consider the branch cut structure in the complex K_r-plane associated with spherical wave reflection from a higher velocity half-space, as shown in Fig. 4.21. Recall that the physical sheets $\mathrm{Im}K_z > 0$ and $\mathrm{Im}K_{1z} > 0$ are displayed in the figure. Determine the sign of $\mathrm{Re}K_z$ in each of the eight regions shown. In doing so, demonstrate that:

(a) The Sommerfeld radiation condition of outgoing, radiating waves at infinity, corresponding to $\mathrm{Re}K_z > 0$, is satisfied only in the second and fourth quadrants (regions III, IV, VII, and VIII), thereby justifying the choice of integration path in Fig. 4.14.

(b) The function $K_z = \sqrt{1-K_r^2}$ is discontinuous when a branch cut is crossed while remaining on the same Riemann sheet.

Hint: For a particular region, the sign of $\mathrm{Re}K_z$ can be determined by examining the leading order behavior in the binomial expansion [cf. Eq. (4.50)] of K_z as a function of K_r for values of K_r which contain a small imaginary part. For example, in region II, we let

$$K_r = \alpha + i\varepsilon \text{ with } 0 < \alpha < 1,\ \ 0 < \varepsilon << 1 \ , \qquad (4.171)$$

so that

$$K_z = \pm\sqrt{1-K_r^2} = \pm\sqrt{1-(\alpha+i\varepsilon)^2}$$
$$\approx \pm\left[1-\frac{(\alpha+i\varepsilon)^2}{2}\right] \approx \pm\left[1-\frac{\alpha^2}{2}-i\varepsilon\alpha\right] . \qquad (4.172)$$

We choose the negative sign in Eq. (4.172) because of the requirement $\mathrm{Im}K_z > 0$, and therefore we obtain the result $\mathrm{Re}K_z < 0$.

4.13 Show that the poles of the integrand do not contribute to the field in the cases of spherical wave reflection from a lower velocity half-space and a higher velocity half-space.

4.14 Apply the asymptotic analysis described in Sec. 4.8.1 to the case of a line source and $n < 1$. Specifically, calculate the specularly reflected and lateral wave contributions to the field.

4.15 For the case of a spherical impulse incident upon a higher velocity half-space at post-critical angles, determine the value of n below which the lateral wave will arrive at the receiver before the direct wave when $z = z_0$ and $r = 4z$. (Ignore the dispersion of the lateral wave.)

Chapter 5. The Method of Normal Modes

That somber theme had to be given a sinister resonance,
a tonality of its own, a continued vibration that, I
hoped, would hang in the air and dwell on the
ear after the last note had been struck.
Joseph Conrad, 1902

5.1 Introduction

In the previous chapter, we discussed several techniques for the construction of the Green's function for the inhomogeneous Helmholtz equation. There is still another approach which is so significant in terms of its physical and mathematical ramifications that its treatment deserves a separate chapter. This is the method of normal modes, which is equivalent to construction of the Green's function using an eigenfunction expansion. Any wave-mechanical system has associated with it a set of natural oscillations or vibrations which are solutions to the homogeneous wave equation and the imposed boundary conditions. The response of this system to an arbitrary driving force can be expressed in terms of these normal modes or eigenfunctions. The fundamental constituents of the solution to the inhomogeneous wave equation are therefore these free modes of vibration or resonances, and only their amplitudes will vary depending upon the particular form of excitation. The ringing of a bell and the plucking of a violin string are examples of modal excitations which we experience in our everyday lives. Equally significant, though more abstract, are the quantum mechanical *bound states* of an atom or the acoustic modes of the oceanic waveguide excited perhaps by a surface ship or a source located deep in the sea.

First, we develop the theory for a point source in a horizontally stratified, fluid medium, illustrating the dependence of the classical technique on separability and Sturm-Liouville theory. We then apply the method to the case of a point source in a homogeneous fluid layer with a soft top and hard bottom, which is the zeroth order model of the ocean acoustic waveguide discussed in Sec. 4.6.5. This example provides the opportunity to introduce the general concepts of modal phase and group velocities, cutoff frequency, interference wavelength, and cycle distance. We show that in the limit of infinite layer thickness, the discrete spectrum of modes becomes continuous. Proceeding to more complicated examples, we determine the eigenvalue equation for a homogeneous fluid layer bounded by arbitrary horizontally stratified media. We then derive the Green's function for a homogeneous fluid layer bounded above by a pressure-release surface and below by a lower velocity, homogeneous fluid half-space. We show that the field is purely a modal continuum which arises due to imperfect trapping, and can be approximated by a sum of discrete, improper (or leaky) modes. We conclude with the derivation of the Green's function for a homogeneous fluid layer bounded above by a pressure-release surface and below by a higher velocity, homogeneous fluid half-space (the Pekeris waveguide). In this case, we show that the modal field is composed of both discrete and continuous spectra, and that the continuous spectrum can again be approximated by a sum of discrete, improper modes. We also compute the phase and group velocities, cutoff frequency, interference wavelength, and cycle distance for the perfectly trapped, Pekeris waveguide modes.

5.2 A Point Source in a Horizontally Stratified, Fluid Medium

We consider a point source with cylindrical coordinates $\mathbf{r}_0 = (0,0,z_0)$ in a horizontally stratified, fluid medium with density $\rho(z)$ and sound velocity $c(z)$, as shown in Fig. 5.1. Then the inhomogeneous, time-independent wave equation for density and sound velocity stratification [cf. Eq. (1.45)] is

$$\rho(z)\nabla \bullet \left[\frac{1}{\rho(z)}\nabla p(\mathbf{r})\right] + k^2(z)p(\mathbf{r}) = -4\pi\frac{\delta(r)}{r}\delta(\theta)\delta(z-z_0) \quad , \tag{5.1}$$

where $\mathbf{r} = (r,\theta,z)$ and the wavenumber $k(z) = \omega/c(z)$. Equation (5.1) can be rewritten as

$$\nabla^2 p(\mathbf{r}) + \rho(z)\nabla\left[\frac{1}{\rho(z)}\right] \bullet \nabla p(\mathbf{r}) + k^2(z)p(\mathbf{r})$$
$$= -4\pi\frac{\delta(r)}{r}\delta(\theta)\delta(z-z_0) \quad , \tag{5.2}$$

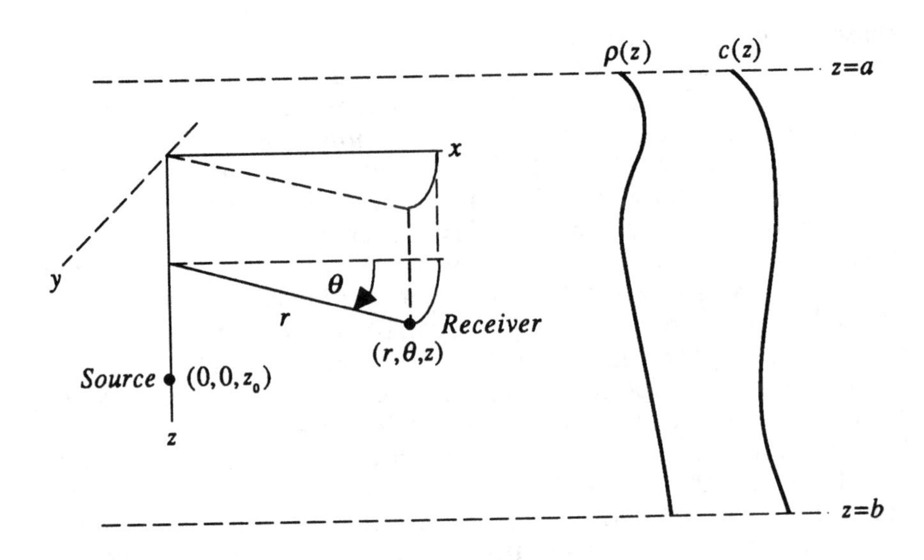

Figure 5.1 Geometry for construction of the normal mode solution.

where

$$\nabla = \mathbf{i}_r \frac{\partial}{\partial r} + \mathbf{i}_\theta \frac{1}{r}\frac{\partial}{\partial \theta} + \mathbf{i}_z \frac{\partial}{\partial z},$$

$$\nabla^2 = \frac{1}{r}\frac{\partial}{\partial r}\left(r\frac{\partial}{\partial r}\right) + \frac{1}{r^2}\frac{\partial^2}{\partial \theta^2} + \frac{\partial^2}{\partial z^2} , \tag{5.3}$$

and $\mathbf{i}_r$, $\mathbf{i}_\theta$, and $\mathbf{i}_z$ are unit vectors in the r, θ, and z directions, respectively. Horizontal stratification and the assumption that the source is at $\mathbf{r}_0 = (0,0,z_0)$ imply cylindrical symmetry in the field about the z axis. Therefore, the θ derivative terms in Eq. (5.2) vanish, and we obtain

$$\frac{1}{r}\frac{\partial}{\partial r}\left[r\frac{\partial p(r,z)}{\partial r}\right] + \frac{\partial^2 p(r,z)}{\partial z^2} + \rho(z)\frac{\partial}{\partial z}\left[\frac{1}{\rho(z)}\right]\frac{\partial p(r,z)}{\partial z}$$

$$+k^2(z)p(r,z) = -4\pi\frac{\delta(r)}{r}\delta(\theta)\delta(z-z_0) . \tag{5.4}$$

Integrating both sides of Eq. (5.4) with respect to θ from 0 to 2π, we find that

$$\frac{1}{r}\frac{\partial}{\partial r}\left[r\frac{\partial p(r,z)}{\partial r}\right]+\frac{\partial^2 p(r,z)}{\partial z^2}+\rho(z)\frac{\partial}{\partial z}\left[\frac{1}{\rho(z)}\right]\frac{\partial p(r,z)}{\partial z}$$

$$+k^2(z)p(r,z)=-2\frac{\delta(r)}{r}\delta(z-z_0) \ . \tag{5.5}$$

We then propose a *separable solution* of Eq. (5.5) in terms of radial and depth *eigenfunctions*, $R_n(r)$ and $u_n(z)$, having amplitudes $a_n(z_0)$:

$$p(r,z)=\sum_n a_n(z_0)u_n(z)R_n(r) \ . \tag{5.6}$$

Here we assume that $u_n(z)$ satisfies the *eigenvalue equation* [cf. Eq. (5.203), which is the more familiar form for constant density]

$$\frac{1}{\rho(z)}\frac{d^2u_n(z)}{dz^2}+\frac{d}{dz}\left[\frac{1}{\rho(z)}\right]\frac{du_n(z)}{dz}+\frac{k_{zn}^2}{\rho(z)}u_n(z)=0 \ , \tag{5.7}$$

where $k_{zn}^2=k^2(z)-k_n^2$, and k_n and k_{zn} are the discrete values of the horizontal and vertical wavenumbers, respectively, associated with the eigenfunction $u_n(z)$. By adding and subtracting the term ω^2/c_m^2 in Eq. (5.7), where c_m is the minimum value of $c(z)$, we obtain

$$\frac{1}{\rho(z)}\frac{d^2u_n(z)}{dz^2}+\frac{d}{dz}\left[\frac{1}{\rho(z)}\right]\frac{du_n(z)}{dz}$$

$$+\left[\frac{k^2(z)-\omega^2/c_m^2}{\rho(z)}+\frac{\omega^2/c_m^2-k_n^2}{\rho(z)}\right]u_n(z)=0 \ . \tag{5.8}$$

Equation (5.8) is an eigenvalue equation of the *Sturm-Liouville type*,

$$\frac{d}{dz}\left[q(z)\frac{du_n(z)}{dz}\right]+\left[r(z)+\lambda_n w(z)\right]u_n(z)=0 \ , \tag{5.9}$$

since Eq. (5.9) can be rewritten as

$$q(z)\frac{d^2u_n(z)}{dz^2}+\frac{dq(z)}{dz}\frac{du_n(z)}{dz}+\left[r(z)+\lambda_n w(z)\right]u_n(z)=0 \ . \tag{5.9a}$$

It is clear that the Sturm-Liouville parameters for Eq. (5.8) are therefore

$$q(z)=w(z)=\frac{1}{\rho(z)},\quad r(z)=\frac{k^2(z)-\omega^2/c_m^2}{\rho(z)},\quad \lambda_n=\frac{\omega^2}{c_m^2}-k_n^2\ . \qquad (5.10)$$

Of particular interest to us are the properties associated with a *proper Sturm-Liouville problem* [Morse and Feshbach, 1953; Arfken, 1985; DeSanto, 1989]:

(i) The eigenfunctions $u_n(z)$ satisfy a Sturm-Liouville equation on the interval $a \le z \le b$. At the endpoints $z=a$ and $z=b$, the $u_n(z)$ satisfy (a) Dirichlet, (b) Neumann, (c) periodic, i.e., $\rho(a)=\rho(b)$, $u_n(a)=u_n(b)$, and $du_n/dz|_{z=a}=du_n/dz|_{z=b}$, or (d) mixed boundary conditions of the type $Au_n+B\,du_n/dz=0$, where A and B are real constants. Note that the latter requirement precludes the use of the impedance boundary condition [cf. Eq. (3.6) and Prob. 3.1] Furthermore, these conditions do not include the Sommerfeld radiation condition, though exponentially decaying fields at infinity are allowed.

(ii) The eigenfunctions $u_n(z)$ are orthonormal with respect to the weighting function w(z):

$$\int_a^b w(z)u_n^*(z)u_m(z)\,dz=\delta_{nm}=\begin{cases}1, & n=m,\\ 0, & n\neq m\end{cases}\ , \qquad (5.11)$$

where δ_{nm} is called the Kronecker delta.

(iii) The eigenfunctions $u_n(z)$ satisfy the closure or completeness relation:

$$\sum_n w(z_0)u_n^*(z_0)u_n(z)=\delta(z-z_0)\ . \qquad (5.12)$$

(iv) The eigenfunctions $u_n(z)$ constitute a complete set in the sense that an arbitrary function f(z) can be expanded in terms of them:

$$f(z)=\sum_n c_n u_n(z)\ , \qquad (5.13)$$

where c_n are the coefficients of the expansion.

(v) The eigenvalues λ_n are discrete, real, and positive.

We now assume that Eq. (5.7), combined with the prescribed boundary conditions, constitute a proper Sturm-Liouville problem. We are then able to use the properties described above to obtain the desired solution. First, we substitute the eigenfunction expansion in Eq. (5.6) into Eq. (5.5) and obtain

$$\sum_n a_n(z_0)\left\{\frac{u_n(z)}{r}\frac{d}{dr}\left[r\frac{dR_n(r)}{dr}\right]\right.$$

$$\left.+R_n(r)\underbrace{\left[\frac{d^2u_n(z)}{dz^2}+\rho(z)\frac{d}{dz}\left[\frac{1}{\rho(z)}\right]\frac{du_n(z)}{dz}+k^2(z)u_n(z)\right]}_{Term\,I}\right\}$$

$$=\sum_n\left\{-2\frac{\delta(r)}{r}\frac{1}{\rho(z_0)}u_n^*(z_0)u_n(z)\right\} , \quad (5.14)$$

where we have applied the closure relation to the right-hand side of Eq. (5.5). From the eigenvalue equation [cf. Eq. (5.7)], it is clear that

$$Term\,I = k_n^2 u_n(z) , \quad (5.15)$$

and therefore Eq. (5.14) becomes

$$\sum_n a_n(z_0)u_n(z)\left\{\frac{1}{r}\frac{d}{dr}\left[r\frac{dR_n(r)}{dr}\right]+k_n^2R_n(r)\right\}$$

$$=\sum_n\left\{-2\frac{\delta(r)}{r}\frac{1}{\rho(z_0)}u_n^*(z_0)u_n(z)\right\} . \quad (5.16)$$

Since Eq. (5.16) must hold true for arbitrary values of r, z, and z_0, we separately equate functions of r, z, and z_0 on the left- and right-hand sides of the equation and obtain

$$a_n(z_0) = u_n^*(z_0)/\rho(z_0) , \quad (5.17)$$

and

$$\frac{1}{r}\frac{d}{dr}\left[r\frac{dR_n(r)}{dr}\right]+k_n^2R_n(r)=-2\frac{\delta(r)}{r} \quad . \tag{5.18}$$

But Eq. (5.18) is simply Bessel's equation with a line source driving term and has the solution [cf. Eqs. (2.21)-(2.23) and (4.165)]

$$R_n(r)=i\pi H_0^{(1)}(k_n r) \quad . \tag{5.19}$$

The complete solution of Eq. (5.5) is therefore

$$p(r,z)=i\pi\sum_{n=1}^{\infty}a_n(z_0)p_n(r,z)=\frac{i\pi}{\rho(z_0)}\sum_{n=1}^{\infty}u_n^*(z_0)u_n(z)H_0^{(1)}(k_n r) \quad , \tag{5.20}$$

where the $p_n(r,z)$ are called the *normal modes* of the waveguide:

$$p_n(r,z)=u_n(z)H_0^{(1)}(k_n r) \quad . \tag{5.21}$$

In the far field ($k_n r >> 1$), we can substitute the asymptotic form for the Hankel function into Eq. (5.20) and obtain

$$p(r,z)\sim\frac{\sqrt{2\pi}e^{i\pi/4}}{\rho(z_0)}\sum_{n=1}^{\infty}u_n^*(z_0)u_n(z)\frac{e^{ik_n r}}{\sqrt{k_n r}} \quad , \tag{5.22}$$

which consists of a sum of cylindrically spreading, outwardly radiating waves in the radial direction. Finally, we note that each term in the modal sum individually satisfies both the homogeneous wave equation and the boundary conditions. This situation is to be contrasted with the image method solution (cf. Sec. 4.6) for which the elements in the sum individually satisfy the wave equation but only collectively satisfy the boundary conditions.

5.2.1 Normal Modes for a Homogeneous Fluid Layer with a Soft Top and Hard Bottom

As an illustrative example, we consider the normal mode solution for the zeroth order model of the ocean acoustic waveguide shown in Fig. 4.12, i.e., a homogeneous fluid layer with a pressure-release surface and hard bottom. For the case of constant density, the eigenvalue equation in Eq. (5.7) becomes

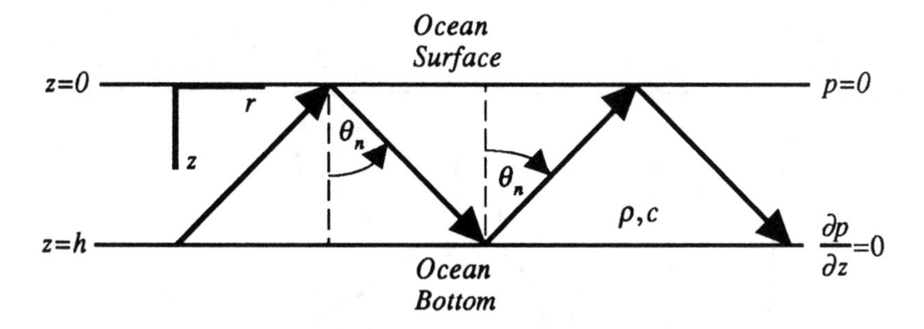

Figure 5.2 Plane wave interpretation of the modal field for a homogeneous fluid layer.

$$\frac{d^2u_n(z)}{dz^2}+k_{zn}^2u_n(z)=0\quad, \tag{5.23}$$

which has the solution

$$u_n(z)=\sqrt{\frac{2\rho}{h}}\sin k_{zn}z,\;\; k_{zn}=\frac{(n-1/2)\pi}{h},\;\; n=1,2,3,\ldots\;. \tag{5.24}$$

The normal mode sum is therefore

$$p(r,z)=\frac{2i\pi}{h}\sum_{n=1}^{\infty}\sin k_{zn}z_0\sin k_{zn}z\,H_0^{(1)}(k_nr)\quad, \tag{5.25}$$

which has the asymptotic behavior ($k_nr >> 1$)

$$\begin{aligned}p(r,z)&\sim\frac{2\sqrt{2\pi}e^{i\pi/4}}{h}\sum_{n=1}^{\infty}\sin k_{zn}z_0\sin k_{zn}z\frac{e^{ik_nr}}{\sqrt{k_nr}}\\&\sim\frac{\sqrt{2\pi}e^{-i\pi/4}}{h}\sum_{n=1}^{\infty}\frac{\sin k_{zn}z_0}{\sqrt{k_nr}}\left[e^{i(k_{zn}z+k_nr)}-e^{-i(k_{zn}z-k_nr)}\right]\quad,\end{aligned} \tag{5.26}$$

where the first and second terms in the square brackets can be interpreted as down- and up-going plane waves, respectively, propagating along the waveguide with angles of inclination θ_n such that (cf. Fig. 5.2)

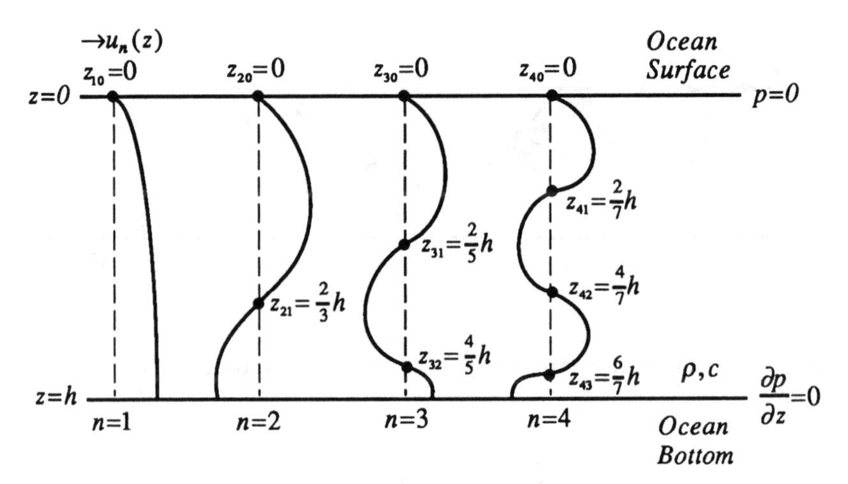

Figure 5.3 Schematic illustration of the first four modes as a function of depth for a homogeneous fluid layer with a soft top and hard bottom.

$$k_n = k \sin \theta_n, \;\; k_{zn} = k \cos \theta_n \;\; . \tag{5.27}$$

As the mode number n increases, the angle of inclination decreases. Thus, the modes correspond to specific, preferred directions of propagation within the waveguide for which constructive interference occurs between up- and down-going plane waves, giving rise to standing waves in depth.

Nodes of the Modal Depth Eigenfunctions

It is of interest to examine the detailed behavior of these standing waves. The nth depth eigenfunction $u_n(z)$ will have nodes at depths z_{nm} which satisfy

$$\sin k_{zn} z_{nm} = 0 \;\; , \tag{5.28}$$

so that

$$k_{zn} z_{nm} = m\pi, \;\; m = 0,1,2,\ldots \;\; , \tag{5.29}$$

and therefore

$$z_{nm} = \frac{mh}{(n - 1/2)}, \;\; m = 0,1,2,\ldots, \;\; z_{nm} < h \;\; . \tag{5.30}$$

The first four eigenfunctions are sketched in Fig. 5.3, and we can see that the mode number n also indicates the number of nodes (including the one at the surface) associated with a given mode. From Eq. (5.26), it is clear that if the source or receiver is placed at one of these nodes, the contribution of the nth mode will vanish. In addition, we note that the pressure and its normal derivative vanish, as required, at the surface and bottom, respectively.

Modal Phase Velocity

Each mode in Eq. (5.26) travels along the channel in the radial direction with a *phase velocity* C_n (cf. Sec. 2.2), corresponding to the speed at which surfaces of constant phase propagate:

$$C_n = \frac{\omega}{k_n} = \frac{\omega}{\sqrt{k^2 - k_{zn}^2}} = \frac{\omega}{\sqrt{k^2 - [(n-1/2)\pi/h]^2}}$$
$$= \frac{c}{\sqrt{1 - [(n-1/2)\lambda/(2h)]^2}} \quad . \tag{5.31}$$

Here we have used the relations $\omega = ck$ and $k = 2\pi/\lambda$. We note that the modal phase velocity is a function of wavelength (or frequency), a phenomenon which is called *geometric dispersion* because this frequency-dependent behavior arises due to the waveguide geometry. *Intrinsic dispersion*, on the other hand, is associated with the intrinsic, frequency-dependent properties of a particular material; we recall the dispersion of white light through a prism as being an example of this effect.

By examining the argument of the square root in Eq. (5.31), we can identify three regimes of characteristic modal behavior:

$(n-1/2)\lambda/(2h) < 1$: This regime is composed of radially propagating modes with phase velocities $C_n > c$.

$(n-1/2)\lambda/(2h) = 1$: In this case, $k_n \to 0$, $C_n \to \infty$, and only pure standing waves in depth occur, with no propagation in range.

$(n-1/2)\lambda/(2h) > 1$: In this region, both k_n and C_n are imaginary, and the modes are exponentially decaying (evanescent or inhomogeneous) in range.

In practice, for long-range propagation, we can obtain a good approximation to the total field by retaining the number of terms $n_{\max}$ in the modal sum which corresponds to the propagating modes only:

$$n_{\max} = \frac{2h}{\lambda} + \frac{1}{2} \approx \frac{2h}{\lambda} \quad \text{for } \frac{2h}{\lambda} >> \frac{1}{2} \ . \tag{5.32}$$

Thus, from the point of view of computation as well as simple physical interpretation, it is clear that the modal representation of the field is most desirable for thin (shallow water in the ocean acoustic case) waveguides for which $n_{\max}$ is small.

Modal Cutoff Frequency

From the discussion above, we can see that the nth mode will not propagate when the frequency is less than the *cutoff frequency* ω_n given by

$$\omega_n = k_{zn} c = \frac{(n - 1/2)\pi c}{h} \ . \tag{5.33}$$

Modal Group Velocity

Group velocity is a measure of the rate of energy transport associated with a dispersive wave system. For the modal wave field, the group velocity V_n is defined as

$$V_n = \frac{d\omega}{dk_n} = \frac{d(C_n k_n)}{dk_n} = C_n + k_n \frac{dC_n}{dk_n} \ , \tag{5.34}$$

where $V_n < c$ by the principle of causality. Perhaps the simplest way to understand this definition is to examine the superposition of two one-dimensional plane waves with equal amplitudes but slightly differing frequencies and wavenumbers:

$$p(x,t) = e^{i(k_1 x - \omega_1 t)} + e^{i(k_2 x - \omega_2 t)} \ , \tag{5.35}$$

where

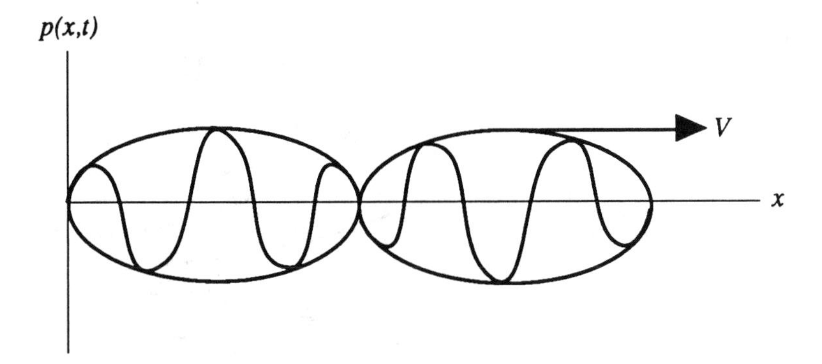

Figure 5.4 Wave packet at time t, propagating with group velocity V.

$$k_1 = k + \Delta k, \quad \omega_1 = \omega + \Delta\omega, \quad \Delta k << k,$$
$$k_2 = k - \Delta k, \quad \omega_2 = \omega - \Delta\omega, \quad \Delta\omega << \omega \ . \tag{5.36}$$

Equation (5.35) can then be rewritten as

$$\begin{aligned} p(x,t) &= e^{i(kx+\Delta kx-\omega t-\Delta\omega t)} + e^{i(kx-\Delta kx-\omega t+\Delta\omega t)} \\ &= e^{i(kx-\omega t)}\left[e^{i(\Delta kx-\Delta\omega t)} + e^{-i(\Delta kx-\Delta\omega t)}\right] \\ &= 2\cos(\Delta kx - \Delta\omega t)e^{i(kx-\omega t)} \ . \end{aligned} \tag{5.37}$$

Thus, the harmonic *wave train* with *carrier frequency* ω and phase velocity c is modulated by a cosine envelope (cf. Fig. 5.4) which propagates with velocity $\Delta\omega/\Delta k$. Examining the limiting behavior of this *wave packet* as $\Delta\omega$ and Δk approach zero, we find that it propagates with group velocity V:

$$\lim_{\substack{\Delta\omega\to 0 \\ \Delta k\to 0}} \frac{\Delta\omega}{\Delta k} = \frac{d\omega}{dk} = V \ . \tag{5.38}$$

For the case of the zeroth order ocean acoustic waveguide shown in Fig. 4.12, it is clear from Eqs. (5.31) and (5.33) that the phase velocity of the nth mode is

$$C_n = \frac{\omega}{\sqrt{k^2 - \omega_n^2/c^2}} = \frac{c}{\sqrt{1 - \omega_n^2/\omega^2}} \ , \tag{5.39}$$

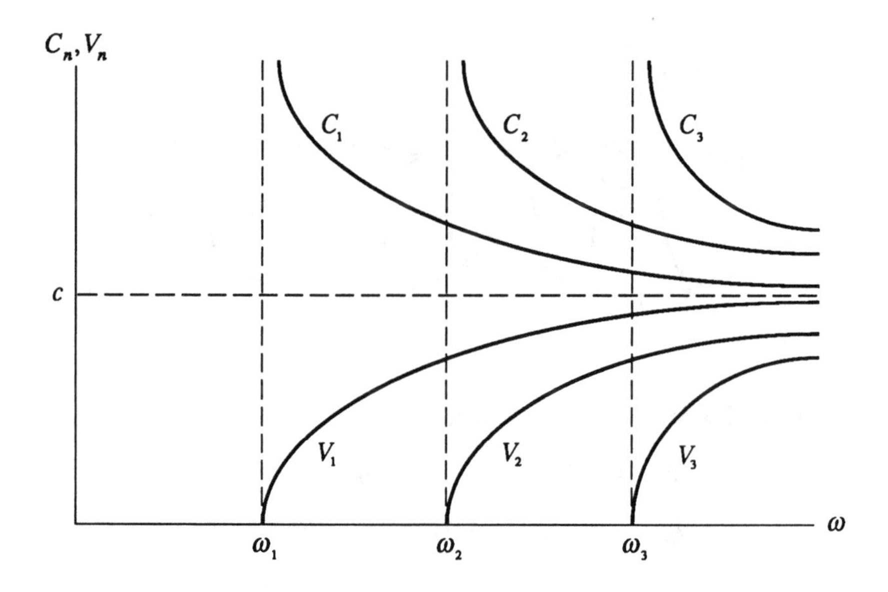

Figure 5.5 Phase and group velocity dispersion curves for the first three modes of a homogeneous fluid layer with a soft top and hard bottom.

which has the limiting behavior

$$C_n \to \infty \text{ as } \omega \to \omega_n,$$
$$C_n \to c \text{ as } \omega \to \infty . \tag{5.40}$$

It is then straightforward to show that

$$k_n = \frac{1}{c}\sqrt{\omega^2 - \omega_n^2} \ , \tag{5.41}$$

and therefore the group velocity of the *n*th mode is [cf. Eq. (5.34)]

$$V_n = c\sqrt{1 - \omega_n^2/\omega^2} \ , \tag{5.42}$$

which has the limiting behavior

$$V_n \to 0 \text{ as } \omega \to \omega_n,$$
$$V_n \to c \text{ as } \omega \to \infty . \tag{5.43}$$

The *dispersion curves* for the first three modes are shown in Fig. 5.5.

Modal Intensity, Interference Wavelength, and Cycle Distance

Using Eq. (2.58), we can show that the intensity of the asymptotic modal field in Eq. (5.26) is given by

$$I = \frac{|p(r,z)|^2}{2\rho c} \sim \frac{4\pi}{\rho c h^2} \sum_{n=1}^{\infty} \sin k_{zn} z_0 \sin k_{zn} z \frac{e^{ik_n r}}{\sqrt{k_n r}} \times \sum_{m=1}^{\infty} \sin k_{zm} z_0 \sin k_{zm} z \frac{e^{-ik_m r}}{\sqrt{k_m r}} , \quad (5.44)$$

which can be rewritten as

$$I \sim \frac{4\pi}{\rho c h^2 r} \sum_{n=1}^{\infty} \frac{\sin^2 k_{zn} z_0 \sin^2 k_{zn} z}{k_n} + \frac{8\pi}{\rho c h^2 r} \sum_{\substack{n=1 \\ n<m=2}}^{\infty} \sin k_{zn} z_0 \sin k_{zm} z_0 \sin k_{zn} z \sin k_{zm} z \frac{\cos(k_n - k_m) r}{\sqrt{k_n k_m}} . \quad (5.45)$$

Equation (5.45) shows that the nth and mth modes have an interference pattern associated with them which has an *interference wavelength* λ_{nm} defined as

$$\lambda_{nm} = \frac{2\pi}{k_n - k_m} . \quad (5.46)$$

For example, if there are only two propagating modes, then we obtain the characteristic *bimodal interference pattern*

$$I \sim \frac{4\pi}{\rho c h^2 r} \left[\frac{\sin^2 k_{z1} z_0 \sin^2 k_{z1} z}{k_1} + \frac{\sin^2 k_{z2} z_0 \sin^2 k_{z2} z}{k_2} \right] + \frac{8\pi}{\rho c h^2 r} \sin k_{z1} z_0 \sin k_{z2} z_0 \sin k_{z1} z \sin k_{z2} z \frac{\cos(k_1 - k_2) r}{\sqrt{k_1 k_2}} , \quad (5.47)$$

which has equally spaced, non-zero nulls at the ranges $r = r_n$ such that

$$\cos(k_1 - k_2) r_n = -1 , \quad (5.48)$$

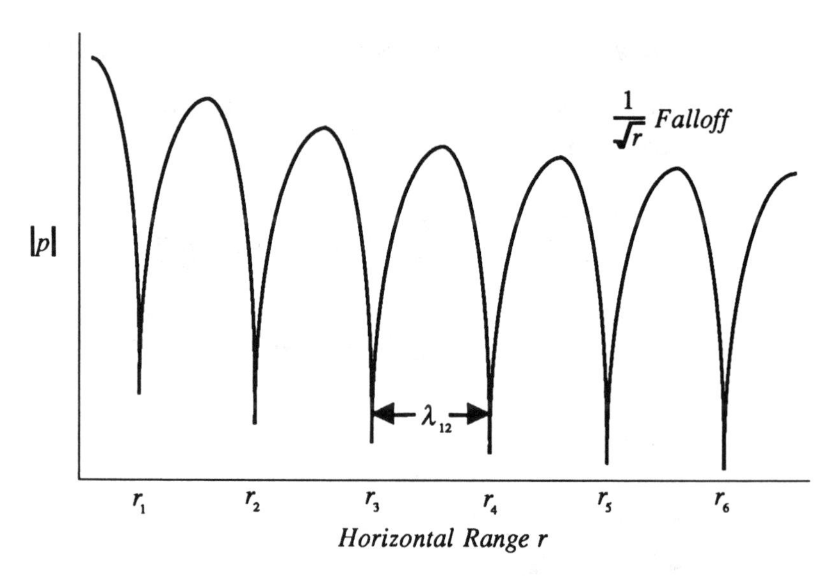

Figure 5.6 Characteristic bimodal interference pattern.

and therefore

$$r_n = \frac{(2n+1)\pi}{k_1 - k_2}, \quad n = 0,1,2,\dots \quad . \tag{5.49}$$

The corresponding pressure magnitude, which has the same null behavior as the intensity, is sketched in Fig. 5.6. It is of interest to compare this interference pattern, which is associated with waveguide propagation, with the Lloyd mirror interference pattern in Fig. 4.7, which arises when there is only one perfectly reflecting boundary. In the latter case, the nulls fall to zero and are unequally spaced in range. Also, in the far field, the Lloyd mirror pattern falls off as $1/r^2$, whereas the modal field falls off as $1/\sqrt{r}$.

Let us examine the interference wavelength for two adjacent modes in more detail. Then we have

$$\lambda_{n,n+1} = \frac{2\pi}{k_n - k_{n+1}} \quad , \tag{5.50}$$

where

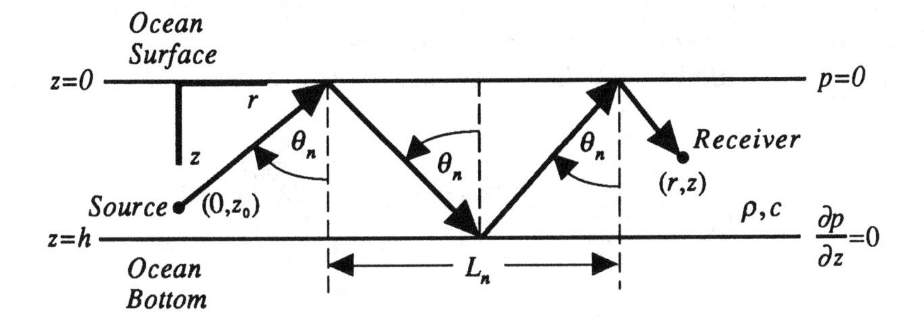

Figure 5.7 Cycle or skip distance L_n associated with the nth mode for a homogeneous fluid layer with a soft top and hard bottom.

$$k_n - k_{n+1} = k\left\{\sqrt{1-\left[\frac{(n-1/2)\pi}{kh}\right]^2} - \sqrt{1-\left[\frac{(n+1/2)\pi}{kh}\right]^2}\right\} . \tag{5.51}$$

Using the first two terms in the binomial expansion [cf. Eq. (4.50)] of the square root terms in Eq. (5.51) for

$$\frac{(n \pm 1/2)\pi}{kh} << 1 \ , \tag{5.52}$$

we obtain

$$k_n - k_{n+1} \approx k\left\{1-\frac{1}{2}\left[\frac{(n-1/2)\pi}{kh}\right]^2 - 1 + \frac{1}{2}\left[\frac{(n+1/2)\pi}{kh}\right]^2\right\} , \tag{5.53}$$

so that

$$k_n - k_{n+1} \approx \frac{n\pi^2}{kh^2} \ . \tag{5.54}$$

Equation (5.50) then becomes

$$\lambda_{n,n+1} \approx \frac{2kh^2}{n\pi} \ . \tag{5.55}$$

But we can also consider a wave, corresponding to the nth mode, which emanates from the source with the *eigenangle* θ_n, propagates along the waveguide, and arrives at the receiver at the same angle. For this path, the horizontal range between successive surface or bottom reflections is called the *cycle* or *skip distance* L_n and is given by (cf. Fig. 5.7)

$$L_n = 2h\tan\theta_n = 2h\frac{k_n}{k_{zn}} = 2h\frac{\sqrt{k^2 - [(n-1/2)\pi/h]^2}}{[(n-1/2)\pi/h]} \approx \frac{2kh^2}{n\pi}, \quad (5.56)$$

where we have again used the binomial expansion under the condition in Eq. (5.52) with the additional requirement that $n >> 1/2$. Thus, we obtain the interesting result that, for thick waveguides, the cycle distance for a particular mode is equal to the interference wavelength between that mode and its nearest neighbor.

Thus, in our discussion, two distinctly different interpretations of the modal field have emerged. On the one hand, we saw that an individual mode can be synthesized from a pair of constructively interfering, down- and up-going plane waves which do not emanate from the source point [cf. Eqs. (5.26) and (5.27) and Fig. 5.2]. On the other hand, Eqs. (5.55) and (5.56) and Fig. 5.7 suggest that the interference of a pair of neighboring modes generates a wave originating at the source and propagating to the receiver along a geometrical acoustics path. This relationship between modes and rays emanating from the source will be explored further in Secs. 5.5.1, 6.3.1, and 7.6.4.

The Modal Continuum

In the discussion so far, we have considered a discrete set of modes, and, in fact, we recall that rigorous Sturm-Liouville theory requires that the *modal spectrum* be discrete (cf. Sec. 5.2). Yet, Eq. (5.54) suggests that as h increases, the spacing between the eigenvalues decreases, and that in the limit of infinite waveguide thickness, the discrete spectrum becomes a continuous one. We shall show that this notion is indeed correct by implementing this limiting procedure in several situations, always retaining a high degree of rigor by posing a proper Sturm-Liouville problem before taking the limit. In this manner, we will be able to clarify the role of the continuous spectrum in the overall modal framework.

We begin by examining the behavior of the modal solution for the zeroth order ocean acoustic waveguide [cf. Eq. (5.25)] in the limit of infinite h. First, we rewrite the trigonometric depth eigenfunctions as exponentials and obtain

$$p(r,z) = \frac{\pi}{2ih}\sum_{n=1}^{\infty}\left[e^{ik_{zn}(z+z_0)} - e^{ik_{zn}(z-z_0)} + e^{-ik_{zn}(z+z_0)} - e^{-ik_{zn}(z-z_0)}\right]H_0^{(1)}(k_n r). \tag{5.57}$$

Then for $n >> 1/2$, we can make the approximations

$$k_{zn} \approx \frac{n\pi}{h}, \; k_n \approx \sqrt{k^2 - \left(\frac{n\pi}{h}\right)^2} \; , \tag{5.58}$$

so that Eq. (5.57) becomes

$$p(r,z) \approx \frac{\pi}{2ih}\sum_{n=1}^{\infty}\left[e^{i\frac{n\pi}{h}(z+z_0)} - e^{i\frac{n\pi}{h}(z-z_0)} + e^{-i\frac{n\pi}{h}(z+z_0)} - e^{-i\frac{n\pi}{h}(z-z_0)}\right]H_0^{(1)}(k_n r). \tag{5.59}$$

But as h tends to infinity, the vertical and horizontal wavenumber spacings [cf. Eq. (5.54)],

$$\Delta k_{zn} = k_{z,n+1} - k_{zn} = \frac{\pi}{h}, \; \Delta k_n = k_n - k_{n+1} \approx \frac{n\pi^2}{kh^2} \; , \tag{5.60}$$

tend to zero, and therefore we can approximate the sum in Eq. (5.59) by an integral, which we can rewrite as two integrals:

$$\begin{aligned} p(r,z) &\approx I_1 + I_2 \\ &\approx \frac{\pi}{2ih}\int_0^{\infty}\left[e^{i\frac{n\pi}{h}(z+z_0)} - e^{i\frac{n\pi}{h}(z-z_0)}\right]H_0^{(1)}(k_n r)\,dn \\ &+ \frac{\pi}{2ih}\int_0^{\infty}\left[e^{-i\frac{n\pi}{h}(z+z_0)} - e^{-i\frac{n\pi}{h}(z-z_0)}\right]H_0^{(1)}(k_n r)\,dn \; . \end{aligned} \tag{5.61}$$

In the second integral, we change variables, letting $n' = -n$, so that

$$I_2 \approx \frac{\pi}{2ih}\int_{-\infty}^{0}\left[e^{i\frac{n'\pi}{h}(z+z_0)} - e^{i\frac{n'\pi}{h}(z-z_0)}\right]H_0^{(1)}(k_{n'} r)\,dn' \; , \tag{5.62}$$

and therefore Eq. (5.61) reduces to

$$p(r,z) \approx \frac{\pi}{2ih}\int_{-\infty}^{\infty}\left[e^{i\frac{n\pi}{h}(z+z_0)} - e^{i\frac{n\pi}{h}(z-z_0)}\right]H_0^{(1)}(k_n r)\,dn \quad . \qquad (5.63)$$

Letting

$$k_z = \frac{n\pi}{h} \text{ and } k_r = \sqrt{k^2 - k_z^2} \quad , \qquad (5.64)$$

we find that Eq. (5.63) becomes

$$p(r,z) \approx \frac{1}{2i}\int_{-\infty}^{\infty}\left[e^{ik_z(z+z_0)} - e^{ik_z(z-z_0)}\right]H_0^{(1)}(k_r r)\,dk_z \quad . \qquad (5.65)$$

But from Gradshteyn and Ryzhik [1965], we know that

$$\frac{1}{2i}\int_{-\infty}^{\infty} e^{ik_z\alpha} H_0^{(1)}\left(\sqrt{k^2 - k_z^2}\, r\right)dk_z = -\frac{e^{ik\sqrt{r^2+\alpha^2}}}{\sqrt{r^2+\alpha^2}} \quad , \qquad (5.66)$$

and therefore, evaluating Eq. (5.65), we obtain

$$p(r,z) \approx \frac{e^{ik\sqrt{r^2+(z-z_0)^2}}}{\sqrt{r^2+(z-z_0)^2}} - \frac{e^{ik\sqrt{r^2+(z+z_0)^2}}}{\sqrt{r^2+(z+z_0)^2}} \quad , \qquad (5.67)$$

which is simply the Lloyd mirror result in Eq. (4.48). Thus, when the lower boundary of the waveguide is removed to infinity, the discrete modes coalesce into a continuum which corresponds to the image method solution for spherical wave reflection from the upper boundary. The removal of the lower boundary, and its associated boundary condition, to infinity has eliminated the mechanism for trapping the waves at discrete angles; all possible angles of propagation are now allowed. This example demonstrates that the limiting behavior of solutions to properly posed Sturm-Liouville problems can be used to obtain solutions of improper Sturm-Liouville problems. We shall exploit this fact in the solution of more complicated problems in subsequent sections, where the role of the continuous spectrum and its relationship to the discrete spectrum become more subtle.

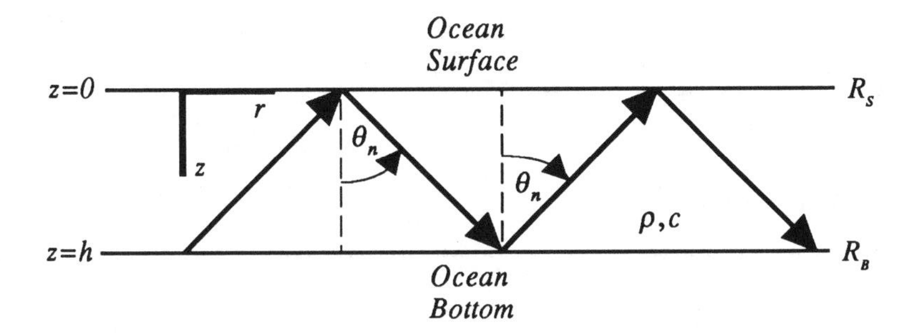

Figure 5.8 Homogeneous fluid layer bounded by arbitrary horizontally stratified media with plane wave reflection coefficients R_S and R_B.

5.3 Eigenvalue Equation for a Homogeneous Fluid Layer Bounded by Arbitrary Horizontally Stratified Media

Suppose that we have a homogeneous fluid layer bounded above by a horizontally stratified medium with plane wave reflection coefficient R_S and below by a horizontally stratified medium with reflection coefficient R_B, where we note that, in general, the bounding media may be elastic solids (cf. Fig. 5.8). Tolstoy and Clay [1987] and Clay and Medwin [1977] describe a useful method for determining the *characteristic equation* for the eigenvalues of this configuration. The depth-dependent part $u(z)$ of the total solution in the layer, with no sources, can be written as the sum of a down-going plane wave u_d and an up-going plane wave u_u,

$$u(z) = u_d(z) + u_u(z) = Ae^{ik_z z} + Be^{-ik_z z} \quad , \tag{5.68}$$

where A and B are arbitrary constants to be determined by the boundary conditions. From the definition of the reflection coefficient as the ratio of the reflected and the incident fields evaluated on the boundary (cf. Sec. 3.3), we know that at the surface,

$$R_S = \left.\frac{u_d(z)}{u_u(z)}\right|_{z=0} = \frac{A}{B} \quad , \tag{5.69}$$

and at the bottom,

$$R_B = \left.\frac{u_u(z)}{u_d(z)}\right|_{z=h} = \frac{B}{A}e^{-2ik_z h} \quad . \tag{5.70}$$

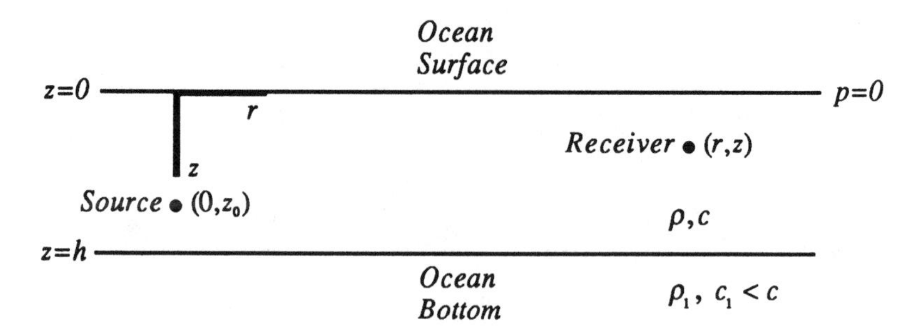

Figure 5.9 A point source in a homogeneous fluid layer with a pressure-release top and a homogeneous, lower velocity fluid bottom.

Combining Eqs. (5.69) and (5.70), we obtain the eigenvalue equation

$$1-R_S R_B e^{2ik_z h}=0 \ , \tag{5.71}$$

a result which we will find to be extremely useful in our subsequent discussion. The parameterization of the eigenvalues in terms of the reflection coefficients of the bounding media is a particularly desirable feature, in that reflection coefficients are commonly encountered as fundamental acoustic characterizations of horizontally stratified media.

5.4 Normal Modes for a Homogeneous Fluid Layer Bounded Above by a Pressure-Release Surface and Below by a Lower Velocity, Homogeneous Fluid Half-Space

5.4.1 An Improper Sturm-Liouville Problem

We now address the problem of constructing the Green's function, using the method of eigenfunction expansion, for the waveguide configuration shown in Fig. 5.9. Here the pressure-release top is characterized by the reflection coefficient $R_s=-1$, and the lower velocity, homogeneous fluid bottom is characterized by the Rayleigh reflection coefficient (cf. Sec. 3.3.3). The eigenvalue equation [cf. Eq. (5.71)] is therefore

$$|R_B|e^{2i(k_z h-\pi/2+\varphi/2)}=1 \ , \tag{5.72}$$

where φ is the phase of R_B, and, for convenience, we have written $R_S = e^{-i\pi}$. But $|R_B(k_z)|$ can be expressed in terms of a function $\alpha(k_z)$,

$$|R_B(k_z)| = e^{-\alpha(k_z)} \Rightarrow \alpha(k_z) = -\ln|R_B(k_z)| \ , \tag{5.73}$$

so that Eq. (5.72) becomes

$$e^{2i(k_z h - \pi/2 + \varphi/2 + i\alpha/2)} = 1 \ , \tag{5.74}$$

which implies that

$$2(k_{zn}h - \pi/2 + \varphi_n/2 + i\alpha_n/2) = 2(n-1)\pi, \ \ n = 1,2,3,\ldots \ , \tag{5.75}$$

and therefore

$$k_{zn}h = (n-1/2)\pi - \varphi_n/2 - i\alpha_n/2, \ \ n = 1,2,3,\ldots \ . \tag{5.76}$$

The solutions of Eq. (5.76) are, in general, complex, and we are thus confronted with an improper Sturm-Liouville problem, since condition *(v)* on p. 115 is violated. This circumstance is a direct consequence of the fact that for a lower velocity half-space (cf. Sec. 3.3.3),

$$|R_B(k_z)| < 1 \ \text{ for } \ k_z < k \ , \tag{5.77}$$

so that total internal reflection cannot occur for plane waves incident upon the lower boundary, thereby also violating condition *(i)* on p. 114. As a result, perfect trapping of the modes does not occur, and energy continually radiates out of the waveguide as the field propagates down the channel. This effect manifests itself through the appearance of complex eigenvalues, the imaginary part thereof reflecting the leakage of energy out of the waveguide. In order to place these statements within the proper theoretical framework, we will examine a closely related, proper Sturm-Liouville problem in the next section.

5.4.2 A Proper Sturm-Liouville Problem

We consider the solution of the properly posed Sturm-Liouville problem shown in Fig. 5.10. By introducing a hard boundary at a finite distance from the surface, we have guaranteed that perfect trapping will occur and that the eigenvalue spectrum will be discrete.

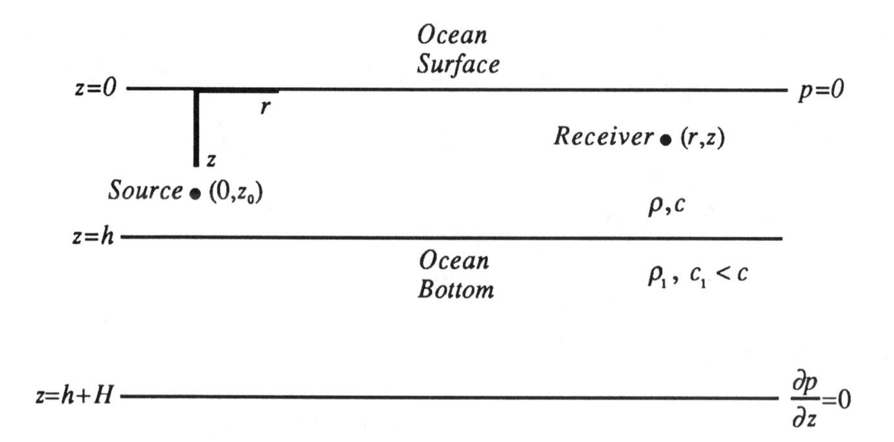

Figure 5.10 The properly posed Sturm-Liouville problem which, in the limit of infinite H, leads to the solution of the improper Sturm-Liouville problem shown in Fig. 5.9.

We will first solve this problem and then examine the limit of the solution as the thickness of the lower layer becomes infinite ($H \to \infty$). We will thereby arrive at the solution of the original, improperly posed Sturm-Liouville problem described in the previous section. This procedure is reminiscent of the use of *box normalization* in quantum mechanics, in which continuous eigenstates are normalized within a large box so that they can fit within the same theoretical structure as the discrete bound states [Schiff, 1968; Merzbacher, 1970]. Our approach is related to the work of Tindle, Stamp, and Guthrie [1976].

The modal depth eigenfunctions for the configuration in Fig. 5.10 are

$$u_n(z) = \begin{cases} A_n \sin k_{zn} z, \quad 0 \le z \le h, \\ B_n\left[e^{ik_{1zn}z} + e^{2ik_{1zn}(h+H)}e^{-ik_{1zn}z}\right] \\ = 2B_n e^{ik_{1zn}(h+H)} \cos\{k_{1zn}[z-(h+H)]\}, \; h \le z \le h+H \end{cases}, \quad (5.78)$$

where $k_{1zn} = \sqrt{k_1^2 - k_n^2}$, and A_n and B_n are arbitrary constants to be determined by the continuity conditions at $z = h$. Continuity of pressure implies that

$$\frac{A_n}{B_n} = 2e^{ik_{1zn}(h+H)} \frac{\cos k_{1zn} H}{\sin k_{zn} h} , \quad (5.79)$$

while continuity of the normal component of particle velocity leads to

$$\frac{A_n}{B_n} = 2\frac{\rho k_{1zn}}{\rho_1 k_{zn}} e^{ik_{1zn}(h+H)} \frac{\sin k_{1zn} H}{\cos k_{zn} h} \quad . \tag{5.80}$$

Combining Eqs. (5.79) and (5.80) yields the eigenvalue equation

$$\tan k_{zn} h \tan k_{1zn} H = \frac{\rho_1 k_{zn}}{\rho k_{1zn}} \quad . \tag{5.81}$$

Using Eq. (5.79) and the orthonormality relation [cf. Eq. (5.11)], we obtain

$$\frac{A_n^2}{\rho}\int_0^h \sin^2 k_{zn} z\, dz + \frac{A_n^2}{\rho_1}\frac{\sin^2 k_{zn} h}{\cos^2 k_{1zn} H}\int_h^{h+H} \cos^2\{k_{1zn}[z-(h+H)]\}dz = 1, \tag{5.82}$$

from which it is straightforward to show that

$$A_n = \left[\frac{1}{\rho}\left(\frac{h}{2} - \frac{\sin 2k_{zn} h}{4k_{zn}}\right) + \frac{1}{\rho_1}\frac{\sin^2 k_{zn} h}{\cos^2 k_{1zn} H}\left(\frac{H}{2} + \frac{\sin 2k_{1zn} h}{4k_{1zn}}\right)\right]^{-1/2} . \tag{5.83}$$

Having obtained the normalization constant, we can write the normal mode solution in the upper layer as

$$p(r,z) = \frac{i\pi}{\rho}\sum_{n=1}^{\infty} A_n^2 \sin k_{zn} z_0 \sin k_{zn} z\, H_0^{(1)}(k_n r), \quad 0 \le z, z_0 \le h. \tag{5.84}$$

We note that in the limit $\rho_1 \to \rho$ and $c_1 \to c$, Eq. (5.83) reduces to

$$A_n = \sqrt{\frac{2\rho}{h+H}} \quad , \tag{5.85}$$

and Eq. (5.84) becomes the solution for a homogeneous fluid layer of thickness $(h+H)$ [cf. Eq. (5.25)], as required.

For $H >> h$, we find that

$$A_n^2 \approx \frac{2\rho_1}{H} \frac{1}{\sin^2 k_{zn}h / \cos^2 k_{1zn}H} \ . \tag{5.86}$$

We can rewrite the trigonometric term in the denominator of Eq. (5.86) as

$$\begin{aligned}
\frac{\sin^2 k_{zn}h}{\cos^2 k_{1zn}H} &= \frac{\sin^2 k_{zn}h}{\cos^2 k_{1zn}H}\left(\sin^2 k_{1zn}H + \cos^2 k_{1zn}H\right) \\
&= \sin^2 k_{zn}h\left(1 + \tan^2 k_{1zn}H\right) \\
&= \sin^2 k_{zn}h\left(1 + \frac{\rho_1^2 k_{zn}^2}{\rho^2 k_{1zn}^2} \frac{\cos^2 k_{zn}h}{\sin^2 k_{zn}h}\right) \\
&= \sin^2 k_{zn}h + \frac{\rho_1^2 k_{zn}^2}{\rho^2 k_{1zn}^2} \cos^2 k_{zn}h \ .
\end{aligned} \tag{5.87}$$

The modal sum in Eq. (5.84) therefore becomes

$$p(r,z) \approx \frac{2i\pi\rho_1}{\rho H} \sum_{n=1}^{\infty} \frac{\sin k_{zn}z_0 \sin k_{zn}z}{\left[\sin^2 k_{zn}h + \frac{\rho_1^2 k_{zn}^2}{\rho^2 k_{1zn}^2}\cos^2 k_{zn}h\right]} H_0^{(1)}(k_n r), \tag{5.88}$$

which, as H tends to infinity, goes over to an integral,

$$p(r,z) \approx \frac{2i\pi\rho_1}{\rho H} \int_0^{\infty} \frac{\sin k_{zn}z_0 \sin k_{zn}z}{\left[\sin^2 k_{zn}h + \frac{\rho_1^2 k_{zn}^2}{\rho^2 k_{1zn}^2}\cos^2 k_{zn}h\right]} H_0^{(1)}(k_n r)\, dn. \tag{5.89}$$

In order to proceed further, we must determine the relationship between n and k_z for large H. Here it is helpful to use the form of the eigenvalue equation in Eq. (5.71), combined with $R_s = e^{-i\pi}$ and Eq. (3.77) for R_B. We then obtain

$$1 - e^{-i\pi}\left[\frac{R_{01} + e^{2ik_{1z}H}}{1 + R_{01}e^{2ik_{1z}H}}\right] e^{2ik_z h} = 0 \ , \tag{5.90}$$

where R_{01} is the Rayleigh reflection coefficient for a plane wave incident from a half-space with acoustic properties ρ, c upon a half-space with properties ρ_1, c_1. Equation (5.90) can be rewritten as

$$e^{-i\pi} e^{2i(k_{1z}H + k_z h)} \left\{1 - R_{01}\left[e^{i\pi} e^{-2ik_z h} - e^{-2ik_{1z}H}\right]\right\} = 1 \ , \qquad (5.90a)$$

which implies that

$$2i(k_{1zn}H + k_{zn}h) + \ln\left\{1 - R_{01}\left[e^{i\pi} e^{-2ik_{zn}h} - e^{-2ik_{1zn}H}\right]\right\}$$
$$= 2i(n - 1/2)\pi, \quad n = 1,2,3,\ldots \ . \qquad (5.91)$$

Equation (5.91) can be rewritten as

$$2i(k_{1zn} + k_{zn}h/H) + \ln\left\{1 - R_{01}\left[e^{i\pi} e^{-2ik_{zn}h} - e^{-2ik_{1zn}H}\right]\right\}/H$$
$$= 2i(n - 1/2)\pi/H, \quad n = 1,2,3,\ldots \ . \qquad (5.91a)$$

For $H \to \infty$ and $n >> 1/2$, Eq. (5.91a) becomes

$$k_{1zn} \approx \frac{n\pi}{H} \ . \qquad (5.92)$$

Therefore, changing variables in Eq. (5.89),

$$k_{1z} = \frac{n\pi}{H} \Rightarrow k_z = \sqrt{k^2 - k_1^2 + \left(\frac{n\pi}{H}\right)^2} \Rightarrow dk_z = \frac{n\pi^2}{k_z H^2} dn = \frac{k_{1z}\pi}{k_z H} dn,$$
$$k_r = \sqrt{k^2 - k_z^2} = \sqrt{k_1^2 - k_{1z}^2} \ , \qquad (5.93)$$

we obtain

$$p(r,z) \approx \frac{2i\rho_1}{\rho} \int_{C_B} \frac{k_z}{k_{1z}} \frac{\sin k_z z_0 \sin k_z z}{\left[\sin^2 k_z h + \dfrac{\rho_1^2 k_z^2}{\rho^2 k_{1z}^2} \cos^2 k_z h\right]} H_0^{(1)}(k_r r)\, dk_z, \quad (5.94)$$

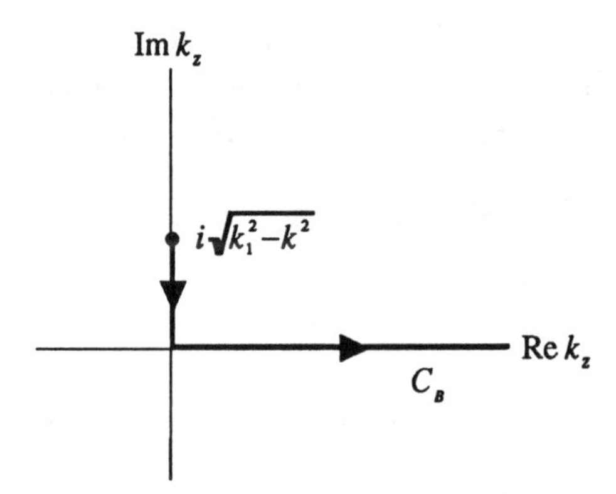

Figure 5.11 Path of integration C_B in the complex k_z-plane associated with the continuous modal solution of the problem shown in Fig. 5.9.

where the path of integration C_B is shown in Fig. 5.11. We note that in the limit $\rho_1 \to \rho$ and $c_1 \to c$, Eq. (5.94) reduces to the Lloyd mirror result [cf. Eq. 5.65)], as expected.

Thus, we see that just as in the Lloyd mirror case (cf. pp. 126-128), the field is composed entirely of a continuous spectrum of modes, reflecting the fact that in both instances, no mechanism exists for perfectly trapping the modes. We recall that, in the Lloyd mirror example, this behavior was due to the absence of a lower reflecting boundary. In the present case, although a lower reflecting surface exists, the fact that it is not a perfect reflector prevents the discretization of the modal field.

5.4.3 Improper Modes

The integral in Eq. (5.94) can be expressed in the form

$$p(r,z) \approx \frac{2i\rho_1}{\rho} \int_{C_B} \frac{k_z}{k_{1z}} F(k_z) H_0^{(1)}(k_r r)\, dk_z \quad , \tag{5.95}$$

where $F(k_z) = N(k_z)/D(k_z)$ and

$$N(k_z) = \sin k_z z_0 \sin k_z z \quad , \tag{5.96a}$$

$$D(k_z) = \sin^2 k_z h + \frac{\rho_1^2 k_z^2}{\rho^2 k_{1z}^2} \cos^2 k_z h \quad . \tag{5.96b}$$

In approaching the evaluation of Eq. (5.95), we note that obvious features of the integrand are the first-order poles of $F(k_z)$, corresponding to the zeros of the denominator, $D(k_z) = 0$, and satisfying the equation

$$\tan^2 k_z h = -\frac{\rho_1^2 k_z^2}{\rho^2 k_{1z}^2} \quad . \tag{5.97}$$

We will show that Eq. (5.97) is simply a different form of the eigenvalue equation for the problem posed in Fig. 5.9, and introduces mode-like behavior to the field in Eq. (5.95). Specifically, the solutions of Eq. (5.97) play a crucial role in the development of an approximation to Eq. (5.95) which consists of a decomposition into a sum of *improper*, or *virtual, modes*. These *leaky modes* are thus associated with the improper Sturm-Liouville problem described in Sec. 5.4.1, and therefore their contributions can be determined within a modal framework only by the careful path through a related proper Sturm-Liouville problem which we have been following. Their effects will be calculated by the asymptotic evaluation of Eq. (5.95) in the vicinity of each pole.

After demonstrating that Eq. (5.97) is indeed the eigenvalue equation, we will estimate approximate pole positions for a specific case. Then we will expand $F(k_z)$ in a Mittag-Leffler expansion and evaluate each term in the sum by asymptotic methods. Our development is strongly influenced by the work of Stickler and Ammicht [1980, 1984] and Tindle, Stamp, and Guthrie [1976].

An Equivalent Eigenvalue Equation

For the configuration shown in Fig. 5.9, we can derive an eigenvalue equation which is equivalent to the reflection coefficient result in Eq. (5.76). First, we write the modal depth eigenfunctions as

$$u_n(z) = \begin{cases} A_n \sin k_{zn} z, & 0 \le z \le h, \\ B_n e^{ik_{1zn} z}, & z \ge h \ , \end{cases} \tag{5.98}$$

where A_n and B_n are arbitrary constants to be determined by the continuity conditions at $z = h$. Continuity of pressure implies that

$$\frac{A_n}{B_n} = \frac{e^{ik_{1zn}h}}{\sin k_{zn}h} , \tag{5.98a}$$

while continuity of the normal component of particle velocity leads to

$$\frac{A_n}{B_n} = i\frac{\rho k_{1zn}}{\rho_1 k_{zn}} \frac{e^{ik_{1zn}h}}{\cos k_{zn}h} . \tag{5.98b}$$

Combining Eqs. (5.98a) and (5.98b) yields

$$\tan k_{zn}h = -i\frac{\rho_1 k_{zn}}{\rho\, k_{1zn}} , \tag{5.99}$$

which, when squared, corresponds precisely to Eq. (5.97). Thus, we anticipate that a modal interpretation may be ascribed to the continuum field in Eq. (5.95), but unorthodox behavior is already suggested by the nature of the solution for $z \geq h$ in Eq. (5.98), which does not satisfy any of the Sturm-Liouville boundary conditions *(i)* on p. 114.

Poles of the Integrand

For simplicity, we will consider the specific case where $\rho_1 < \rho$ in Fig. 5.9 and determine the approximate pole locations using an iterative perturbation technique. It is clear that for the parameters of the waveguide that we have prescribed,

$$\frac{\rho_1 k_{zn}}{\rho\, k_{1zn}} < 1 , \tag{5.100}$$

and therefore we can obtain the zeroth order solution $k_{zn}^{(0)}$ for the pole positions by assuming that the right-hand side of Eq. (5.99) is zero, so that

$$k_{zn}^{(0)} = \frac{(n-1)\pi}{h}, \quad n = 1,2,3,\ldots . \tag{5.101}$$

These are simply the eigenvalues for a pressure-release bottom which we now substitute into the right-hand side of Eq. (5.99), while substituting a sum of zeroth and first order terms into the left-hand side:

$$k_{zn} \approx k_{zn}^{(0)} + k_{zn}^{(1)} \ . \tag{5.102}$$

Using the identity

$$\tan\left(k_{zn}^{(0)} + k_{zn}^{(1)}\right)h = \frac{\sin k_{zn}^{(0)}h \cos k_{zn}^{(1)}h + \cos k_{zn}^{(0)}h \sin k_{zn}^{(1)}h}{\cos k_{zn}^{(0)}h \cos k_{zn}^{(1)}h - \sin k_{zn}^{(0)}h \sin k_{zn}^{(1)}h}, \tag{5.103}$$

we obtain

$$\tan k_{zn}^{(1)}h \approx -i\frac{\rho_1 k_{zn}^{(0)}}{\rho\, k_{1zn}^{(0)}} \ , \tag{5.104}$$

where

$$k_{1zn}^{(0)} = \sqrt{k_1^2 - k^2 + k_{zn}^{(0)^2}} \ . \tag{5.105}$$

But we know that

$$\tan k_{zn}^{(1)}h \approx k_{zn}^{(1)}h \ \text{ for } \left|k_{zn}^{(1)}h\right| << 1 \ , \tag{5.106}$$

and therefore

$$k_{zn}^{(1)} \approx -i\frac{\rho_1}{\rho h}\frac{k_{zn}^{(0)}}{k_{1zn}^{(0)}} \ , \tag{5.107}$$

so that we can write

$$k_{zn} \approx k_{zn}^{(0)} - i\alpha_n \ \text{ with } \ \alpha_n = \frac{\rho_1}{\rho h}\frac{k_{zn}^{(0)}}{k_{1zn}^{(0)}} > 0 \ . \tag{5.108}$$

Using Eq. (5.108) and the relation $k_n = \sqrt{k^2 - k_{zn}^2}$, we find that

$$k_n \approx \sqrt{k^2 - k_{zn}^{(0)^2} + 2ik_{zn}^{(0)}\alpha_n} \approx \sqrt{k^2 - k_{zn}^{(0)^2}}\left[1 + i\frac{k_{zn}^{(0)}\alpha_n}{k^2 - k_{zn}^{(0)^2}}\right] , \quad (5.109)$$

where we have used the binomial expansion [cf. Eq. (4.50)] in Eq. (5.109) for

$$\left|\frac{2k_{zn}^{(0)}\alpha_n}{k^2 - k_{zn}^{(0)^2}}\right| < 1 . \quad (5.110)$$

Thus, we can also write

$$k_n \approx k_n^{(0)} + i\beta_n \text{ with } \beta_n = \frac{\rho_1}{\rho h}\frac{k_{zn}^{(0)^2}}{k_n^{(0)}k_{1zn}^{(0)}} > 0 , \quad (5.111)$$

and $k_n^{(0)} = \sqrt{k^2 - k_{zn}^{(0)^2}}$. Finally, using the relation $k_{1zn} = \sqrt{k_1^2 - k_n^2}$ and Eq. (5.111), we obtain

$$k_{1zn} \approx \sqrt{k_1^2 - k_n^{(0)^2} - 2ik_n^{(0)}\beta_n} \approx \sqrt{k_1^2 - k_n^{(0)^2}}\left[1 - i\frac{k_n^{(0)}\beta_n}{k_1^2 - k_n^{(0)^2}}\right] , \quad (5.112)$$

where we have used the binomial expansion in Eq. (5.112) for

$$\left|\frac{2k_n^{(0)}\beta_n}{k_1^2 - k_n^{(0)^2}}\right| < 1 , \quad (5.113)$$

and we have the result

$$k_{1zn} \approx k_{1zn}^{(0)} - i\varepsilon_n \text{ with } \varepsilon_n = \frac{\rho_1}{\rho h}\frac{k_{zn}^{(0)^2}}{k_{1zn}^{(0)^2}} > 0 . \quad (5.114)$$

Thus, we see that in addition to the fact that the three types of eigenvalues are complex, the imaginary parts of k_{zn} and k_{1zn} are negative, giving rise to fields which are exponentially growing with depth. A negative imaginary part is not a difficulty in the instance of k_{zn}, as it is the vertical wavenumber associated with a bounded region. The wavenumber k_{1zn}, on the other hand, describes propagation in an unbounded half-space, leading to the physically disturbing, though mathematically sound, feature of exponential growth in the wave field. In this manner, the improper ramifications of a modal approach to this problem manifest themselves through complex eigenvalues and sound fields with atypical behavior.

A Leaky Mode Decomposition

We will now proceed with an asymptotic evaluation of the field in Eq. (5.95). First, we change variables,

$$k_{1z} = \sqrt{k_1^2 - k^2 + k_z^2} \Rightarrow dk_{1z} = \frac{k_z}{k_{1z}} dk_z \ , \tag{5.115}$$

and Eq. (5.95) becomes

$$p(r,z) \approx \frac{2i\rho_1}{\rho} \int_0^\infty F(k_{1z}) H_0^{(1)}(k_r r)\, dk_{1z} \ , \tag{5.116}$$

where $F(k_{1z})$ is defined in Eq. (5.96). Since $F(k_{1z})$ is a *meromorphic function* of k_{1z}, i.e., it contains pole-type singularities only, we can expand it in a *Mittag-Leffler expansion* [Stickler and Ammicht, 1980, 1984]:

$$F(k_{1z}) = \sum_{n=1}^{\infty} \left[\frac{r_n(k_{1zn})}{k_{1z} - k_{1zn}} + \frac{r_n(-k_{1zn})}{k_{1z} + k_{1zn}} \right] \ , \tag{5.117}$$

where $D(\pm k_{1zn}) = 0$, and $r_n(\pm k_{1zn})$ are the residues at the poles $k_{1z} = \pm k_{1zn}$. It is straightforward to show that $r_n(-k_{1zn}) = -r_n(k_{1zn})$, and therefore Eq. (5.117) becomes

$$F(k_{1z}) = 2\sum_{n=1}^{\infty} \frac{k_{1zn} r_n(k_{1zn})}{k_{1z}^2 - k_{1zn}^2} \ , \tag{5.118}$$

where we sum over positive k_{1zn}. Substituting Eq. (5.118) into Eq. (5.116), we obtain

$$p(r,z) \approx \frac{4i\rho_1}{\rho} \int_0^\infty \sum_{n=1}^{\infty} \frac{k_{1zn} r_n(k_{1zn})}{k_{1z}^2 - k_{1zn}^2} H_0^{(1)}(k_r r)\, dk_{1z} \quad . \qquad (5.119)$$

Interchanging the order of integration and summation, and substituting the asymptotic form for the Hankel function [cf. Eq. (4.119)] for $k_r r >> 1$ in Eq. (5.119), we find that

$$p(r,z) \sim 4\sqrt{\frac{2}{\pi r}} e^{i\pi/4} \frac{\rho_1}{\rho} \sum_{n=1}^{\infty} k_{1zn} r_n(k_{1zn}) \int_0^\infty \frac{1}{\left(k_{1z}^2 - k_{1zn}^2\right)} \frac{e^{ik_r r}}{\sqrt{k_r}} dk_{1z} \ . \quad (5.120)$$

Changing variables,

$$K_{1z} = \frac{k_{1z}}{k}, \quad K_{1zn} = \frac{k_{1zn}}{k}, \quad K_r = \frac{k_r}{k} \ , \qquad (5.121)$$

we can rewrite the integral $I(r)$ in Eq. (5.120) as

$$I(r) = \int_0^\infty \frac{1}{\left(k_{1z}^2 - k_{1zn}^2\right)} \frac{e^{ik_r r}}{\sqrt{k_r}} dk_{1z} = \frac{1}{k^{3/2}} \int_0^\infty \frac{1}{\left(K_{1z}^2 - K_{1zn}^2\right)} \frac{e^{ikK_r r}}{\sqrt{K_r}} dK_{1z}, \quad (5.122)$$

in which case it is in a canonical form for the application of asymptotic methods [cf. Eq. (4.123)] for large values of the parameter k:

$$I(r) = \frac{1}{k^{3/2}} \int_0^\infty f(K_{1z}) e^{ikh(K_{1z})}\, dK_{1z} \ , \qquad (5.123)$$

where

$$f(K_{1z}) = \frac{1}{\left(K_1^2 - K_{1z}^2\right)^{1/4}} \frac{1}{\left(K_{1z}^2 - K_{1zn}^2\right)},$$

$$h(K_{1z}) = K_r r = \sqrt{K_1^2 - K_{1z}^2}\, r = i\sqrt{K_{1z}^2 - K_1^2}\, r \ \text{ for } |K_{1z}| > K_1 \ . \quad (5.124)$$

The behavior of the integrand in Eq. (5.123) is dominated by the large values of $f(K_{1z})$ for $K_{1z} \approx K_{1zn}^{(0)} = k_{1zn}^{(0)}/k$ and the exponentially decaying factor introduced by the phase term for $|K_{1z}| > K_1$. It is therefore sensible to proceed with the choice of the critical point as $K_c = K_{1zn}^{(0)}$, and we expand $h(K_{1z})$ in a Taylor series about this point:

$$h(K_{1z}) \approx h(K_c) + h'(K_c)(K_{1z} - K_c)$$
$$\approx \sqrt{K_1^2 - K_c^2}\, r - \frac{K_c r}{\sqrt{K_1^2 - K_c^2}}(K_{1z} - K_c) \ . \qquad (5.125)$$

In evaluating $f(K_{1z})$ close to the critical point, we note that

$$\left.\frac{1}{K_{1z}^2 - K_{1zn}^2}\right|_{K_{1z} \approx K_c} = \left.\frac{1}{(K_{1z} - K_{1zn})(K_{1z} + K_{1zn})}\right|_{K_{1z} \approx K_c} \approx \frac{1}{2K_c(K_{1z} - K_{1zn})} , \qquad (5.126)$$

and therefore

$$f(K_{1z})\Big|_{K_{1z} \approx K_c} \approx \frac{1}{(K_1^2 - K_c^2)^{1/4}} \frac{1}{2K_c(K_{1z} - K_{1zn})} \ . \qquad (5.127)$$

The integral in Eq. (5.123) then becomes

$$I(r) \sim \frac{e^{ik\sqrt{K_1^2 - K_c^2}\, r}}{2k^{3/2}(K_1^2 - K_c^2)^{1/4} K_c} \int_{-\infty}^{\infty} \frac{e^{-ikK_c r(K_{1z} - K_c)/\sqrt{K_1^2 - K_c^2}}}{K_{1z} - K_{1zn}} dK_{1z} \ , \qquad (5.128)$$

where, following Erdelyi [1956], we have extended the lower limit of integration to $-\infty$. Changing variables in Eq. (5.128),

$$\tau = K_{1z} - K_c \Rightarrow d\tau = dK_{1z} \ , \qquad (5.129)$$

we obtain

$$I(r) \sim \frac{e^{ik\sqrt{K_1^2 - K_c^2}\, r}}{2k^{3/2}(K_1^2 - K_c^2)^{1/4} K_c} \int_{-\infty}^{\infty} \frac{e^{-ikK_c r\tau/\sqrt{K_1^2 - K_c^2}}}{\tau + i\varepsilon_n/k} d\tau, \qquad (5.130)$$

where we have used Eq. (5.114). The integrand in Eq. (5.130) has a simple pole at $\tau = -i\varepsilon_n/k$, and therefore we close the contour in the lower half-plane (cf. Fig. 5.12) and apply Cauchy's residue theorem to obtain the result

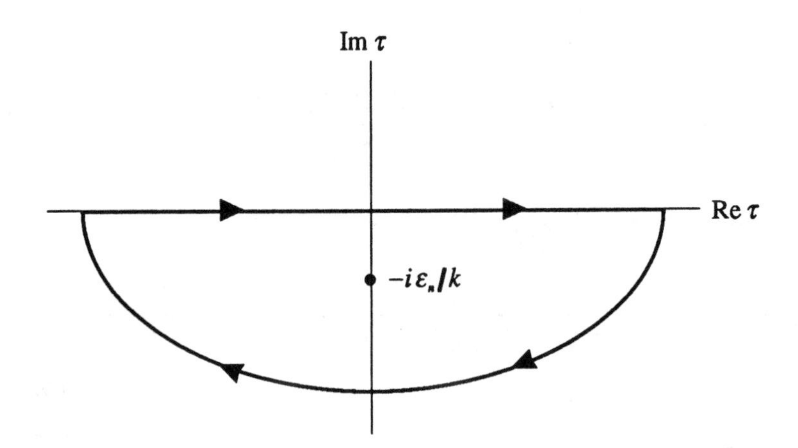

Figure 5.12 Closed contour of integration used in the asymptotic evaluation of the solution to the problem shown in Fig. 5.9 with $\rho_1 < \rho$ and $c_1 < c$.

$$
\begin{aligned}
I(r) &\sim \frac{e^{ik\sqrt{K_1^2-K_c^2}\,r}}{2k^{3/2}\left(K_1^2-K_c^2\right)^{1/4}K_c}\left[-2\pi i\,\mathrm{Res}\left(-i\varepsilon_n/k\right)\right] \\
&\sim \frac{e^{ik\sqrt{K_1^2-K_c^2}\,r}}{2k^{3/2}\left(K_1^2-K_c^2\right)^{1/4}K_c}\left[-2\pi i\frac{e^{-ikK_c r\tau/\sqrt{K_1^2-K_c^2}}}{\partial\left(\tau+i\varepsilon_n/k\right)/\partial\tau}\right]_{\tau=-i\varepsilon_n/k} \\
&\sim -\frac{\pi i e^{ik\sqrt{K_1^2-K_c^2}\,r}}{k^{3/2}\left(K_1^2-K_c^2\right)^{1/4}K_c}e^{-K_c r\varepsilon_n/\sqrt{K_1^2-K_c^2}} \quad . \qquad (5.131)
\end{aligned}
$$

Finally, using Eqs. (5.114) and (5.121) and recalling that $K_c = K_{1zn}^{(0)}$, we can rewrite Eq. (5.131) as

$$
I(r) \sim -\frac{\pi i e^{ik_n^{(0)}r}}{k_n^{(0)1/2}k_{1zn}^{(0)}}\exp\left[-\frac{\rho_1 k_{zn}^{(0)2}}{\rho h k_n^{(0)}k_{1zn}^{(0)}}r\right] \quad . \qquad (5.132)
$$

The only remaining quantity to compute in our evaluation of the field in Eq. (5.120) is the residue

$$
r_n(k_{1zn}) = \left.\frac{N(k_{1z})}{\partial D(k_{1z})/\partial k_{1z}}\right|_{k_{1z}=k_{1zn}} \quad , \qquad (5.133)
$$

where we recall that $N(k_{1z})$ and $D(k_{1z})$ are defined in Eq. (5.96). We make the approximation

$$N(k_{1z})\big|_{k_{1z}=k_{1zn}} \approx N(k_{1z})\big|_{k_{1z}=k_{1zn}^{(0)}} = \sin k_{zn}^{(0)} z_0 \sin k_{zn}^{(0)} z \ , \qquad (5.134)$$

and, using the eigenvalue equation [cf. Eq. (5.99)], determine that to leading order (cf. Prob. 5.7),

$$\left.\frac{\partial D(k_{1z})}{\partial k_{1z}}\right|_{k_{1z}=k_{1zn}} \approx -2ih\frac{\rho_1}{\rho} \ . \qquad (5.135)$$

Finally, substituting the results in Eqs. (5.132)-(5.135) into Eq. (5.120) and letting $k_{1zn} r_n(k_{1zn}) \approx k_{1zn}^{(0)} r_n(k_{1zn})$ in Eq. (5.120), we obtain

$$p(r,z) \sim \frac{2\sqrt{2\pi} e^{i\pi/4}}{h} \sum_{n=1}^{\infty} \sin k_{zn}^{(0)} z_0 \sin k_{zn}^{(0)} z \, e^{-\frac{\rho_1 k_{zn}^{(0)2}}{\rho h k_n^{(0)} k_{1zn}^{(0)}} r} \frac{e^{ik_n^{(0)} r}}{\sqrt{k_n^{(0)} r}} \ . \qquad (5.136)$$

We associate the improper modes in Eq. (5.136) with up- and down-going plane waves that only partly constructively interfere with one another due to partial transmission into the bottom [cf. Eq. (5.98)]. As these modes propagate along the waveguide, they lose energy into the bottom, causing an exponential decay with range as well as an exponential growth in depth. The modal sum in Eq. (5.136) is identical to that for a waveguide with pressure-release top and bottom [cf. Eq. (5.26)] except for the additional exponentially decaying factor associated with leakage into the penetrable bottom in Fig. 5.9. In fact, this leaky mode decomposition reduces to the result for a pressure-release bottom in the limit $\rho_1/\rho \to 0$, as required.

5.5 Normal Modes for a Homogeneous Fluid Layer Bounded Above by a Pressure-Release Surface and Below by a Higher Velocity, Homogeneous Fluid Half-Space

In our discussion of normal modes so far, we have considered extreme cases where, on the one hand, when the waveguide had impenetrable boundaries, the solution consisted totally of a sum of discrete modes (cf. Sec. 5.2.1); on the other hand, when one of the bounding media was penetrable and had lower sound velocity, the solution was composed entirely of a continuum of modes, which could

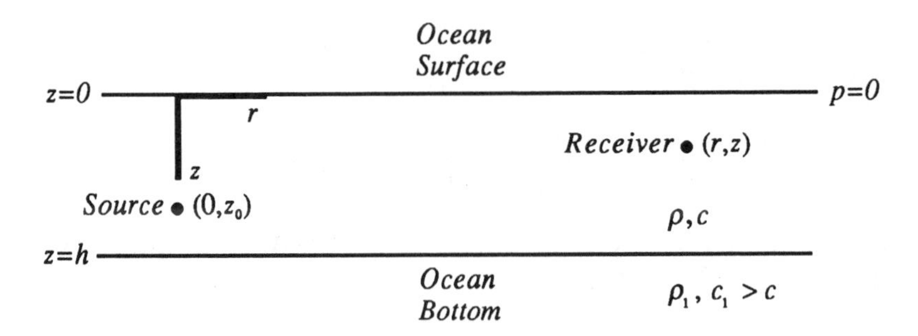

Figure 5.13 A point source in a homogeneous fluid layer with a pressure-release top and a homogeneous, higher velocity fluid bottom.

be approximated by a sum of discrete virtual modes (cf. Sec. 5.4). In this section, we derive the modal solution for the case of a higher velocity bottom, shown in Fig. 5.13. We will find that the phenomenon of total internal reflection which occurs in this instance (cf. pp. 45-51) gives rise to both discrete and continuum sets of modes. In our calculations, we will also assume that $\rho_1 > \rho$, which together with the condition $c_1 > c$ and a pressure-release top, constitute a *Pekeris waveguide* [Pekeris, 1948]. This canonical example is of particular interest in ocean acoustics because it embodies many of the fundamental features of acoustic propagation in shallow water.

We recall that for plane wave reflection from a higher velocity half-space, the critical angle θ_c delineates two distinct regions for which the reflection coefficient has characteristic behavior (cf. Fig. 3.10). For angles of incidence $\theta < \theta_c$, the reflection coefficient has magnitude $|R_B| < 1$, while for $\theta \geq \theta_c$, we have $|R_B| = 1$. As a result, for $\theta < \theta_c$, the behavior of the eigenvalue equation is similar to that described in Sec. 5.4.1, in that complex eigenvalues arise, corresponding to an improperly posed Sturm-Liouville problem. For $\theta \geq \theta_c$, however, Eq. (5.72) becomes

$$e^{2i(k_z h - \pi/2 + \varphi/2)} = 1 \quad , \tag{5.137}$$

where φ is the phase of R_B [cf. Eq. (3.58)]:

$$\varphi = -2\tan^{-1}\left[\frac{\sqrt{\sin^2\theta - c^2/c_1^2}}{m\cos\theta}\right] \quad , \tag{5.138}$$

and we recall that $m = \rho_1/\rho$. Eq. (5.137) implies that

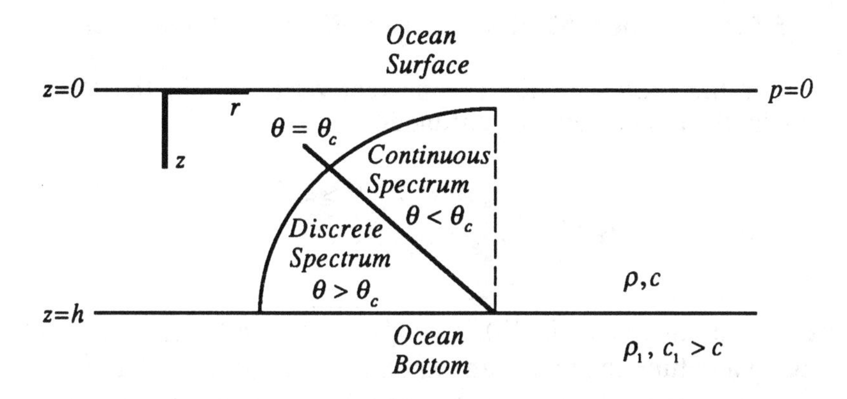

Figure 5.14 Angular regions for the discrete and continuous modal spectra associated with the solution of the problem shown in Fig. 5.13.

$$2(k_{zn}h - \pi/2 + \varphi_n/2) = 2(n-1)\pi, \quad n = 1,2,3,\ldots \quad , \tag{5.139}$$

and therefore

$$k_{zn}h = (n - 1/2)\pi - \varphi_n/2, \quad n = 1,2,3,\ldots \quad . \tag{5.140}$$

The solutions of Eq. (5.140) are real and are thus associated with a proper Sturm-Liouville problem.

We will therefore solve the problem in Fig. 5.13 by first obtaining the modal solutions separately for each of the two angular regions depicted in Fig. 5.14. The total solution will then be constructed as the superposition of the solutions for these two angular regimes. We shall see that for $\theta > \theta_c$, a discrete set of perfectly trapped modes will be generated because $|R_B| = 1$. For $\theta < \theta_c$, a continuum of modes will arise because $|R_B| < 1$; this modal continuum can, in turn, be approximated by a leaky mode decomposition analogous to that described in Sec. 5.4.3. Thus, we write the total field $p(r,z)$ as

$$p(r,z) = p_t(r,z) + p_c(r,z) \quad , \tag{5.140a}$$

where $p_t(r,z)$ and $p_c(r,z)$ are the contributions of the trapped and continuum modes, respectively.

5.5.1 Proper Modes for the Pekeris Waveguide

For the modes which are perfectly trapped in the Pekeris waveguide, the modal depth eigenfunctions are

$$u_n(z) = \begin{cases} A_n \sin k_{zn} z, & 0 \le z \le h, \\ B_n e^{-k_{1zn} z}, & z \ge h \ , \end{cases} \tag{5.141}$$

where k_{zn} satisfies Eq. (5.140), $k_n = \sqrt{k^2 - k_{zn}^2}$, and $k_{1zn} = \sqrt{k_n^2 - k_1^2}$. Here the wave functions have an exponential decay with depth in the bottom due to the effect of total internal reflection. Continuity of pressure at $z = h$ implies that

$$\frac{A_n}{B_n} = \frac{e^{-k_{1zn} h}}{\sin k_{zn} h} \ , \tag{5.142}$$

which, when used in the orthonormality relation [cf. Eq. (5.11)], yields

$$\frac{A_n^2}{\rho} \int_0^h \sin^2 k_{zn} z \, dz + \frac{A_n^2}{\rho_1} e^{2k_{1zn} h} \sin^2 k_{zn} h \int_h^\infty e^{-2k_{1zn} z} \, dz = 1, \tag{5.143}$$

from which it is straightforward to show that

$$A_n = \sqrt{2} \left[\frac{1}{\rho} \left(h - \frac{\sin 2k_{zn} h}{2k_{zn}} \right) + \frac{1}{\rho_1} \frac{\sin^2 k_{zn} h}{k_{1zn}} \right]^{-1/2} . \tag{5.144}$$

The full solution for the trapped modes $p_t(r,z)$ with the source in the water column $(z_0 \le h)$ is then given by

$$p_t(r,z) = \begin{cases} \dfrac{i\pi}{\rho} \displaystyle\sum_{n=1}^{n_{\max}} A_n^2 \sin k_{zn} z_0 \sin k_{zn} z \, H_0^{(1)}(k_n r), & 0 \le z \le h, \\ \dfrac{i\pi}{\rho} \displaystyle\sum_{n=1}^{n_{\max}} A_n^2 \sin k_{zn} z_0 \sin k_{zn} h \, e^{-k_{1zn}(z-h)} H_0^{(1)}(k_n r), & z \ge h, \end{cases} \tag{5.145}$$

which has the asymptotic behavior $(k_n r >> 1)$

$$p_t(r,z) \sim \begin{cases} \dfrac{\sqrt{2\pi}e^{i\pi/4}}{\rho}\displaystyle\sum_{n=1}^{n_{\max}} A_n^2 \sin k_{zn}z_0 \sin k_{zn}z \frac{e^{ik_n r}}{\sqrt{k_n r}}, \; 0 \le z \le h, \\ \dfrac{\sqrt{2\pi}e^{i\pi/4}}{\rho}\displaystyle\sum_{n=1}^{n_{\max}} A_n^2 \sin k_{zn}z_0 \sin k_{zn}h\, e^{-k_{1zn}(z-h)} \frac{e^{ik_n r}}{\sqrt{k_n r}}, \; z \ge h. \end{cases} \quad (5.146)$$

Here $n_{\max}$ is the maximum number of trapped, propagating modes. We will now examine in detail key features of this modal field, indicating relationships to the case with impenetrable boundaries (the zeroth order model of Sec. 5.2.1) where appropriate.

Nodes of the Modal Depth Eigenfunctions

The nth depth eigenfunction $u_n(z)$ will have nodes at depths z'_{nm} which satisfy [cf. Eqs. (5.28)-(5.30)]

$$\sin k_{zn} z'_{nm} = 0 \quad , \qquad (5.147)$$

so that

$$k_{zn} z'_{nm} = m\pi, \;\; m = 0,1,2,\ldots \quad , \qquad (5.148)$$

and therefore

$$z'_{nm} = \frac{m\pi h}{\left[(n - 1/2)\pi - \varphi_n/2\right]}, \;\; m = 0,1,2,\ldots, \;\; z'_{nm} < h \quad , \qquad (5.149)$$

where we have used Eq. (5.140). By comparing Eqs. (5.149) and (5.30), we conclude that $z'_{nm} < z_{nm}$, where z_{nm} are the nodes of the eigenfunctions for the hard bottom model of Sec. 5.2.1, since $-\pi \le \varphi_n \le 0$. It is also clear that

$$z'_{nm} \to z_{nm} \;\text{ for }\; \varphi \to 0 \quad , \qquad (5.150)$$

as expected. Thus, in the Pekeris case, the nodes are compressed toward the surface, and the modes oscillate at a higher spatial frequency than in the hard bottom case. In Fig. 5.15, we have sketched the first four eigenfunctions, which are to be contrasted with the modes for the zeroth order model depicted in Fig. 5.3. The

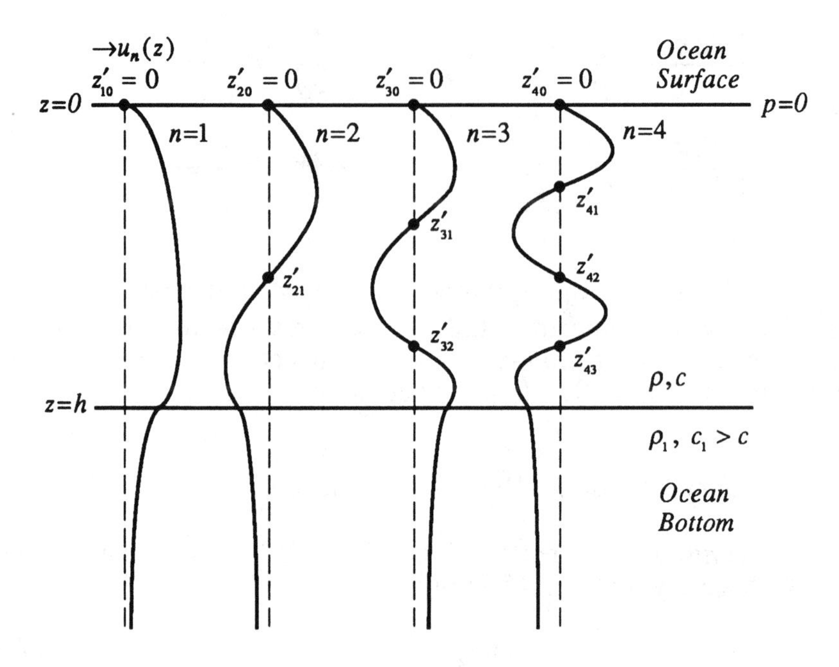

Figure 5.15 Schematic illustration of the first four modes as a function of depth for a homogeneous fluid layer with a soft top and a homogeneous, higher velocity fluid bottom.

continuity of pressure at the water-bottom interface is illustrated as well as the fact that the rate of exponential decay in the bottom increases with decreasing mode number.

Modal Cutoff Frequency

Both the maximum number of propagating, proper modes and their cutoff frequencies can be derived by beginning with the condition for perfect trapping, $\theta_n > \theta_c$, which implies that $\sin\theta_n > \sin\theta_c$, and therefore

$$\sin^2\theta_n > \frac{c^2}{c_1^2} , \tag{5.151}$$

since $\sin\theta_c = c/c_1$. But using Eq. (5.27), we find that

$$\sin^2\theta_n = \frac{k_n^2}{k^2} = 1 - \frac{k_{zn}^2}{k^2} , \tag{5.152}$$

which, when substituted into Eq. (5.151), yields

$$\frac{k_{zn}^2}{k^2} < 1 - \frac{c^2}{c_1^2} \ , \tag{5.153}$$

or

$$k_{zn} < \frac{2\pi}{\lambda}\sqrt{1 - \frac{c^2}{c_1^2}} \ . \tag{5.154}$$

We also know from the eigenvalue equation [cf. Eq. (5.140)] that

$$k_{zn} = \frac{1}{h}\left[(n - 1/2)\pi - \varphi_n/2\right] \ , \tag{5.155}$$

which, when combined with Eq. (5.154), yields

$$n < \frac{2h}{\lambda}\sqrt{1 - \frac{c^2}{c_1^2}} + \frac{\varphi_n}{2\pi} + \frac{1}{2} \ . \tag{5.156}$$

But $-\pi \le \varphi_n \le 0$, and therefore the maximum number n_{max} of propagating, trapped modes is

$$n_{max} = \frac{2h}{\lambda}\sqrt{1 - \frac{c^2}{c_1^2}} + \frac{1}{2} \ , \tag{5.157}$$

which reduces to the hard bottom result in the limit $c_1 \to \infty$. For a given ratio h/λ, the value of n_{max} in Eq. (5.157) is smaller than that in the hard bottom case because of the more constrained angular regime for perfect reflection in the Pekeris waveguide.

We can also determine the cutoff frequency ω_n' below which the nth mode will not propagate by using the relations $\lambda = 2\pi/k$ and $\omega = ck$ in Eq. (5.157):

$$\omega_n' = \frac{\pi c}{h}\frac{(n - 1/2)}{\sqrt{1 - c^2/c_1^2}} = \frac{\omega_n}{\sqrt{1 - c^2/c_1^2}} \ , \tag{5.158}$$

where ω_n is the cutoff frequency for the hard bottom model. Thus, because of the more restrictive propagation conditions, the cutoff frequencies for the Pekeris model are higher than those for the hard bottom case. It is also evident that $\omega_n' = \omega_n$ in the limit $c_1 \to \infty$.

Modal Phase Velocity

The behavior of the modal phase velocity C_n can be determined from the eigenvalue equation [cf. Eqs. (5.138) and (5.140)] which, using Eqs. (5.27) and (5.152), can be rewritten as

$$\frac{\omega h}{c}\sqrt{1-\frac{c^2}{C_n^2}} = (n-1/2)\pi + \tan^{-1}\left[\frac{\rho\sqrt{c^2/C_n^2 - c^2/c_1^2}}{\rho_1\sqrt{1-c^2/C_n^2}}\right] . \quad (5.159)$$

Equation (5.159) can also be expressed in terms of the cutoff frequency by using Eq. (5.158), so that we have

$$\omega = \frac{c}{h\sqrt{1-c^2/C_n^2}}\left\{\frac{\omega_n' h}{c}\sqrt{1-c^2/c_1^2} + \tan^{-1}\left[\frac{\rho\sqrt{c^2/C_n^2 - c^2/c_1^2}}{\rho_1\sqrt{1-c^2/C_n^2}}\right]\right\} . \quad (5.160)$$

It is clear from Eq. (5.160) that $\omega = \omega_n'$ when $C_n = c_1$ and that $\omega \to \infty$ as $C_n \to c$. Thus, the phase velocity is bounded, with $c \le C_n \le c_1$, as shown in Fig. 5.16 for the first mode.

Modal Group Velocity

The modal group velocity V_n can be obtained by first taking the tangent of both sides of the eigenvalue equation [cf. Eq. (5.159)] and using the identity [Abramowitz and Stegun, 1964]

$$\tan(z_1 + z_2) = \frac{\tan z_1 + \tan z_2}{1 - \tan z_1 \tan z_2} , \quad (5.161)$$

to obtain

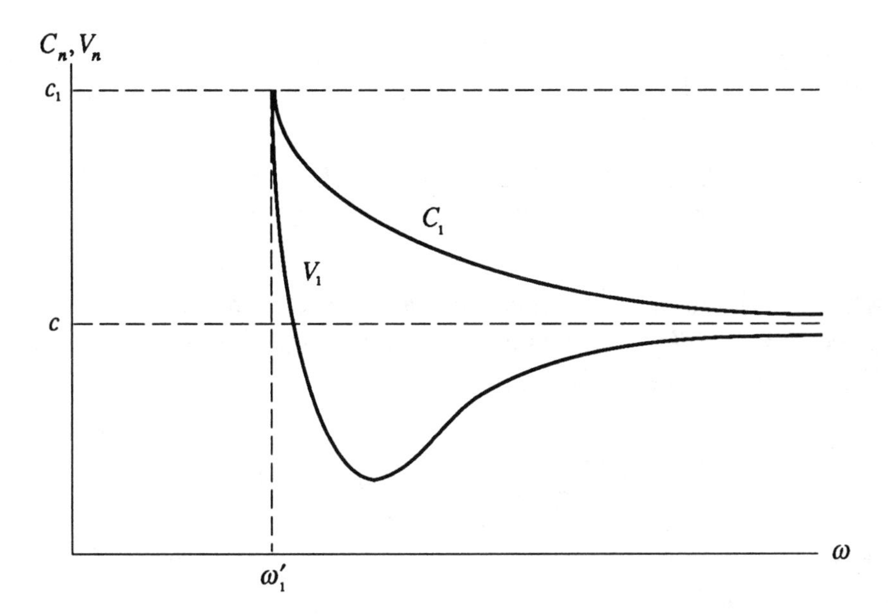

Figure 5.16 Phase and group velocity dispersion curves for the first mode of a homogeneous fluid layer with a soft top and a homogeneous, higher velocity fluid bottom.

$$\tan\left[\frac{\omega h}{c}\sqrt{1-\frac{c^2}{C_n^2}}\right]=-\frac{\rho_1\sqrt{1-c^2/C_n^2}}{\rho\sqrt{c^2/C_n^2-c^2/c_1^2}} \quad . \tag{5.162}$$

Equation (5.162) can be rewritten as

$$\tan\left[h\sqrt{k^2-k_n^2}\right]=-\frac{\rho_1\sqrt{k^2-k_n^2}}{\rho\sqrt{k_n^2-k_1^2}} \quad , \tag{5.163}$$

and is thus a form of the eigenvalue equation equivalent to that presented in Eq. (5.137) [cf. Eq. (5.99) for the case $c_1 < c$]. Differentiating both sides of Eq. (5.163) with respect to ω, we obtain

$$\frac{h}{\cos^2\left[h\left(k^2-k_n^2\right)^{1/2}\right]}\left(\frac{k}{c}-k_n\frac{dk_n}{d\omega}\right)$$

$$=-\frac{\rho_1}{\rho}\left[\frac{1}{\left(k_n^2-k_1^2\right)^{1/2}}\left(\frac{k}{c}-k_n\frac{dk_n}{d\omega}\right)-\frac{\left(k^2-k_n^2\right)}{\left(k_n^2-k_1^2\right)^{3/2}}\left(k_n\frac{dk_n}{d\omega}-\frac{k_1}{c_1}\right)\right], \quad (5.164)$$

where we have used the relations $dk/d\omega=1/c$ and $dk_1/d\omega=1/c_1$. Since $V_n=d\omega/dk_n$, it is then a straightforward procedure to solve Eq. (5.164) for the group velocity

$$V_n=\frac{k_n\left\{\frac{h\left(k_n^2-k_1^2\right)^{3/2}}{\cos^2\left[h\left(k^2-k_n^2\right)^{1/2}\right]}+\frac{\rho_1}{\rho}\left(k^2-k_1^2\right)\right\}}{\frac{k}{c}\left\{\frac{h\left(k_n^2-k_1^2\right)^{3/2}}{\cos^2\left[h\left(k^2-k_n^2\right)^{1/2}\right]}+\frac{\rho_1}{\rho}\left(k_n^2-k_1^2\right)\right\}+\frac{\rho_1 k_1}{\rho c_1}\left(k^2-k_n^2\right)} \quad . \quad (5.165)$$

Recalling the limiting behavior of the phase velocity,

$$\begin{aligned}&\omega\to\infty\Rightarrow C_n\to c\Rightarrow k_n\to k,\\&\omega\to\omega_n'\Rightarrow C_n\to c_1\Rightarrow k_n\to k_1 \quad ,\end{aligned} \quad (5.166)$$

and examining Eq. (5.165), it becomes clear that

$$\lim_{\omega\to\infty}V_n=c \text{ and } \lim_{\omega\to\omega_n'}V_n=c_1 \quad . \quad (5.167)$$

Thus, the phase and group velocities for the Pekeris waveguide have the same limiting behavior. However, the group velocity dispersion curve has an additional feature, namely, a minimum called the *Airy phase* (cf. Prob. 5.11), which is shown in Fig. 5.16. For frequencies greater than the Airy phase frequency, the group velocity tends toward the sound velocity in the fluid layer, corresponding to *water wave* propagation in ocean acoustics; for frequencies less than the Airy phase frequency, the group velocity tends toward the sound velocity in the bottom, corresponding to *ground wave* propagation.

Modal Intensity, Interference Wavelength, and Cycle Distance

The asymptotic behavior of the intensity for the perfectly trapped modes in the Pekeris waveguide for both source and receiver in the water column $(z, z_0 < h)$ is given by [cf. Eq. (5.146)]

$$I = \frac{|p_t(r,z)|^2}{2\rho c} \sim \frac{\pi}{\rho^3 c} \sum_{n=1}^{n_{\max}} A_n^2 \sin k_{zn} z_0 \sin k_{zn} z \frac{e^{ik_n r}}{\sqrt{k_n r}}$$

$$\times \sum_{m=1}^{n_{\max}} A_m^2 \sin k_{zm} z_0 \sin k_{zm} z \frac{e^{-ik_m r}}{\sqrt{k_m r}} \quad , \qquad (5.168)$$

which can be rewritten as

$$I \sim \frac{\pi}{\rho^3 cr} \sum_{n=1}^{n_{\max}} A_n^4 \frac{\sin^2 k_{zn} z_0 \sin^2 k_{zn} z}{k_n}$$

$$+ \frac{2\pi}{\rho^3 cr} \sum_{\substack{n=1 \\ n<m=2}}^{n_{\max}} A_n^2 A_m^2 \sin k_{zn} z_0 \sin k_{zm} z_0 \sin k_{zn} z \sin k_{zm} z \frac{\cos(k_n - k_m) r}{\sqrt{k_n k_m}} \quad , (5.169)$$

where A_n and A_m are defined in Eq. (5.144). Equation (5.169) has a mathematical form similar to the intensity for the zeroth order ocean acoustic waveguide [cf. Eq. (5.45)]. However, the differing eigenvalues in the two cases lead to pronounced differences in the corresponding interference patterns and their physical interpretation. Specifically, let us consider the interference wavelength for two adjacent modes in the Pekeris case. Then, for thick waveguides and sufficiently closely spaced modes, we can approximate the finite difference in Eq. (5.50) by a continuous derivative

$$\lambda_{n,n+1} = \frac{2\pi}{k_n - k_{n+1}} \approx -\frac{2\pi}{dk_n/dn} \quad . \qquad (5.170)$$

But rewriting the eigenvalue equation [cf. Eqs. (5.138) and (5.140)],

$$(n - 1/2)\pi = \sqrt{k^2 - k_n^2}\, h + \varphi_n/2 \quad , \qquad (5.171)$$

and differentiating both sides of Eq. (5.171) with respect to k_n, we obtain

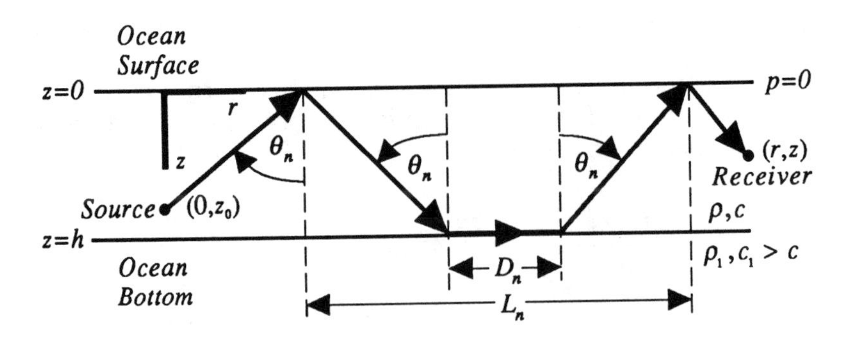

Fig. 5.17 Cycle or skip distance L_n and ray displacement D_n associated with the nth mode for a homogeneous fluid layer with a soft top and a homogeneous, higher velocity fluid bottom.

$$\pi \frac{dn}{dk_n} = -\frac{k_n h}{\sqrt{k^2 - k_n^2}} + \frac{1}{2}\frac{d\varphi_n}{dk_n} \quad . \tag{5.172}$$

Equation (5.170) therefore becomes

$$\lambda_{n,n+1} \approx 2h \tan\theta_n - \frac{d\varphi_n}{dk_n} \quad , \tag{5.173}$$

and we see that the geometrical acoustics term $2h\tan\theta_n$, which was the only contribution in the case of impenetrable boundaries [cf. Eq. (5.56)], has been augmented by a term which depends on the phase of the reflection coefficient for the penetrable bottom. This behavior suggests that the cycle or skip distance L_n associated with the nth mode consists of a geometrical part and a *ray displacement* D_n given by (cf. Fig. 5.17)

$$D_n = -\frac{d\varphi_n}{dk_n} = \frac{2mk_n\left(k^2 - k_1^2\right)}{\sqrt{\left(k^2 - k_n^2\right)\left(k_n^2 - k_1^2\right)}\left[m^2\left(k^2 - k_n^2\right) + k_n^2 - k_1^2\right]} \quad . \tag{5.174}$$

It can be shown that this additional lateral shift is precisely equal to the displacement which an acoustic beam of finite width undergoes when it interacts with the bottom [Brekhovskikh, 1980; Tindle and Weston, 1980; Brekhovskikh and Lysanov, 1991]. The rigorous connection between a group of modes and a geometrical path with lateral

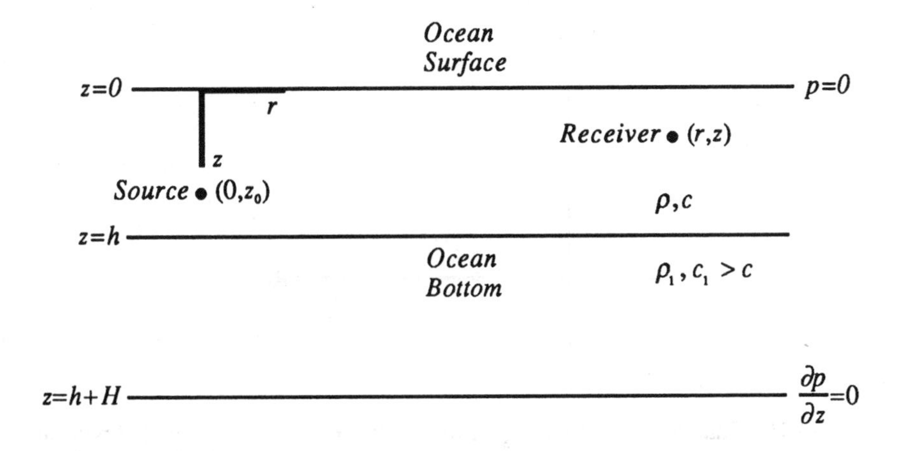

Figure 5.18 The properly posed Sturm-Liouville problem which, in the limit of infinite H, leads to the modal continuum part of the solution to the improper Sturm-Liouville problem shown in Figs. 5.13 and 5.14.

displacement is derived by Kamel and Felsen [1982] using the Poisson sum formula (cf. Prob. 5.6).

5.5.2 Improper Modes for the Pekeris Waveguide

In order to evaluate the contribution of the modal continuum to the field in the Pekeris waveguide, we utilize a procedure analogous to that employed for the solution of the low-velocity bottom problem in Fig. 5.9. That is, we solve the related proper Sturm-Liouville problem shown in Fig. 5.18 and determine the behavior of the solution as $H \to \infty$. In fact, the sequence of steps associated with solving the problem in Fig. 5.18 is identical to that described in Sec. 5.4.2, and the final result for the modal continuum $p_c(r,z)$ has the same form as Eq. (5.94):

$$p_c(r,z) \approx \frac{2i\rho_1}{\rho} \int_{C_B} \frac{k_z}{k_{1z}} \frac{\sin k_z z_0 \sin k_z z}{\left[\sin^2 k_z h + \dfrac{\rho_1^2 k_z^2}{\rho^2 k_{1z}^2} \cos^2 k_z h\right]} H_0^{(1)}(k_r r)\, dk_z. \quad (5.175)$$

The only difference between the two results lies in the path of integration C_B, which is now the one shown in Fig. 5.19 rather than the one given in Fig. 5.11. This difference arises from the change-of-variable step in Eq. (5.93).

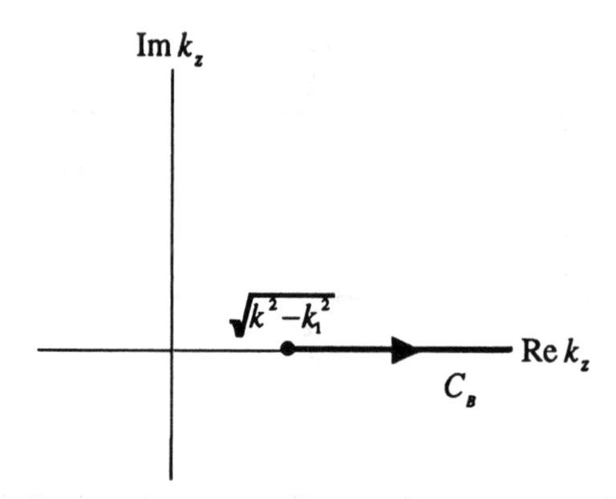

Figure 5.19 Path of integration C_B in the complex k_z-plane associated with the modal continuum part of the solution to the problem shown in Figs. 5.13 and 5.14.

A Leaky Mode Decomposition

Here, as in Sec. 5.4.3, a leaky mode decomposition of the continuum field in Eq. (5.175) can be developed by the use of a Mittag-Leffler expansion and asymptotic methods. Specifically, the eigenvalue equation is the same as Eq. (5.99), which can be rewritten as

$$\cot k_{zn}h = i\frac{\rho k_{1zn}}{\rho_1 k_{zn}} \quad . \tag{5.176}$$

It is clear that for the parameters of the waveguide that we have prescribed ($\rho_1 > \rho$, $c_1 > c$),

$$\frac{\rho k_{1zn}}{\rho_1 k_{zn}} < 1 \quad , \tag{5.177}$$

and therefore we can obtain the zeroth order solution $k_{zn}^{(0)}$ for the pole positions by assuming that the right-hand side of Eq. (5.176) is zero, so that

$$k_{zn}^{(0)} = \frac{(n-1/2)\pi}{h}, \quad n = 1,2,3,\ldots \quad . \tag{5.178}$$

These are simply the eigenvalues for a hard bottom which we now substitute into the right-hand side of Eq. (5.176), while substituting a sum of zeroth and first order terms into the left-hand side:

$$k_{zn} \approx k_{zn}^{(0)} + k_{zn}^{(1)} \quad . \tag{5.179}$$

Using the identity

$$\cot\left(k_{zn}^{(0)} + k_{zn}^{(1)}\right)h = \frac{\cos k_{zn}^{(0)}h \cos k_{zn}^{(1)}h - \sin k_{zn}^{(0)}h \sin k_{zn}^{(1)}h}{\sin k_{zn}^{(0)}h \cos k_{zn}^{(1)}h + \cos k_{zn}^{(0)}h \sin k_{zn}^{(1)}h}, \tag{5.180}$$

we obtain

$$\tan k_{zn}^{(1)}h \approx -i\frac{\rho k_{1zn}^{(0)}}{\rho_1 k_{zn}^{(0)}} \quad , \tag{5.181}$$

where

$$k_{1zn}^{(0)} = \sqrt{k_1^2 - k^2 + k_{zn}^{(0)^2}} \quad . \tag{5.182}$$

But we know that

$$\tan k_{zn}^{(1)}h \approx k_{zn}^{(1)}h \ \text{ for } \left|k_{zn}^{(1)}h\right| << 1 \quad , \tag{5.183}$$

and therefore

$$k_{zn}^{(1)} \approx -i\frac{\rho}{\rho_1 h}\frac{k_{1zn}^{(0)}}{k_{zn}^{(0)}} \quad , \tag{5.184}$$

so that we can write

$$k_{zn} \approx k_{zn}^{(0)} - i\alpha_n \ \text{ with } \ \alpha_n = \frac{\rho}{\rho_1 h}\frac{k_{1zn}^{(0)}}{k_{zn}^{(0)}} > 0 \quad . \tag{5.185}$$

Using Eq. (5.185) and the relation $k_n = \sqrt{k^2 - k_{zn}^2}$, we find that

$$k_n \approx \sqrt{k^2 - k_{zn}^{(0)^2} + 2ik_{zn}^{(0)}\alpha_n} \approx \sqrt{k^2 - k_{zn}^{(0)^2}}\left[1 + i\frac{k_{zn}^{(0)}\alpha_n}{k^2 - k_{zn}^{(0)^2}}\right] , \quad (5.186)$$

where we have used the binomial expansion [cf. Eq. (4.50)] in Eq. (5.186) for

$$\left|\frac{2k_{zn}^{(0)}\alpha_n}{k^2 - k_{zn}^{(0)^2}}\right| < 1 . \quad (5.187)$$

Thus, we can also write

$$k_n \approx k_n^{(0)} + i\beta_n \text{ with } \beta_n = \frac{\rho}{\rho_1 h}\frac{k_{1zn}^{(0)}}{k_n^{(0)}} > 0 , \quad (5.188)$$

and $k_n^{(0)} = \sqrt{k^2 - k_{zn}^{(0)^2}}$. Finally, using the relation $k_{1zn} = \sqrt{k_1^2 - k_n^2}$ and Eq. (5.188), we obtain

$$k_{1zn} \approx \sqrt{k_1^2 - k_n^{(0)^2} - 2ik_n^{(0)}\beta_n} \approx \sqrt{k_1^2 - k_n^{(0)^2}}\left[1 - i\frac{k_n^{(0)}\beta_n}{k_1^2 - k_n^{(0)^2}}\right] , \quad (5.189)$$

where we have used the binomial expansion in Eq. (5.189) for

$$\left|\frac{2k_n^{(0)}\beta_n}{k_1^2 - k_n^{(0)^2}}\right| < 1 , \quad (5.190)$$

and we have the result

$$k_{1zn} \approx k_{1zn}^{(0)} - i\varepsilon_n \text{ with } \varepsilon_n = \frac{\rho}{\rho_1 h} > 0 \cdot \quad (5.191)$$

Here it is interesting to note that ε_n is independent of mode number.

Having obtained the leaky pole positions, we proceed with an asymptotic evaluation of the field in Eq. (5.175). Again, we change variables,

$$k_{1z} = \sqrt{k_1^2 - k^2 + k_z^2} \Rightarrow dk_{1z} = \frac{k_z}{k_{1z}} dk_z \ , \tag{5.192}$$

and Eq. (5.175) becomes

$$p_c(r,z) \approx \frac{2i\rho_1}{\rho} \int_0^\infty F(k_{1z}) H_0^{(1)}(k_r r)\, dk_{1z} \ , \tag{5.193}$$

where $F(k_{1z}) = N(k_{1z})/D(k_{1z})$, and $N(k_{1z})$ and $D(k_{1z})$ are defined in Eq. (5.96). Since Eq. (5.193) is identical to Eq. (5.95), the steps involved in the application of the Mittag-Leffler expansion and its asymptotic evaluation are identical to those described in Sec. 5.4.3, with one subtle difference. In order that the transmitted wave field [cf. Eq. (5.98)] satisfy the Sommerfeld radiation condition, we require that

$$\operatorname{Re} k_{1zn}^{(0)} > 0 \Rightarrow \operatorname{Re} \sqrt{k_1^2 - k^2 + \left[\frac{(n - 1/2)\pi}{h}\right]^2} > 0 \ , \tag{5.194}$$

and therefore the Mittag-Leffler expansion consists only of terms which satisfy $n > n_{\max}$, where $n_{\max}$ is defined in Eq. (5.157). As a result, we obtain [cf. Eqs. (5.120) and (5.131)]

$$p_c(r,z) \sim 4\sqrt{\frac{2}{\pi r}} e^{i\pi/4} \frac{\rho_1}{\rho} \sum_{n=n_{\max}+1}^{\infty} k_{1zn} r_n(k_{1zn}) \int_0^\infty \frac{1}{\left(k_{1z}^2 - k_{1zn}^2\right)} \frac{e^{ik_r r}}{\sqrt{k_r}} dk_{1z} . \tag{5.195}$$

where

$$\int_0^\infty \frac{1}{\left(k_{1z}^2 - k_{1zn}^2\right)} \frac{e^{ik_r r}}{\sqrt{k_r}} dk_{1z} \sim -\frac{\pi i e^{ik_n^{(0)} r}}{k_n^{(0)1/2} k_{1zn}^{(0)}} \exp\left[-\frac{\varepsilon_n k_{1zn}^{(0)}}{k_n^{(0)}} r\right] . \tag{5.196}$$

and ε_n is now given by Eq. (5.191). We approximate the residue,

$$r_n(k_{1zn}) = \left. \frac{N(k_{1z})}{\partial D(k_{1z})/\partial k_{1z}} \right|_{k_{1z} = k_{1zn}} \ , \tag{5.197}$$

in a manner analogous to that used in Sec. 5.4.3 and obtain the same results:

$$N(k_{1z})\big|_{k_{1z}=k_{1zn}} \approx N(k_{1z})\big|_{k_{1z}=k_{1zn}^{(0)}} = \sin k_{zn}^{(0)} z_0 \sin k_{zn}^{(0)} z \ , \qquad (5.198)$$

$$\left.\frac{\partial D(k_{1z})}{\partial k_{1z}}\right|_{k_{1z}=k_{1zn}} \approx -2ih\frac{\rho_1}{\rho} \ . \qquad (5.199)$$

Finally, substituting the results in Eqs. (5.196)-(5.199) into Eq. (5.195) and letting $k_{1zn} r_n(k_{1zn}) \approx k_{1zn}^{(0)} r_n(k_{1zn})$ in Eq. (5.195), we obtain

$$p_c(r,z) \sim \frac{2\sqrt{2\pi} e^{i\pi/4}}{h} \sum_{n=n_{\max}+1}^{\infty} \sin k_{zn}^{(0)} z_0 \sin k_{zn}^{(0)} z \, e^{-\frac{\rho k_{1zn}^{(0)}}{\rho_1 h k_n^{(0)}} r} \frac{e^{ik_n^{(0)} r}}{\sqrt{k_n^{(0)} r}} \ . \qquad (5.200)$$

Just as in the case of Eq. (5.136), we associate the improper modes in Eq. (5.200) with up- and down-going plane waves that only partly constructively interfere with one another due to partial transmission into the bottom. As these modes propagate along the waveguide, they lose energy into the bottom, causing an exponential decay with range as well as an exponential growth in depth. We also note that Eq. (5.200) agrees with the results of Tindle, Stamp, and Guthrie [1976] to leading order in the parameter $\rho k_{1zn}^{(0)} / \left(\rho_1 k_{zn}^{(0)}\right)$.

5.5.3 The Total Field in the Pekeris Waveguide

Summarizing then, the total field in the Pekeris waveguide is composed of discrete and continuum modal contributions which, for source and receiver in the water column $(z, z_0 \le h)$, are given by

$$\begin{aligned} p(r,z) &= p_t(r,z) + p_c(r,z) \\ &= \frac{i\pi}{\rho} \sum_{n=1}^{n_{\max}} A_n^2 \sin k_{zn} z_0 \sin k_{zn} z \, H_0^{(1)}(k_n r) \\ &+ \frac{2i\rho_1}{\rho} \int_{C_B} \frac{k_z}{k_{1z}} \frac{\sin k_z z_0 \sin k_z z}{\left[\sin^2 k_z h + \dfrac{\rho_1^2 k_z^2}{\rho^2 k_{1z}^2} \cos^2 k_z h\right]} H_0^{(1)}(k_r r) \, dk_z . \qquad (5.201) \end{aligned}$$

Asymptotically, the continuum field can be approximated by a sum of discrete, improper modes, and Eq. (5.201) becomes

$$p(r,z) \sim \frac{\sqrt{2\pi}e^{i\pi/4}}{\rho} \sum_{n=1}^{n_{max}} A_n^2 \sin k_{zn} z_0 \sin k_{zn} z \frac{e^{ik_n r}}{\sqrt{k_n r}}$$

$$+ \frac{2\sqrt{2\pi}e^{i\pi/4}}{h} \sum_{n=n_{max}+1}^{\infty} \sin k_{zn}^{(0)} z_0 \sin k_{zn}^{(0)} z \, e^{-\frac{\rho k_{1zn}^{(0)}}{\rho_1 h k_n^{(0)}} r} \frac{e^{ik_n^{(0)} r}}{\sqrt{k_n^{(0)} r}} \ . \quad (5.202)$$

Equation (5.202) illustrates several interesting modal characteristics. First, with increasing range, the perfectly trapped modes tend to dominate the total field because of the exponentially decaying factor in the leaky mode field. Second, both the proper and improper modes satisfy the same eigenvalue equation [cf. Eqs. (5.99) and (5.163)], and, as kh increases [cf. Eq. (5.157)], virtual modes migrate from the continuum to the perfectly trapped part of the spectrum. However, we note that Eq. (5.202) is accurate only for modes which lie outside the region $\theta \approx \theta_c$ (cf. Fig. 5.14). For modes which lie in a transition zone close to the critical angle, other methods of analysis, such as uniform asymptotic expansions, must be used in order to properly describe modal transitions between the discrete and continuous spectra [Williams, 1978; Stickler and Ammicht, 1980]. Third, in the limit $\rho/\rho_1 \to 0$ and $c/c_1 \to 0$, Eq. (5.202) goes over to the modal solution for a waveguide with pressure-release top and hard bottom [cf. Eq. (5.26)], as required (cf. Prob. 5.12).

Finally, in Chap. 6, we will derive Eq. (5.201) by an entirely different approach, namely, the method of Hankel transforms. That discussion will further elucidate the relationship among the various field constituents, their alternative representations, and their physical interpretation.

PROBLEMS

5.1 Show that for constant density, Eq. (5.8) takes on the form of a one-dimensional, time-independent Schrodinger equation (cf. Sec. 1.4),

$$\frac{d^2 u_n(z)}{dz^2} + \frac{2m}{\hbar^2}[E_n - V(z)]u_n(z) = 0 \ . \quad (5.203)$$

What are the acoustic quantities corresponding to the energy eigenvalues E_n and potential energy $V(z)$? Provide a physical interpretation for the results in the context of this quantum mechanical analogy.

5.2 Using the orthonormality relation in Eq. (5.11), derive an integral expression for the coefficients c_n in the eigenfunction expansion in Eq. (5.13). Show that the derived expression and Eq. (5.13) imply the closure relation in Eq. (5.12).

5.3 Use the method of eigenfunction expansion and the endpoint method to obtain the following representation for the two-dimensional free-space Green's function in cylindrical coordinates:

$$G_0(\mathbf{r},\mathbf{r}_0) = i\pi H_0^{(1)}\left(k\left|\mathbf{r}-\mathbf{r}_0\right|\right) = i\pi \sum_{n=-\infty}^{\infty} J_n(kr_<)H_n^{(1)}(kr_>)e^{in(\theta-\theta_0)} \quad , \qquad (5.204)$$

where $\mathbf{r} = (r,\theta)$, $\mathbf{r}_0 = (r_0,\theta_0)$, $r_<$ ($r_>$) is the lesser (greater) of r and r_0, and J_n and $H_n^{(1)}$ are nth order Bessel and Hankel functions of the first kind, respectively.

Hint: The Wronskian for J_n and $H_n^{(1)}$ is [Abramowitz and Stegun, 1964]

$$W[J_n(x),H_n^{(1)}(x)] = \frac{2i}{\pi x} \quad . \qquad (5.205)$$

5.4 Consider a point source at $(0,z_0)$ in a homogeneous fluid layer with a soft top and bottom.

(a) What is the normalized normal mode solution for the field $p(r,z)$, including the eigenvalues?

(b) What are the cutoff frequencies ω_n for the two lowest modes?

(c) Calculate and sketch the phase and group velocities for the two lowest modes as a function of frequency ω.

(d) Sketch the depth eigenfunctions for the two lowest modes, indicating the positions of the nodes in each case.

5.5 Consider a point source at $(0, z_0)$ in a homogeneous fluid layer with a hard top and bottom.

(a) What is the normalized normal mode solution for the field $p(r,z)$, including the eigenvalues?

(b) What are the cutoff frequencies ω_n for the two lowest modes?

(c) Calculate and sketch the phase and group velocities for the two lowest modes as a function of frequency ω.

(d) Sketch the depth eigenfunctions for the two lowest modes, indicating the positions of the nodes in each case.

5.6 For the zeroth order ocean acoustic waveguide, show that the mode expansion in Eq. (5.25) and the image method solution in Eq. (4.82) are equivalent.

Hint: Use the *Poisson sum formula* [Brekhovskikh and Lysanov, 1982]

$$\sum_{n=-\infty}^{\infty} F(n) = \sum_{n=-\infty}^{\infty} f(2\pi n) \quad , \tag{5.206}$$

where

$$f(x) = \int_{-\infty}^{\infty} F(n) e^{-inx} dn \quad , \tag{5.207}$$

and Eq. (5.66).

5.7 Derive Eq. (5.135).

5.8 Consider the characteristics of normal mode propagation for the waveguide configuration shown in Fig. 5.20.

(a) For what range of values of the propagation angle θ in the layer $0 \le z \le h$ will perfectly trapped modes be excited?

(b) Derive the eigenvalue equation for the perfectly trapped modes.

(c) Derive an expression for the maximum number of propagating, perfectly trapped modes.

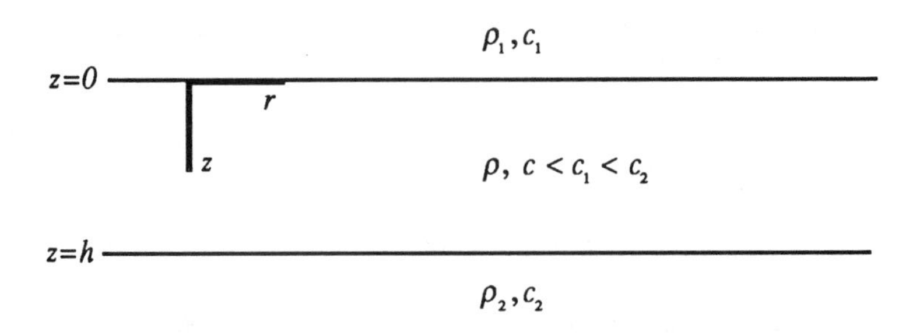

Figure 5.20 Waveguide configuration for Prob. 5.8.

5.9 Consider the characteristics of normal mode propagation for the waveguide configuration shown in Fig. 5.21.

(a) What is the eigenvalue equation for the perfectly trapped modes which lie in the angular regime $\sin^{-1}(c/c_2) < \theta < \sin^{-1}(c/c_1)$, where θ is the propagation angle in the layer with sound speed c?

(b) What are the normalized modal depth eigenfunctions in each of the three regions for the angular regime $\sin^{-1}(c/c_2) < \theta < \sin^{-1}(c/c_1)$?

(c) What is the functional form of the modal depth eigenfunctions in each of the three regions for $\theta > \sin^{-1}(c/c_1)$? In this case, it is not necessary to calculate any normalization or multiplicative constants.

5.10 Suppose the sound source in the Pekeris waveguide were located in the ocean bottom $(z_0 > h)$. What is the normal mode solution for the field $p(r,z)$ for the perfectly trapped modes in this case?

5.11 Show that there is a minimum in the group velocity dispersion curve (the Airy phase) for the Pekeris waveguide.

5.12 Show that in the limit $\rho/\rho_1 \to 0$ and $c/c_1 \to 0$, the asymptotic modal solution for the Pekeris waveguide in Eq. (5.202) goes over to the asymptotic modal solution for a waveguide with pressure-release top and hard bottom [cf. Eq. (5.26)].

5.13 Use *first-order perturbation theory* [Schiff, 1968; Merzbacher, 1970] to obtain an approximate expression for the attenuation δ_n of the nth mode due to an attenuation profile $\alpha(z)$ in the bottom (cf. Prob. 3.4).

First assume that we know the normalized eigenfunctions $u_n^{(0)}(z)$ and discrete horizontal wavenumbers $k_n^{(0)}$ for a waveguide with arbitrary density and sound velocity profiles, $\rho(z)$ and $c(z)$, and zero

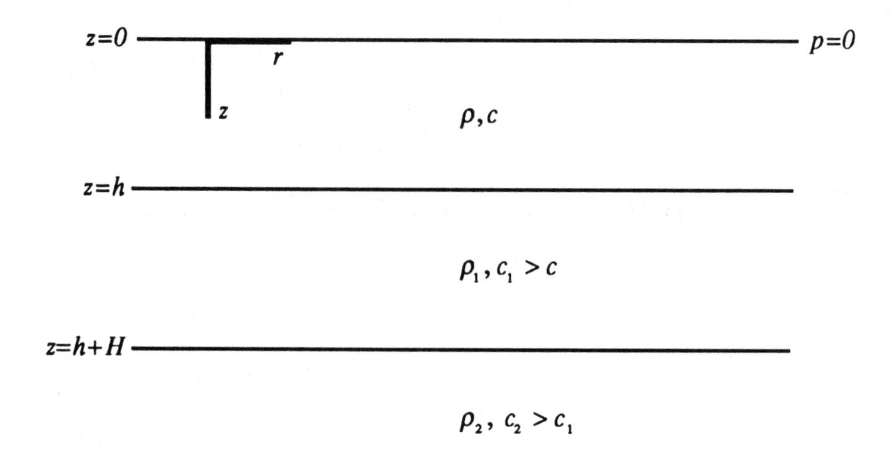

Figure 5.21 Waveguide configuration for Prob. 5.9.

attenuation (cf. Fig. 5.1 with $b=\infty$). Then add an attenuation profile $\alpha(z)$ in the bottom ($z \geq h$) such that

$$k(z)=k_0(z)+i\varepsilon\alpha(z), \quad z \geq h \ , \tag{5.208}$$

where $k_0(z)=\omega/c(z)$, and we introduce a parameter ε purely for bookkeeping purposes. In Eq. (5.208), $\alpha(z)$ is treated as a first-order perturbation to the wavenumber $k_0(z)$ because of the requirement that $\alpha(z) << k_0(z)$. Then assume that the perturbed wavenumber causes first-order perturbations, $u_n^{(1)}(z)$ and $k_n^{(1)}$, in the eigenfunctions $u_n^{(0)}(z)$ and horizontal wavenumbers $k_n^{(0)}$, so that we can write

$$\begin{aligned} u_n(z) &\approx u_n^{(0)}(z)+\varepsilon u_n^{(1)}(z), \\ k_n &\approx k_n^{(0)}+\varepsilon k_n^{(1)} \ , \end{aligned} \tag{5.209}$$

where $u_n(z)$ and k_n are the eigenfunctions and horizontal wavenumbers for the problem *with* attenuation. Substitute Eq. (5.209) into the eigenvalue Eq. (5.7) and group terms of equal power in ε to obtain zeroth and first-order eigenvalue equations. In the first-order equation, use the fact that the $u_n^{(0)}(z)$ constitute a complete, orthonormal set to obtain the result

$$\delta_n \approx \frac{1}{k_n^{(0)}}\int_h^{\infty}\frac{\alpha(z)}{\rho(z)}k_0(z)\,u_n^{(0)^2}(z)\,dz \ . \tag{5.210}$$

Chapter 6. The Hankel Transform: A Unified Approach to Wave Propagation in Horizontally Stratified Media

There is always a better way...
your challenge is to find it.
Anonymous

6.1 Introduction

In Chap. 4, we introduced the Hankel transform in our discussion of spherical wave reflection from an arbitrary horizontally stratified medium (cf. Sec. 4.8). In that context, we saw that the description of the acoustic field in terms of a two-dimensional, spatial/wavenumber Fourier transform in Cartesian coordinates was equivalent to a one-dimensional characterization in terms of the Hankel transform in cylindrical coordinates. This equivalence was a consequence of the cylindrically symmetric nature of the sound field in horizontally stratified media. In fact, the Hankel transform, combined with the endpoint method, provides a general, unified approach for constructing the Green's function in arbitrary horizontally stratified media. Intuitively, the generality of this method stems from the fact that the sound field is expressed as an integral over all possible horizontal wavenumbers, and, as such, embraces all possible wave types propagating through and interacting with the medium. As a result, all of the field representations described in previous chapters can be obtained from the Hankel transform representation, while, in general, the converse is not true. Furthermore, the Hankel transform readily lends itself to numerical implementations.

First, we derive the solution for a point source in a horizontally stratified, fluid layer bounded by arbitrary horizontally stratified media. We then examine in detail the behavior of this result for the case where the source is located in a homogeneous fluid layer and provide a geometrical acoustics interpretation of the solution. For the zeroth order ocean acoustic waveguide, we demonstrate that it is equivalent to

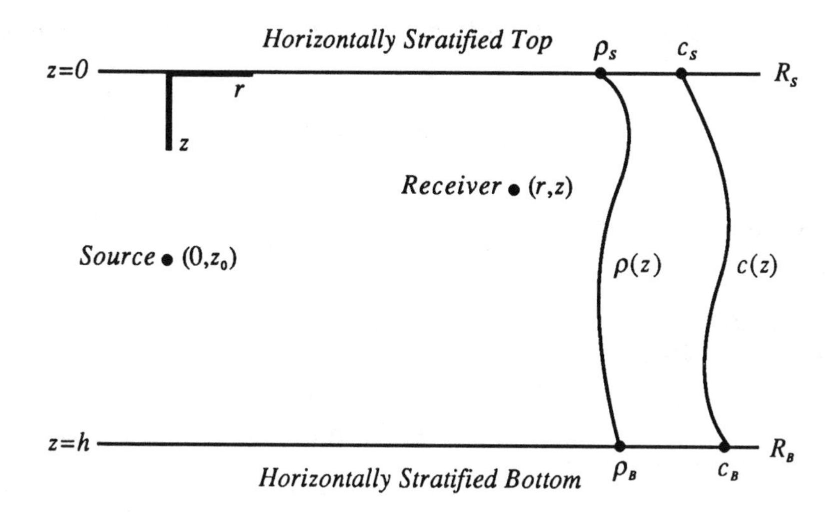

Figure 6.1 Geometry for construction of the Hankel transform solution.

the image method result. For general penetrable boundaries, however, we show that a physical interpretation that includes geometrical acoustic as well as diffraction contributions can be achieved only through approximate asymptotic analysis. Finally, we discuss the relationship between the normal mode and Hankel transform representations and derive the normal mode solution from the Hankel transform result for the Pekeris waveguide.

6.2 A Point Source in a Horizontally Stratified, Fluid Layer Bounded by Arbitrary Horizontally Stratified Media

We consider a point source with cylindrical coordinates $(0, z_0)$ in a horizontally stratified, fluid layer with density $\rho(z)$ and sound velocity $c(z)$ bounded by arbitrary horizontally stratified media (cf. Fig. 6.1). Then the inhomogeneous, time-independent wave equation for density and sound velocity stratification [cf. Eqs. (1.45) and (5.5)] and $0 \le z, z_0 \le h$ is

$$\frac{1}{r}\frac{\partial}{\partial r}\left[r\frac{\partial p(r;z,z_0)}{\partial r}\right]+\frac{\partial^2 p(r;z,z_0)}{\partial z^2}+\rho(z)\frac{\partial}{\partial z}\left[\frac{1}{\rho(z)}\right]\frac{\partial p(r;z,z_0)}{\partial z}$$

$$+k^2(z)p(r;z,z_0)=-2\frac{\delta(r)}{r}\delta(z-z_0) \ , \tag{6.1}$$

with the wavenumber $k(z) = \omega/c(z)$. Applying the inverse zero order Hankel transform operator,

$$I.H.T.\{\bullet\} = \int_0^\infty \{\bullet\} J_0(k_r r) r\, dr \quad , \tag{6.2}$$

to both sides of Eq. (6.1), we obtain

$$\left\{ \frac{d^2}{dz^2} + \rho(z)\frac{d}{dz}\left[\frac{1}{\rho(z)}\right]\frac{d}{dz} + k^2(z) - k_r^2 \right\} g(k_r; z, z_0) = -2\delta(z - z_0), \tag{6.3}$$

where $p(r; z, z_0)$ and $g(k_r; z, z_0)$ are conjugate zero order Hankel transform pairs (cf. Sec. 4.8)

$$p(r; z, z_0) = \int_0^\infty g(k_r; z, z_0) J_0(k_r r)\, k_r\, dk_r \quad , \tag{6.4}$$

$$g(k_r; z, z_0) = \int_0^\infty p(r; z, z_0) J_0(k_r r)\, r\, dr \quad , \tag{6.5}$$

and we have used the relation

$$I.H.T.\left\{ \frac{1}{r}\frac{\partial}{\partial r}\left[r \frac{\partial p(r; z, z_0)}{\partial r} \right] \right\} = -k_r^2 g(k_r; z, z_0) \quad . \tag{6.6}$$

Thus, the conjugate Hankel transform variables are the horizontal wavenumber k_r and the horizontal range r. The quantity $g(k_r; z, z_0)$ is a Green's function in the variables z and z_0, and is therefore called the depth-dependent Green's function (cf. Sec. 4.8). We solve Eq. (6.3) using the endpoint method (cf. Sec. 4.7) and obtain the result

$$g(k_r; z, z_0) = \begin{cases} -\dfrac{2}{W(z_0)} p_S(k_r; z) p_B(k_r; z_0), & 0 \le z \le z_0, \\ -\dfrac{2}{W(z_0)} p_S(k_r; z_0) p_B(k_r; z), & z_0 \le z \le h \quad , \end{cases} \tag{6.7}$$

where the Wronskian $W(z_0)$ is defined as [cf. Eq. (4.85)]

$$W(z_0)=p_S(z_0)p_B'(z_0)-p_S'(z_0)p_B(z_0) \ . \tag{6.8}$$

The functions $p_S(k_r;z)$ and $p_B(k_r;z)$ are linearly independent solutions of the homogeneous version of Eq. (6.3) which satisfy homogeneous boundary conditions at $z=0$ and $z=h$, respectively. These conditions can be expressed as the impedance relations

$$\frac{i\omega\rho_S p_S(k_r;z)}{\partial p_S(k_r;z)/\partial z}=\xi_S(k_r) \text{ at } z=0 \ , \tag{6.9a}$$

$$\frac{i\omega\rho_B p_B(k_r;z)}{\partial p_B(k_r;z)/\partial z}=\xi_B(k_r) \text{ at } z=h \ , \tag{6.9b}$$

where ρ_S and ρ_B are the values of $\rho(z)$ at $z=0$ and $z=h$, respectively (cf. Fig. 6.1). The required information about the bounding media in Eq. (6.9) is also contained in the plane wave reflection coefficients $R_S(k_r)$ and $R_B(k_r)$, even in the case when the layer $0\le z\le h$ is an *inhomogeneous* fluid medium [Bucker, 1970; Frisk, Oppenheim, and Martinez, 1980]. Then $R_S(k_r)$ is the reflection coefficient associated with a plane wave impinging upon the region $z<0$ from a half-space with sound velocity c_S, while $R_B(k_r)$ is the reflection coefficient associated with a plane wave impinging upon the region $z>h$ from a half-space with sound velocity c_B. Furthermore, the bounding media may, in general, be elastic solids.

Once $g(k_r;z,z_0)$ is determined, the pressure field can be calculated by executing the Hankel transform in Eq. (6.4). Although we will emphasize analytic evaluations of the transform, it is important to point out that Eq. (6.4) is particularly well suited for the application of numerical algorithms. For example, it is clear from Sec. 4.8.1 that we can rewrite Eq. (6.4) as

$$p(r;z,z_0)=\frac{1}{2}\int_{-\infty}^{\infty}g(k_r;z,z_0)H_0^{(1)}(k_r r)k_r\,dk_r \ . \tag{6.10}$$

When we approximate the Hankel function by its asymptotic form [cf. Eq. (4.119)] for $k_r r>>1$, Eq. (6.10) takes on the form of a Fourier transform,

$$p(r;z,z_0) \sim \frac{e^{-i\pi/4}}{\sqrt{2\pi r}} \int_{-\infty}^{\infty} g(k_r;z,z_0)\sqrt{k_r}\, e^{ik_r r}\, dk_r \ , \tag{6.11}$$

which can be evaluated by numerical methods such as the Fast Fourier Transform [Brigham, 1988]. This approach is the basis for rapid computational algorithms known as Fast Field Programs [DiNapoli and Deavenport, 1980; Schmidt and Jensen, 1985]. The inverse transform,

$$g(k_r;z,z_0) \sim \frac{e^{i\pi/4}}{\sqrt{2\pi k_r}} \int_{-\infty}^{\infty} p(r;z,z_0)\sqrt{r}\, e^{-ik_r r}\, dr \ , \tag{6.12}$$

can be used in waveguide characterization techniques and inversion methods for inferring the acoustic properties of the seabed from measurements of the pressure field as a function of range [Frisk and Lynch, 1984; Frisk, 1990; Rajan, Frisk, and Lynch, 1992].

6.3 A Point Source in a Homogeneous Fluid Layer Bounded by Arbitrary Horizontally Stratified Media

Let us examine the behavior of the acoustic field in the case where the layer $0 \le z \le h$ has both constant density ρ and constant sound velocity c. The solution $p_S(k_r;z)$ and its derivative $p_S'(k_r;z)$ are

$$p_S(k_r;z) = A\left[e^{-ik_z z} + R_S e^{ik_z z}\right] \ , \tag{6.13}$$

$$p_S'(k_r;z) = -ik_z A\left[e^{-ik_z z} - R_S e^{ik_z z}\right] \ , \tag{6.14}$$

where A is an arbitrary constant, R_S is the reflection coefficient for the upper bounding medium, and $k_z = \sqrt{k^2 - k_r^2}$. Similarly, $p_B(k_r;z)$ and its derivative $p_B'(k_r;z)$ are

$$p_B(k_r;z) = B\left[e^{ik_z z} + R_B e^{2ik_z h} e^{-ik_z z}\right] \ , \tag{6.15}$$

$$p_B'(k_r;z) = ik_z B\left[e^{ik_z z} - R_B e^{2ik_z h} e^{-ik_z z}\right] \ , \tag{6.16}$$

where B is an arbitrary constant and R_B is the reflection coefficient for the lower bounding medium. The Wronskian W is given by

$$W = 2ik_z AB\left(1 - R_S R_B e^{2ik_z h}\right) \ , \tag{6.17}$$

and it is then straightforward to show that

$$g(k_r) = \frac{i\left\{e^{ik_z|z-z_0|} + R_S e^{ik_z(z+z_0)} + R_B e^{2ik_z h}\left[e^{-ik_z(z+z_0)} + R_S e^{-ik_z|z-z_0|}\right]\right\}}{k_z\left[1 - R_S R_B e^{2ik_z h}\right]} . \tag{6.18}$$

The full solution for the pressure field is therefore

$$p = \int_0^\infty \frac{i\left\{e^{ik_z|z-z_0|} + R_S e^{ik_z(z+z_0)} + R_B e^{2ik_z h}\left[e^{-ik_z(z+z_0)} + R_S e^{-ik_z|z-z_0|}\right]\right\}}{k_z\left[1 - R_S R_B e^{2ik_z h}\right]}$$

$$\times J_0(k_r r) k_r \, dk_r \ . \tag{6.19}$$

At this point, we can pursue two analytic approaches to the evaluation of the integral in Eq. (6.19). In the first case, we expand the denominator of the integrand and obtain a geometrical acoustics interpretation of the field. In the second case, we examine the singularities of the integrand and use complex contour integration methods to obtain a normal mode decomposition of the field.

6.3.1 A Geometrical Acoustics Decomposition

The geometrical acoustics result is based on the expansion [cf. Eq. (3.40)]

$$\frac{1}{1 - R_S R_B e^{2ik_z h}} = \sum_{n=0}^{\infty} (R_S R_B)^n e^{2ink_z h} \ , \tag{6.20}$$

which is valid for $|R_S R_B| < 1$. The latter condition is satisfied as long as at least one of the bounding media has even a small amount of attenuation associated with it. Substituting Eq. (6.20) into Eq. (6.19), we obtain

$$p(r) = \sum_{n=0}^{\infty} \int_0^{\infty} \frac{i}{k_z} \left\{ e^{ik_z|z-z_0|} + R_S e^{ik_z(z+z_0)} + R_B e^{2ik_z h} \left[e^{-ik_z(z+z_0)} + R_S e^{-ik_z|z-z_0|} \right] \right\}$$

$$\times (R_S R_B)^n e^{2ink_z h} J_0(k_r r) k_r \, dk_r \ . \tag{6.21}$$

In general, Eq. (6.21) must be evaluated using approximate methods. Before addressing the general case, however, we will discuss the application of Eq. (6.21) to the zeroth order model of the ocean acoustic waveguide, for which a closed form result can be obtained.

Zeroth Order Model of the Ocean Acoustic Waveguide

In the zeroth order model (cf. Fig. 4.12), $R_S = -1$ and $R_B = 1$, and Eq. (6.21) therefore becomes

$$p = \sum_{n=0}^{\infty} (-1)^n \int_0^{\infty} \frac{i}{k_z} \left\{ e^{ik_z[|z-z_0|+2nh]} + e^{ik_z[-(z+z_0)+2(n+1)h]} - e^{ik_z[z+z_0+2nh]} \right.$$

$$\left. - e^{ik_z[-|z-z_0|+2(n+1)h]} \right\} J_0(k_r r) k_r \, dk_r \ . \tag{6.22}$$

The four integrals in Eq. (6.22) can be evaluated using the Hankel transform relation [cf. Eq. (4.101) and Keller and Papadakis, 1977]

$$\int_0^{\infty} \frac{i}{k_z} e^{ik_z a} J_0(k_r r) k_r \, dk_r = \frac{e^{ik\sqrt{r^2+a^2}}}{\sqrt{r^2+a^2}} \ , \tag{6.23}$$

where a is a positive or negative real number. For $z < z_0$, the result is

$$p(r) = \sum_{n=0}^{\infty} (-1)^n \left[\frac{e^{ikR_{n1}}}{R_{n1}} + \frac{e^{ikR_{n2}}}{R_{n2}} - \frac{e^{ikR_{n3}}}{R_{n3}} - \frac{e^{ikR_{n4}}}{R_{n4}} \right] , \tag{6.24}$$

$$R_{n1} = \sqrt{r^2 + (z - z_0 - 2nh)^2}, \quad R_{n2} = \sqrt{r^2 + [z + z_0 - 2(n+1)h]^2},$$

$$R_{n3} = \sqrt{r^2 + (z + z_0 + 2nh)^2}, \quad R_{n4} = \sqrt{r^2 + [z - z_0 + 2(n+1)h]^2} \ . \tag{6.25}$$

Equations (6.24) and (6.25) are identical to the image method results in Eqs. (4.80) and (4.82).

For $z > z_0$, demonstrating the equivalence between the Hankel transform and image method results is more subtle. In this case, when we apply Eq. (6.23) to Eq. (6.22), we obtain

$$p(r) = \sum_{n=0}^{\infty}(-1)^n\left[\frac{e^{ikR'_{n1}}}{R'_{n1}} + \frac{e^{ikR_{n2}}}{R_{n2}} - \frac{e^{ikR_{n3}}}{R_{n3}} - \frac{e^{ikR'_{n4}}}{R'_{n4}}\right]$$
$$= Term\,I + Term\,II + Term\,III + Term\,IV \quad , \tag{6.26}$$

$$R'_{n1} = \sqrt{r^2 + (z - z_0 + 2nh)^2}, \; R_{n2} = \sqrt{r^2 + [z + z_0 - 2(n+1)h]^2},$$
$$R_{n3} = \sqrt{r^2 + (z + z_0 + 2nh)^2}, \; R'_{n4} = \sqrt{r^2 + [z - z_0 - 2(n+1)h]^2} \; . \tag{6.27}$$

We see that *Terms I* and *IV* in Eq. (6.26) differ from the first and fourth terms in the Hankel transform result for $z < z_0$ and the image method result. In order to resolve this inconsistency, we change variables, letting $n = m + 1$ in *Term I* and $m = n + 1$ in *Term IV*, so that

$$Term\,I = \sum_{m=-1}^{\infty}(-1)^{m+1}\frac{e^{ik\sqrt{r^2+[z-z_0+2(m+1)h]^2}}}{\sqrt{r^2 + [z - z_0 + 2(m+1)h]^2}} \quad , \tag{6.28}$$

$$Term\,IV = \sum_{m=1}^{\infty}(-1)^m\frac{e^{ik\sqrt{r^2+(z-z_0-2mh)^2}}}{\sqrt{r^2 + (z - z_0 - 2mh)^2}} \quad . \tag{6.29}$$

But the $m = -1$ term in *Term I* can be written as the $m = 0$ term in *Term IV*, and we find that

$$Term\,I + Term\,IV = \sum_{m=0}^{\infty}\left\{(-1)^{m+1}\frac{e^{ik\sqrt{r^2+[z-z_0+2(m+1)h]^2}}}{\sqrt{r^2 + [z - z_0 + 2(m+1)h]^2}} + (-1)^m\frac{e^{ik\sqrt{r^2+(z-z_0-2mh)^2}}}{\sqrt{r^2 + (z - z_0 - 2mh)^2}}\right\}$$
$$= \sum_{n=0}^{\infty}(-1)^n\left[-\frac{e^{ikR_{n4}}}{R_{n4}} + \frac{e^{ikR_{n1}}}{R_{n1}}\right] \; , \tag{6.30}$$

where R_{n1} and R_{n4} are defined in Eq. (6.25). Thus, it is clear that when we substitute Eq. (6.30) into Eq. (6.26), we recover the image method result in Eq. (6.24). Finally, we recall that each of the four fundamental terms corresponding to $n=0$ in Eq. (6.24) is associated with a specific path connecting the source and receiver. The R_{01} term corresponds to the direct arrival, the R_{02} term to a bottom reflection, the R_{03} term to the surface-reflected arrival, and the R_{04} term to a path which reflects off the bottom and then the surface before reaching the receiver (cf. Fig. 4.11). The higher order terms ($n>0$) correspond to multiple reflections of these basic field constituents within the waveguide (cf. Prob. 4.7). This interpretation will help guide us through the subtleties of the boundary interactions in the situation when the boundaries are not perfectly reflecting.

The General Case of Penetrable Boundaries

The equivalence, in closed form, between the Hankel transform and image method results can be demonstrated for any set of boundary conditions corresponding to impenetrable boundaries, for which $|R_S|=|R_B|=1$ for all k_r. When one or both bounding media are penetrable, however, so that R_S and/or R_B are functions of k_r, we must use approximate asymptotic methods to arrive at a geometrical acoustic interpretation which, in general, will also have additional diffracted components.

First, it is important to observe that, in the general case, there are also four basic wave types which underlie the field decomposition. These can be seen by rewriting Eq. (6.21):

$$
\begin{aligned}
p(r) = &\sum_{n=0}^{\infty}\int_0^{\infty}\frac{i}{k_z}e^{ik_z|z-z_0|}\left(R_S R_B\right)^n e^{2ink_z h}J_0(k_r r)k_r\,dk_r && \textit{Term I}\\
&+\sum_{n=0}^{\infty}\int_0^{\infty}\frac{i}{k_z}R_B e^{ik_z[2h-(z+z_0)]}\left(R_S R_B\right)^n e^{2ink_z h}J_0(k_r r)k_r\,dk_r && \textit{Term II}\\
&+\sum_{n=0}^{\infty}\int_0^{\infty}\frac{i}{k_z}R_S e^{ik_z(z+z_0)}\left(R_S R_B\right)^n e^{2ink_z h}J_0(k_r r)k_r\,dk_r && \textit{Term III}\\
&+\sum_{n=0}^{\infty}\int_0^{\infty}\frac{i}{k_z}R_B R_S e^{ik_z[2h-|z-z_0|]}\left(R_S R_B\right)^n e^{2ink_z h}J_0(k_r r)k_r\,dk_r. && \textit{Term IV}
\end{aligned}
\tag{6.31}
$$

The $n = 0$ term in each of the four expressions corresponds to the plane wave decomposition of a particular wave species emanating from a specific real or image source plane. Let us explore this interpretation for the case $z_0 > z$ (cf. Sec. 4.8 and Fig. 4.11). Then *Term I* consists of a spectrum of plane waves leaving the real source plane $z = z_0$ and propagating to the receiver. For $n = 0$, it is the only term in Eq. (6.31) which can be evaluated in closed form to yield the direct spherical wave corresponding to the R_{01} term in Eq. (6.24). *Term II* is composed of plane waves which emanate from the image source plane $z = 2h - z_0$ and are multiplied by R_B; they correspond to plane waves which originate at the real source plane $z = z_0$ and reflect off the bottom interface. *Term III* is a synthesis of plane waves which leave the image source plane $z = -z_0$ and are multiplied by R_S; they are equivalent to plane waves which emerge from the real source plane $z = z_0$ and reflect off the top boundary. *Term IV* is a spectrum of plane waves which emanate from the image source plane $z = -2h + z_0$ and are multiplied by R_B and R_S; they correspond to plane waves which leave the real source plane $z = z_0$ and reflect off the bottom and the top boundaries. Again, the higher order terms ($n > 0$), which are associated with higher order image source planes (cf. Fig. 4.11), correspond to multiple reflections of these fundamental wave types within the waveguide.

In order to proceed further in our analysis of the general case, we must use approximate methods. We begin by executing a sequence of steps analogous to those described in Eqs. (4.112)-(4.122) for the asymptotic analysis of spherical wave reflection from a homogeneous fluid half-space. For $k_r r >> 1$, the result is

$$
\begin{aligned}
p \sim & \sum_{n=0}^{\infty} \frac{\sqrt{k} e^{i\pi/4}}{\sqrt{2\pi r}} \int_{-\infty}^{\infty} \frac{\sqrt{K_r}}{K_z} (R_S R_B)^n e^{ik\{K_z[|z-z_0|+2nh]+K_r r\}} \, dK_r && I \\
& + \sum_{n=0}^{\infty} \frac{\sqrt{k} e^{i\pi/4}}{\sqrt{2\pi r}} \int_{-\infty}^{\infty} \frac{\sqrt{K_r}}{K_z} R_B (R_S R_B)^n e^{ik\{K_z[-(z+z_0)+2(n+1)h]+K_r r\}} \, dK_r && II \\
& + \sum_{n=0}^{\infty} \frac{\sqrt{k} e^{i\pi/4}}{\sqrt{2\pi r}} \int_{-\infty}^{\infty} \frac{\sqrt{K_r}}{K_z} R_S (R_S R_B)^n e^{ik\{K_z[z+z_0+2nh]+K_r r\}} \, dK_r && III \\
& + \sum_{n=0}^{\infty} \frac{\sqrt{k} e^{i\pi/4}}{\sqrt{2\pi r}} \int_{-\infty}^{\infty} \frac{\sqrt{K_r}}{K_z} R_B R_S (R_S R_B)^n e^{ik\{K_z[-|z-z_0|+2(n+1)h]+K_r r\}} \, dK_r, && IV
\end{aligned}
\tag{6.32}
$$

where we recall that

$$K_r = \frac{k_r}{k} = \sin\theta, \;\; K_z = \frac{k_z}{k} = \cos\theta \;\; , \tag{6.33}$$

and θ is the angle of propagation of a particular plane wave component within the waveguide. Each of the four types of integrals in Eq. (6.32) is in a canonical form for the application of asymptotic methods for large values of the parameter k [cf. Eq. (4.123)]. By analogy with Sec. 4.8.1, we determine the saddle points associated with the phase factor in each integral and evaluate the integral using the modified method of stationary phase. For *Term I* with $z_0 > z$, the saddle points occur at values $K_r = K_{n1} = \sin\theta_{n1}$ which satisfy the equation

$$\frac{d}{dK_r}\{K_z[z_0 - z + 2nh] + K_r r\} = 0 \;\; , \tag{6.34}$$

so that

$$-\frac{K_{n1}}{\sqrt{1-K_{n1}^2}}(z_0 - z + 2nh) + r = -(z_0 - z + 2nh)\tan\theta_{n1} + r = 0 \;\; . \tag{6.35}$$

Equation (6.35) has the solution

$$\tan\theta_{n1} = r/(z_0 - z + 2nh), \;\; \sin\theta_{n1} = r/R_{n1},$$
$$\cos\theta_{n1} = (z_0 - z + 2nh)/R_{n1}, \;\; R_{n1} = \sqrt{r^2 + (z - z_0 - 2nh)^2} \, . \tag{6.36}$$

Similarly, we obtain for *Term II*, $K_{n2} = \sin\theta_{n2}$, with

$$\tan\theta_{n2} = r/[2(n+1)h - (z+z_0)], \;\; \sin\theta_{n2} = r/R_{n2},$$
$$\cos\theta_{n2} = [2(n+1)h - (z+z_0)]/R_{n2}, \;\; R_{n2} = \sqrt{r^2 + [z + z_0 - 2(n+1)h]^2}, \tag{6.37}$$

for *Term III*, $K_{n3} = \sin\theta_{n3}$, with

$$\tan\theta_{n3} = r/(z + z_0 + 2nh), \;\; \sin\theta_{n3} = r/R_{n3},$$
$$\cos\theta_{n3} = (z + z_0 + 2nh)/R_{n3}, \;\; R_{n3} = \sqrt{r^2 + (z + z_0 + 2nh)^2} \, , \tag{6.38}$$

and for *Term IV*, $K_{n4} = \sin\theta_{n4}$, with

$$\tan\theta_{n4} = r/\left[z - z_0 + 2(n+1)h\right], \quad \sin\theta_{n4} = r/R_{n4},$$
$$\cos\theta_{n4} = \left[z - z_0 + 2(n+1)h\right]/R_{n4}, \quad R_{n4} = \sqrt{r^2 + \left[z - z_0 + 2(n+1)h\right]^2}. \tag{6.39}$$

The specular angles of propagation θ_{nj} $(j = 1,\dots 4)$, corresponding to the saddle points K_{nj} $(j = 1,\dots 4)$, are shown in Fig. 6.2 for $n = 0$.

We proceed with a straightforward evaluation of the integrals in Eq. (6.32) using the modified method of stationary phase described in Sec. 4.8.1 [cf. Eqs. (4.129)-(4.137)]. We obtain the result

$$\begin{aligned} p(r) \sim \sum_{n=0}^{\infty} \Bigg\{ & \left[R_S(\theta_{n1}) R_B(\theta_{n1})\right]^n \frac{e^{ikR_{n1}}}{R_{n1}} \\ & + R_B(\theta_{n2})\left[R_S(\theta_{n2}) R_B(\theta_{n2})\right]^n \frac{e^{ikR_{n2}}}{R_{n2}} \\ & + R_S(\theta_{n3})\left[R_S(\theta_{n3}) R_B(\theta_{n3})\right]^n \frac{e^{ikR_{n3}}}{R_{n3}} \\ & + R_B(\theta_{n4}) R_S(\theta_{n4})\left[R_S(\theta_{n4}) R_B(\theta_{n4})\right]^n \frac{e^{ikR_{n4}}}{R_{n4}} \Bigg\} \\ & + \textit{Diffraction Terms}\ , \end{aligned} \tag{6.40}$$

where the R_{nj} and θ_{nj} $(j = 1,\dots 4)$ are defined in Eq. (6.25) and Eqs. (6.36)-(6.39). Equation (6.40) is valid as long as the saddle points are sufficiently well separated from one another and from any singularities of the integrands [Felsen and Marcuvitz, 1973]. The four types of terms in the sum in Eq. (6.40) have a geometrical acoustic interpretation which is analogous to that associated with the solution for the zeroth order ocean acoustic waveguide in Eq. (6.24). For $n = 0$, the R_{01} term corresponds to the direct path joining the source and receiver at the angle θ_{01}, the R_{02} term to a specular reflection from the bottom at the angle θ_{02}, the R_{03} term to a specular reflection off the surface at the angle θ_{03}, and the R_{04} term to a path which is specularly reflected from the bottom and then the surface at the angle θ_{04} (cf. Fig. 6.2). Again, the higher order terms $(n > 0)$ correspond to multiple reflections of these basic field components at the specular angles θ_{nj}

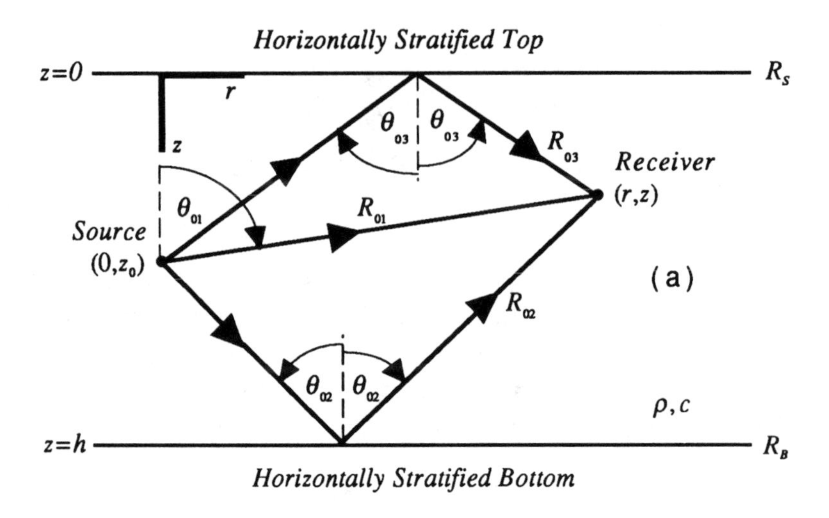

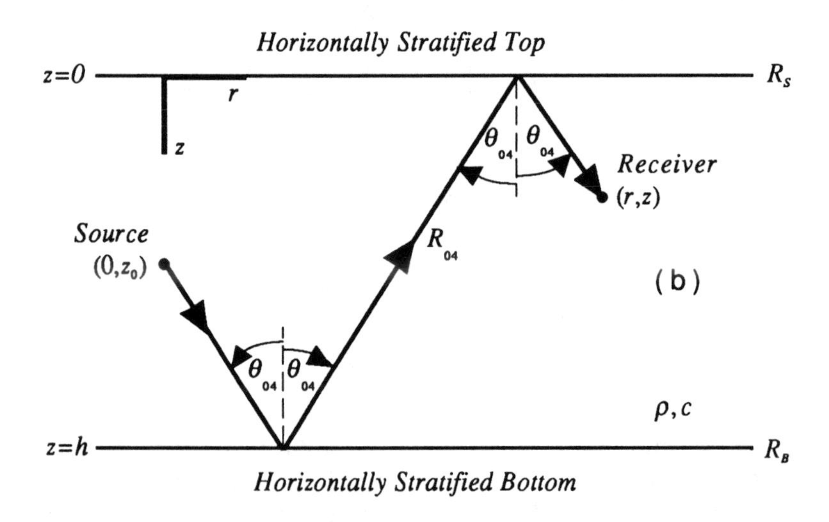

Figure 6.2 The four fundamental field constituents in the geometrical acoustics approximation for a point source in a homogeneous fluid layer bounded by arbitrary horizontally stratified media: (a) the direct, bottom-reflected, and surface-reflected arrivals; (b) the bottom-reflected, then surface-reflected path.

$(j = 1, \ldots 4)$ within the waveguide. With each boundary interaction, a particular field constituent is multiplied by the appropriate plane wave reflection coefficient evaluated at the specular angle.

The diffraction terms in Eq. (6.40) arise due to the contributions of any singularities in the integrands encountered in the process of deforming the original contour (cf. Fig. 4.14) into the saddle point contour. These singularities depend on the detailed characteristics of the bounding media, which manifest themselves in the analytic properties of the reflection coefficients. For example, for the Pekeris waveguide, just as in the case of spherical wave reflection from a fast, homogeneous fluid half-space, the analytic properties of the bottom reflection coefficient are the dominant factor. In that case, it is the branch line associated with the branch point $K_r = c/c_1$ that plays a critical role in the calculations, and Eq. (6.40) becomes

$$p(r) \sim \sum_{n=0}^{\infty}(-1)^n\left\{\left[R_B(\theta_{n1})\right]^n \frac{e^{ikR_{n1}}}{R_{n1}} + \left[R_B(\theta_{n2})\right]^{n+1} \frac{e^{ikR_{n2}}}{R_{n2}} - \left[R_B(\theta_{n3})\right]^n \frac{e^{ikR_{n3}}}{R_{n3}} - \left[R_B(\theta_{n4})\right]^{n+1} \frac{e^{ikR_{n4}}}{R_{n4}}\right\} + \sum_{m=1}^{M} \textit{Branch Line Integrals}\ , \tag{6.41}$$

where a particular branch line integral contribution will arise only when $\theta_{nj} > \sin^{-1}(c/c_1)$. These branch line integrals may also have lateral wave interpretations [Tamir and Felsen, 1965].

An Alternative Geometrical Acoustics Decomposition for the Pekeris Waveguide

An alternative geometrical acoustics decomposition for the Pekeris waveguide has been described by Tindle [1983]. It is useful in clarifying the relationship between groups of modes and geometrical acoustic paths with lateral displacement (cf. pp. 155-156).

The alternative approach is based on the assumption that, at long ranges, the pressure field is dominated by acoustic paths which have been totally internally reflected from the bottom. The reflection coefficient R_B in Eq. (6.32) is therefore approximated by its behavior for total internal reflection [cf. Eq. (5.138)]:

$$R_B \approx e^{i\varphi}, \text{ with } \varphi = -2\tan^{-1}\left[\frac{\sqrt{K_r^2 - c^2/c_1^2}}{mK_z}\right] , \tag{6.42}$$

where we recall that $m = \rho_1/\rho$. Equation (6.32), with $R_S = -1$, then becomes

$$
\begin{aligned}
p \approx & \sum_{n=0}^{\infty}(-1)^n \frac{\sqrt{k}e^{i\pi/4}}{\sqrt{2\pi r}} \int_{-\infty}^{\infty} \frac{\sqrt{K_r}}{K_z} e^{ik\{K_z[|z-z_0|+2nh]+n\varphi/k+K_r r\}} dK_r & I \\
& +\sum_{n=0}^{\infty}(-1)^n \frac{\sqrt{k}e^{i\pi/4}}{\sqrt{2\pi r}} \int_{-\infty}^{\infty} \frac{\sqrt{K_r}}{K_z} e^{ik\{K_z[-(z+z_0)+2(n+1)h]+(n+1)\varphi/k+K_r r\}} dK_r & II \\
& -\sum_{n=0}^{\infty}(-1)^n \frac{\sqrt{k}e^{i\pi/4}}{\sqrt{2\pi r}} \int_{-\infty}^{\infty} \frac{\sqrt{K_r}}{K_z} e^{ik\{K_z[z+z_0+2nh]+n\varphi/k+K_r r\}} dK_r & III \\
& -\sum_{n=0}^{\infty}(-1)^n \frac{\sqrt{k}e^{i\pi/4}}{\sqrt{2\pi r}} \int_{-\infty}^{\infty} \frac{\sqrt{K_r}}{K_z} e^{ik\{K_z[-|z-z_0|+2(n+1)h]+(n+1)\varphi/k+K_r r\}} dK_r, & IV
\end{aligned}
\tag{6.43}
$$

where we see that the bottom reflection coefficient now appears in the phase terms associated with the four types of integrands. Analogous to the situation discussed in the previous section, each of the four types of integrals in Eq. (6.43) is in a canonical form for the application of asymptotic methods for large values of the parameter k [cf. Eq. (4.123)]. We therefore determine the saddle points associated with the phase factor in each integral and evaluate the integral using the modified method of stationary phase. In this case, however, the saddle point positions are influenced by the phase of the bottom reflection coefficient. For *Term I* with $z_0 > z$, the saddle points occur at values $K_r = K_{n1} = \sin\theta_{n1}$, which satisfy the equation

$$
\frac{d}{dK_r}\{K_z[z_0 - z + 2nh] + n\varphi/k + K_r r\} = 0 \ , \tag{6.44}
$$

so that

$$
\begin{aligned}
& -\frac{K_{n1}}{\sqrt{1-K_{n1}^2}}(z_0 - z + 2nh) - nD_{n1} + r \\
& = -(z_0 - z + 2nh)\tan\theta_{n1} - nD_{n1} + r = 0 \ ,
\end{aligned}
\tag{6.45}
$$

where D_{nj} $(j = 1,\ldots 4)$ is the ray displacement associated with the nth geometrical acoustic path of type j [cf. Eq. (5.174)]:

$$D_{nj} = -\frac{1}{k}\frac{d\varphi_{nj}}{dK_r}\bigg|_{K_r = K_{nj}} = -\frac{1}{k}\frac{d\varphi_{nj}}{dK_{nj}} = -\frac{d\varphi_{nj}}{dk_{nj}} \ . \tag{6.46}$$

Similarly, we obtain for *Term II*, $K_{n2} = \sin\theta_{n2}$, with

$$-\frac{K_{n2}}{\sqrt{1-K_{n2}^2}}\left[-(z+z_0)+2(n+1)h\right]-(n+1)D_{n2}+r$$
$$= -\left[-(z+z_0)+2(n+1)h\right]\tan\theta_{n2}-(n+1)D_{n2}+r=0 \ , \tag{6.47}$$

for *Term III*, $K_{n3} = \sin\theta_{n3}$, with

$$-\frac{K_{n3}}{\sqrt{1-K_{n3}^2}}(z+z_0+2nh)-nD_{n3}+r$$
$$= -(z+z_0+2nh)\tan\theta_{n3}-nD_{n3}+r=0 \ , \tag{6.48}$$

and for *Term IV*, $K_{n4} = \sin\theta_{n4}$, with

$$-\frac{K_{n4}}{\sqrt{1-K_{n4}^2}}\left[z-z_0+2(n+1)h\right]-(n+1)D_{n4}+r$$
$$= -\left[z-z_0+2(n+1)h\right]\tan\theta_{n4}-(n+1)D_{n4}+r=0 \ . \tag{6.49}$$

We proceed with an evaluation of the integrals in Eq. (6.43) using the modified method of stationary phase described in Sec. 4.8.1 [cf. Eqs. (4.129)-(4.133)]. We find that

$$p \approx \sum_{n=0}^{\infty}(-1)^n e^{i\pi/2}\left[\sum_{j=1}^{2}\frac{\sqrt{\sin\theta_{nj}}}{\cos\theta_{nj}}\frac{e^{ikH(K_{nj})}}{\sqrt{rH''(K_{nj})}}-\sum_{j=3}^{4}\frac{\sqrt{\sin\theta_{nj}}}{\cos\theta_{nj}}\frac{e^{ikH(K_{nj})}}{\sqrt{rH''(K_{nj})}}\right], \tag{6.50}$$

where, for example,

$$kH(K_{n1}) = (z_0-z+2nh)k\cos\theta_{n1}+n\varphi_{n1}+rk\sin\theta_{n1}$$
$$= k_{zn1}(z_0-z+2nh)+k_{n1}r+n\varphi_{n1} \ . \tag{6.51}$$

with $k_{znj} = k\cos\theta_{nj}$ and $k_{nj} = k\sin\theta_{nj}$ $(j = 1,\ldots 4)$. But from Eq. (6.45), we know that

$$r = \left(z_0 - z + 2nh\right)\tan\theta_{n1} + nD_{n1} \ , \tag{6.52}$$

which we substitute into Eq. (6.51) to obtain

$$kH\left(K_{n1}\right) = \overbrace{k_{zn1}\left(z_0 - z + 2nh\right)}^{\textit{Vertical Geometrical Phase Shift}} + \overbrace{k_{n1}\left(z_0 - z + 2nh\right)\tan\theta_{n1}}^{\textit{Horizontal Geometrical Phase Shift}}$$
$$+ \overbrace{nk_{n1}D_{n1}}^{\substack{\textit{Phase Shift Due to} \\ \textit{Horizontal Ray Displacement}}} + \overbrace{n\varphi_{n1}}^{\substack{\textit{Phase Shift Due to} \\ \textit{Bottom Reflection}}} \ . \tag{6.53}$$

Equation (6.53) describes the total phase accumulated by a direct path $(n = 0)$ joining source and receiver and its multiple reflections $(n > 0)$. In addition to the vertical and horizontal geometrical components of the phase, there are ray displacement and reflection coefficient phase contributions associated with the bottom interaction of the sound waves. Since Eq. (6.53) describes the total phase for a path which has undergone n surface and n bottom reflections, we can calculate the phase accumulated in the lateral direction between successive surface or bottom reflections:

$$\begin{aligned} &\left\{k_{n1}\left(z_0 - z + 2nh\right)\tan\theta_{n1} + nk_{n1}D_{n1}\right\} \\ &-\left\{k_{n1}\left[z_0 - z + 2(n-1)h\right]\tan\theta_{n1} + (n-1)k_{n1}D_{n1}\right\} \\ &= k_{n1}\left[2h\tan\theta_{n1} + D_{n1}\right] = k_{n1}L_n \ . \end{aligned} \tag{6.54}$$

Here L_n is precisely the cycle distance for the nth normal mode discussed on pp. 155-156. This result solidifies the connection between a group of modes and a geometrical acoustic path which undergoes ray displacement.

In order to determine the amplitude of the $j = 1$ term in Eq. (6.50), we calculate

$$H''\left(K_{n1}\right) = \left.\frac{\partial^2 H\left(K_r\right)}{\partial^2 K_r}\right|_{K_r = K_{n1}} = -\frac{\left(z_0 - z + 2nh\right)}{K_{zn1}^3} - n\frac{\partial D_{n1}}{\partial K_{n1}} . \tag{6.55}$$

But from Eq. (6.45), we have

$$\frac{\partial r}{\partial K_{n1}} = \frac{\left(z_0 - z + 2nh\right)}{K_{zn1}^3} + n\frac{\partial D_{n1}}{\partial K_{n1}} = -H''\left(K_{n1}\right) \ , \qquad (6.56)$$

and furthermore, we know that

$$\frac{\partial r}{\partial \theta_{n1}} = \frac{\partial r}{\partial K_{n1}} \frac{\partial K_{n1}}{\partial \theta_{n1}} = -H''\left(K_{n1}\right)\cos\theta_{n1} \ , \qquad (6.57)$$

and therefore

$$H''\left(K_{n1}\right) = -\frac{1}{\cos\theta_{n1}} \frac{\partial r}{\partial \theta_{n1}} \ . \qquad (6.58)$$

The $j = 1$ term p_{n1} in Eq. (6.50) therefore becomes

$$p_{n1} \approx \sum_{n=0}^{\infty} (-1)^n A_{n1} e^{i\left\{\left(k_{zn1} + k_{n1}\tan\theta_{n1}\right)\left[z_0 - z + 2nh\right] + n\left(k_{n1}D_{n1} + \varphi_{n1}\right)\right\}} \ , \qquad (6.59)$$

where A_{nj} $(j = 1, \ldots 4)$ is given by

$$A_{nj} = \left[\frac{\sin\theta_{nj}}{r\cos\theta_{nj}\left(\partial r / \partial\theta_{nj}\right)}\right]^{1/2} \ . \qquad (6.60)$$

The analysis of terms $j = 2, \ldots 4$ in Eq. (6.50) proceeds in a manner analogous to that used for $j = 1$, and we obtain the final result

$$\begin{aligned}
p \approx &\sum_{n=0}^{\infty} (-1)^n A_{n1} e^{i\left\{\left(k_{zn1} + k_{n1}\tan\theta_{n1}\right)\left[z_0 - z + 2nh\right] + n\left(k_{n1}D_{n1} + \varphi_{n1}\right)\right\}} && I \\
&+ \sum_{n=0}^{\infty} (-1)^n A_{n2} e^{i\left\{\left(k_{zn2} + k_{n2}\tan\theta_{n2}\right)\left[-(z + z_0) + 2(n+1)h\right] + (n+1)\left(k_{n2}D_{n2} + \varphi_{n2}\right)\right\}} && II \\
&- \sum_{n=0}^{\infty} (-1)^n A_{n3} e^{i\left\{\left(k_{zn3} + k_{n3}\tan\theta_{n3}\right)\left[z + z_0 + 2nh\right] + n\left(k_{n3}D_{n3} + \varphi_{n3}\right)\right\}} && III \\
&- \sum_{n=0}^{\infty} (-1)^n A_{n4} e^{i\left\{\left(k_{zn4} + k_{n4}\tan\theta_{n4}\right)\left[z - z_0 + 2(n+1)h\right] + (n+1)\left(k_{n4}D_{n4} + \varphi_{n4}\right)\right\}} \ . && IV
\end{aligned}$$

$$(6.61)$$

Equation (6.61) is similar in several ways to the conventional geometrical acoustics decompositions described in the previous section. For example, *Terms II*, *III*, and *IV* correspond to bottom-reflected, surface-reflected, and bottom-reflected/then surface-reflected paths, respectively. As before, the $n=0$ term is the basic field constituent in each case, and the $n>0$ terms correspond to multiple reflections within the waveguide. Again, with each boundary interaction, a particular field component is multiplied by the appropriate plane wave reflection coefficient. These similarities can be further confirmed by showing that, in the case of a rigid bottom, Eq. (6.61) reduces to Eq. (6.24) (cf. Prob. 6.4). As might be expected, the differences between the alternative and conventional field decompositions stem entirely from the approximations used for the bottom reflection coefficient in Eq. (6.42) and the pressure field in Eq. (6.43). As a result, explicit diffraction terms do not arise, and geometrical acoustic paths with ray displacement become the basic field constituents. These types of paths appear to be key links in connecting geometrical acoustic and modal representations.

6.3.2 A Normal Mode Decomposition

We consider now the derivation of a normal mode solution from the Hankel transform representation of the field due to a point source in a homogeneous fluid layer bounded by arbitrary horizontally stratified media. We recall that the pressure field is given by [cf. Eq. (6.19)]

$$p=\int_0^\infty \frac{i\left\{e^{ik_z|z-z_0|}+R_S e^{ik_z(z+z_0)}+R_B e^{2ik_z h}\left[e^{-ik_z(z+z_0)}+R_S e^{-ik_z|z-z_0|}\right]\right\}}{k_z\left[1-R_S R_B e^{2ik_z h}\right]}$$

$$\times J_0(k_r r)k_r\, dk_r \ . \tag{6.62}$$

The normal mode decomposition is obtained through a careful examination of the singularities of the integrand in Eq. (6.62) and the use of complex contour integration techniques. Perhaps the most striking feature of the integrand is its first-order poles (zeros of the denominator), which satisfy the equation

$$1-R_S R_B e^{2ik_z h}=0 \ . \tag{6.63}$$

But Eq. (6.63) corresponds exactly to the characteristic equation for the normal modes of propagation in the waveguide [cf. Eq. (5.71)]. As a result, we shall see that the normal mode description is intimately connected with the contributions of pole-type singularities in the Hankel transform representation of the field. In order to understand this connection further, we will examine in detail the case of the Pekeris waveguide.

The Pekeris Waveguide

In the Pekeris case, $R_S = -1$, and Eq. (6.62) becomes

$$p = \int_0^\infty \frac{i\left\{e^{ik_z|z-z_0|} - e^{ik_z(z+z_0)} + R_B e^{2ik_z h}\left[e^{-ik_z(z+z_0)} - e^{-ik_z|z-z_0|}\right]\right\}}{k_z\left[1+R_B e^{2ik_z h}\right]} \times J_0(k_r r) k_r \, dk_r \ , \tag{6.64}$$

which we write as

$$p(r) = \int_0^\infty g(k_r) J_0(k_r r) k_r \, dk_r \ , \tag{6.65}$$

where $g(k_r) = N(k_r)/D(k_r)$ and

$$N(k_r) = \frac{i}{k_z}\left\{e^{ik_z|z-z_0|} - e^{ik_z(z+z_0)} + R_B e^{2ik_z h}\left[e^{-ik_z(z+z_0)} - e^{-ik_z|z-z_0|}\right]\right\}, \tag{6.66}$$

$$D(k_r) = 1 + R_B e^{2ik_z h} \ . \tag{6.67}$$

Before discussing the pole-type singularities, it is important to determine the locations of any branch points in the integrand of Eq. (6.64). It is left as an exercise for the reader (cf. Prob. 6.5) to show that branch points occur only at the locations $k_r = \pm k_1 = \pm\omega/c_1$. This behavior is to be contrasted with the two half-space problem discussed in Sec. 4.8.1, in which meaningful branch points occur at $k_r = \pm k$ and $k_r = \pm k_1$. The difference between the two cases occurs because branch points arise in spatial regions for which the field may become unbounded as $z \to \pm\infty$. In that case, the signs of certain square root functions of k_r, which are associated with the vertical wavenumbers of propagation in the particular regions, must be chosen so that the field remains bounded as $z \to \pm\infty$. In the two half-space problem, there are two regions in which the field may become unbounded as $z \to \pm\infty$, and therefore there are two pairs of branch points (cf. p. 93). In the Pekeris waveguide, on the other hand, the finite extent of the homogeneous layer of thickness h guarantees that the field in the layer remains bounded, and therefore the branch points at $k_r = \pm k$ do not appear. Only the branch points $k_r = \pm k_1$, associated with the underlying half-space, arise. For these, as in the case of the two half-space problem, we select EJP branch cuts emanating from the branch points along the lines $\mathrm{Im}\sqrt{k_1^2 - k_r^2} = 0$ and define the physical sheet as $\mathrm{Im}\sqrt{k_1^2 - k_r^2} > 0$ (cf. Fig. 6.3).

The poles $k_r = \pm k_n$ $(\mathrm{Re}\, k_n > 0)$ of the integrand in Eq. (6.65) are solutions of the modal eigenvalue equation

$$D(k_r) = 1 + R_B e^{2ik_z h} = 0 \ , \tag{6.68}$$

and fall into two categories. For $k_1 \leq |\mathrm{Re} k_r| \leq k$, total internal reflection from the bottom occurs, and we have

$$R_B = e^{i\varphi}, \ \text{with} \ \varphi = -2\tan^{-1}\left[\frac{\sqrt{k_r^2 - k_1^2}}{mk_z}\right] \ , \tag{6.69}$$

so that Eq. (6.68) becomes

$$D(k_r) = 1 + e^{2i(k_z h + \varphi/2)} = 0 \ . \tag{6.70}$$

Equation (6.70) is identical to the eigenvalue equation for the perfectly trapped (proper) modes discussed in Sec. 5.5.1 [cf. Eqs. (5.137) and (5.138)]. There is a finite number of these poles $(n = 1, \ldots n_{\max})$ lying on the real axis, as shown in Fig. 6.3. For $0 \leq |\mathrm{Re} k_r| < k_1$, we have

$$R_B = e^{-\alpha(k_z)}, \ \text{with} \ \alpha(k_z) = -\ln R_B \ , \tag{6.71}$$

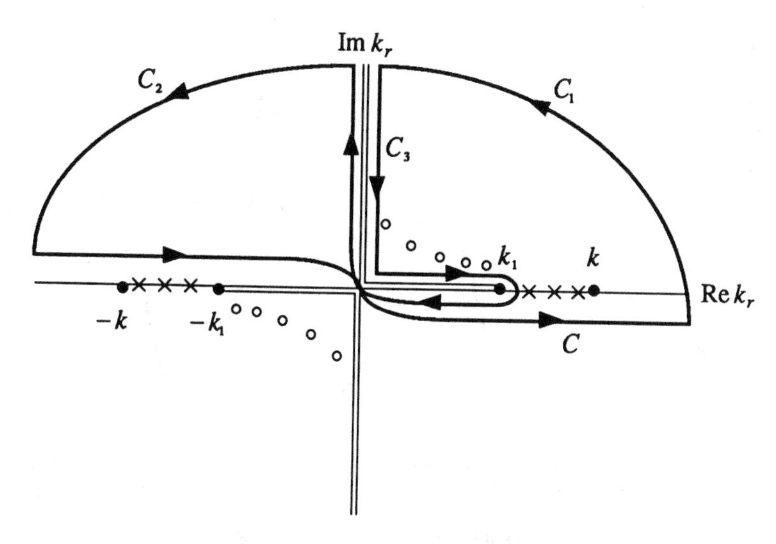

Figure 6.3 Contour of integration in the complex k_r-plane used in obtaining the normal mode solution from the Hankel transform representation for the Pekeris waveguide. Proper modes are indicated by ×, and improper modes by ∘.

and Eq. (6.68) becomes

$$D(k_r) = 1 + e^{2i(k_z h + i\alpha/2)} = 0 \quad . \tag{6.72}$$

Equation (6.72) is identical to the eigenvalue equation for the improper modes discussed in Sec. 5.5.2 [cf. Eqs. (5.73) and (5.74)]. There is an infinite number of these complex poles ($n = n_{\max} + 1, \ldots \infty$) having non-zero imaginary parts, as shown in Fig. 6.3. We recall that these poles have the behavior $\mathrm{Im}\sqrt{k_1^2 - k_n^2} < 0$ and therefore lie on an unphysical Riemann sheet in the complex k_r-plane. They may, however, contribute indirectly to the total field, depending upon their proximity to the branch cut $\mathrm{Im}\sqrt{k_1^2 - k_r^2} = 0$. In order to clarify this point, we have sketched the typical behavior of $|g(k_r)|$ for $\mathrm{Re}\, k_r \geq 0$ in Fig. 6.4. Here, the proper modes appear as perfect resonances in the region $k_1 < k_r < k$, whereas the improper modes manifest themselves as imperfect resonances in the region $0 < k_r < k_1$. Thus, the improper modes may introduce weaker, mode-like contributions to the Hankel transform integral in Eq. (6.65). An approximate decomposition in

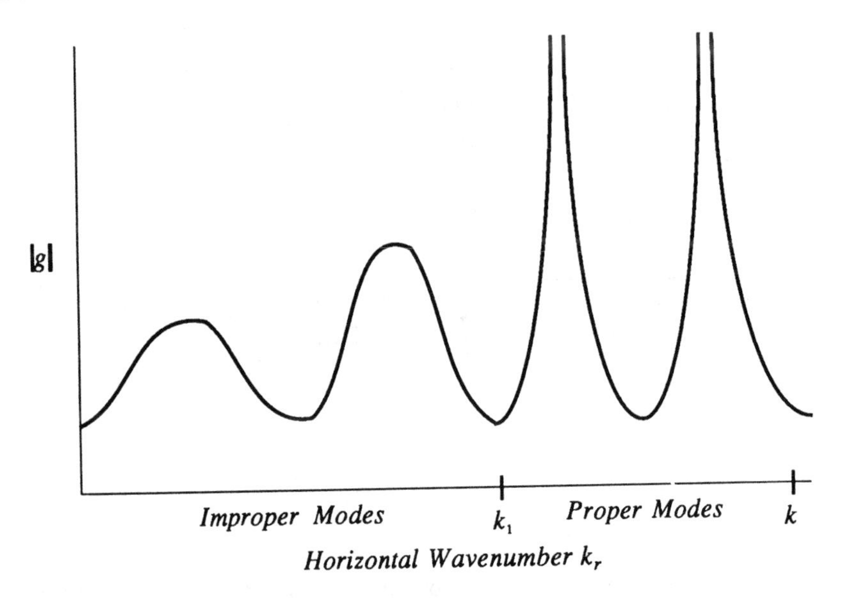

Figure 6.4 Schematic illustration of the influence of the proper and improper modes on the depth-dependent Green's function magnitude $|g|$ for the Pekeris waveguide.

terms of these leaky modes was discussed in Sec. 5.5.2. Finally, we reiterate the fact that, as kh increases, improper poles migrate into the proper pole region.

Having identified the nature of the singularities in the integrand, we now proceed with an evaluation of the integral in Eqs. (6.65)-(6.67) using complex contour integration methods. First, Eq. (6.10) enables us to rewrite Eq. (6.65) as

$$p(r) = \frac{1}{2}\int_{-\infty}^{\infty} g(k_r) H_0^{(1)}(k_r r)\, k_r\, dk_r \ , \qquad (6.73)$$

where the original path of integration C is shown in Fig. 6.3 (cf. Fig. 4.14). We then close the contour of integration in the upper half-plane on the physical Riemann sheet, as shown in Fig. 6.3, thereby enclosing only the pole-type singularities, associated with the proper modes, with $k_1 < \operatorname{Re} k_n < k$. From Cauchy's residue theorem [Churchill and Brown, 1984], we can then conclude that

$$2\pi i\sum_{n=1}^{n_{\max}}\text{Res}\left[\frac{1}{2}g(k_r)H_0^{(1)}(k_r r)k_r\right]_{k_r=k_n}$$
$$=\frac{1}{2}\int_{-\infty}^{\infty}g(k_r)H_0^{(1)}(k_r r)k_r\,dk_r+\frac{1}{2}\int_{C_1}g(k_r)H_0^{(1)}(k_r r)k_r\,dk_r$$
$$+\frac{1}{2}\int_{C_2}g(k_r)H_0^{(1)}(k_r r)k_r\,dk_r+\frac{1}{2}\int_{C_3}g(k_r)H_0^{(1)}(k_r r)k_r\,dk_r\,, \quad (6.74)$$

where the path C_3 runs around the branch line. It can be shown that the integrals along the semi-circular paths C_1 and C_2 vanish as the radius of the semi-circle tends to ∞ (cf. Prob. 6.6), and therefore Eq. (6.74) becomes

$$2\pi i\sum_{n=1}^{n_{\max}}\text{Res}\left[\frac{1}{2}g(k_r)H_0^{(1)}(k_r r)k_r\right]_{k_r=k_n}$$
$$=\frac{1}{2}\int_{-\infty}^{\infty}g(k_r)H_0^{(1)}(k_r r)k_r\,dk_r+\frac{1}{2}\int_{C_3}g(k_r)H_0^{(1)}(k_r r)k_r\,dk_r\,. \quad (6.75)$$

Evaluating the residues, we find that

$$p=\pi i\sum_{n=1}^{n_{\max}}\frac{k_n N(k_n)H_0^{(1)}(k_n r)}{\partial D(k_r)/\partial k_r\big|_{k_r=k_n}}-\frac{1}{2}\int_{C_3}g(k_r)H_0^{(1)}(k_r r)k_r\,dk_r\,, \quad (6.76)$$

where $N(k_r)$ and $D(k_r)$ are defined in Eqs. (6.66) and (6.67). We shall see that the residue series in Eq. (6.76) yields the discrete part of the modal spectrum, while the branch line integral gives rise to the modal continuum.

Residue Series: For the perfectly trapped modes, we know that

$$R_B=e^{i\varphi_n},\ \text{with}\ \varphi_n=-2\tan^{-1}\left[\frac{\sqrt{k_n^2-k_1^2}}{mk_{zn}}\right], \quad (6.77)$$

and therefore Eq. (6.66) becomes

$$N=\frac{i}{k_{zn}}\left\{e^{ik_{zn}|z-z_0|}-e^{ik_{zn}(z+z_0)}+e^{i\varphi_n}e^{2ik_{zn}h}\left[e^{-ik_{zn}(z+z_0)}-e^{-ik_{zn}|z-z_0|}\right]\right\}, \quad (6.78)$$

where $k_{zn}=\sqrt{k^2-k_n^2}$. But the eigenvalue equation is $e^{i\varphi_n}e^{2ik_{zn}h}=-1$ [cf. Eq. (6.70)], so that Eq. (6.78) becomes

$$N(k_n)=\frac{i}{k_{zn}}\left\{e^{ik_{zn}|z-z_0|}-e^{ik_{zn}(z+z_0)}-\left[e^{-ik_{zn}(z+z_0)}-e^{-ik_{zn}|z-z_0|}\right]\right\} \quad , \tag{6.79}$$

which reduces in a straightforward manner to (cf. Prob. 6.7)

$$N(k_n)=\frac{4i}{k_{zn}}\sin k_{zn}z\sin k_{zn}z_0 \quad . \tag{6.80}$$

Examining the denominator of the residue series terms in Eq. (6.76), we obtain

$$\left.\frac{\partial D(k_r)}{\partial k_r}\right|_{k_r=k_n}=i\left(\frac{\partial\varphi}{\partial k_r}+2h\frac{\partial k_z}{\partial k_r}\right)e^{i\varphi}e^{2ik_zh}\bigg|_{k_r=k_n}=i\left(\frac{d\varphi_n}{dk_n}-2h\frac{k_n}{k_{zn}}\right)e^{i\varphi_n}e^{2ik_{zn}h}$$

$$=i\left(2h\tan\theta_n-\frac{d\varphi_n}{dk_n}\right)=iL_n \quad , \tag{6.81}$$

where θ_n is the angle of propagation, and, interestingly, L_n is the cycle distance of the nth mode [cf. Eq. (5.173)]. At this point, then, the residue series can be written as

$$\pi i\sum_{n=1}^{n_{\max}}\frac{k_nN(k_n)H_0^{(1)}(k_nr)}{\left.\frac{\partial D(k_r)}{\partial k_r}\right|_{k_r=k_n}}=\pi i\sum_{n=1}^{n_{\max}}\frac{4\sin k_{zn}z\sin k_{zn}z_0}{\left(2h-\frac{k_{zn}}{k_n}\frac{d\varphi_n}{dk_n}\right)}H_0^{(1)}(k_nr). \tag{6.82}$$

From Eq. (5.174), we determine that

$$-\frac{k_{zn}}{k_n}\frac{d\varphi_n}{dk_n}=\frac{2m\left(k^2-k_1^2\right)}{k_{1zn}\left(m^2k_{zn}^2+k_{1zn}^2\right)} \quad , \tag{6.83}$$

where $k_{1zn}^2=\sqrt{k_n^2-k_1^2}$. Squaring the eigenvalue equation in Eq. (5.163),

$$k_{1zn}^2 = m^2 k_{zn}^2 / \tan^2 k_{zn} h \quad , \tag{6.84}$$

we substitute Eq. (6.84) into Eq. (6.83) to obtain

$$\begin{aligned}
-\frac{k_{zn}}{k_n}\frac{d\varphi_n}{dk_n} &= \frac{2\left(k^2 - k_1^2\right)}{mk_{1zn}k_{zn}^2\left(1+\cot^2 k_{zn}h\right)} = \frac{2\left(k^2 - k_1^2\right)\sin^2 k_{zn}h}{mk_{1zn}k_{zn}^2} \\
&= \frac{2\left(k^2 - k_n^2 - k_1^2 + k_n^2\right)\sin^2 k_{zn}h}{mk_{1zn}k_{zn}^2} = \frac{2\left(k_{zn}^2 + k_{1zn}^2\right)\sin^2 k_{zn}h}{mk_{1zn}k_{zn}^2} \\
&= \frac{2}{mk_{1zn}}\sin^2 k_{zn}h + \frac{2k_{1zn}}{mk_{zn}^2}\sin^2 k_{zn}h \ .
\end{aligned} \tag{6.85}$$

Again using the eigenvalue equation in Eq. (5.163),

$$k_{1zn} = -mk_{zn}\cot k_{zn}h \quad , \tag{6.86}$$

we substitute Eq. (6.86) into Eq. (6.85) and find that

$$-\frac{k_{zn}}{k_n}\frac{d\varphi_n}{dk_n} = \frac{2}{mk_{1zn}}\sin^2 k_{zn}h - \frac{1}{k_{zn}}\sin 2k_{zn}h \quad , \tag{6.87}$$

where we have used the identity

$$\sin 2\vartheta = 2\sin\vartheta\cos\vartheta \ . \tag{6.88}$$

As a result, the residue series in Eq. (6.82) becomes

$$\pi i \sum_{n=1}^{n_{\max}} \frac{k_n N(k_n) H_0^{(1)}(k_n r)}{\left.\dfrac{\partial D(k_r)}{\partial k_r}\right|_{k_r = k_n}} = \frac{\pi i}{\rho}\sum_{n=1}^{n_{\max}} \frac{2\sin k_{zn}z_0 \sin k_{zn}z H_0^{(1)}(k_n r)}{\left[\dfrac{1}{\rho}\left(h - \dfrac{\sin 2k_{zn}h}{2k_{zn}}\right) + \dfrac{\sin^2 k_{zn}h}{\rho_1 k_{1zn}}\right]}, \tag{6.89}$$

which is identical to the discrete sum of trapped modes in Eq. (5.145), for $0 \le z, z_0 \le h$, obtained using the method of normal modes.

Branch Line Integral: In order to evaluate the branch line integral in Eq. (6.76), we first rewrite it as

$$-\frac{1}{2}\int_{C_3} g(k_r)H_0^{(1)}(k_r r)k_r\,dk_r = -\frac{1}{2}\left[\int_{i\infty}^{0}+\int_{0}^{k_1}\right]g_-(k_r)H_0^{(1)}(k_r r)k_r\,dk_r$$

$$-\frac{1}{2}\left[\int_{k_1}^{0}+\int_{0}^{i\infty}\right]g_+(k_r)H_0^{(1)}(k_r r)k_r\,dk_r$$

$$=-\frac{1}{2}\left[\int_{k_1}^{0}+\int_{0}^{i\infty}\right]\left[g_+(k_r)-g_-(k_r)\right]H_0^{(1)}(k_r r)k_r\,dk_r, \quad (6.90)$$

where $g_-(k_r)$ is the value of $g(k_r)$ along the portion of the path C_3 in the first quadrant, and $g_+(k_r)$ is the value of $g(k_r)$ along the part of the path C_3 in the fourth and second quadrants. Specifically, we have

$$g_\pm(k_r)=\frac{i\left\{e^{ik_z|z-z_0|}-e^{ik_z(z+z_0)}+R_\pm e^{2ik_z h}\left[e^{-ik_z(z+z_0)}-e^{-ik_z|z-z_0|}\right]\right\}}{k_z\left[1+R_\pm e^{2ik_z h}\right]}, \quad (6.91)$$

where [cf. Eq. (4.141)]

$$R_\pm = R_B^\pm = \frac{m\sqrt{k^2-k_r^2}\mp\sqrt{k_1^2-k_r^2}}{m\sqrt{k^2-k_r^2}\pm\sqrt{k_1^2-k_r^2}}. \quad (6.92)$$

The appropriate signs of the square root $\sqrt{k_1^2-k_r^2}$ were determined by imposing the physical requirement $\operatorname{Im}\sqrt{k_1^2-k_r^2}>0$. For example, in the region $0<\operatorname{Re}k_r<k_1$ of the first quadrant, we let (cf. Prob. 4.12)

$$k_r=\alpha+i\varepsilon \text{ with } 0<\alpha<k_1,\ 0<\varepsilon<<k_1\ , \quad (6.93)$$

so that

$$k_{1z}=\pm\sqrt{k_1^2-k_r^2}=\pm\sqrt{k_1^2-(\alpha+i\varepsilon)^2}=\pm k_1\sqrt{1-\frac{(\alpha+i\varepsilon)^2}{k_1^2}}$$

$$\approx\pm k_1\left[1-\frac{(\alpha+i\varepsilon)^2}{2k_1^2}\right]\approx\pm k_1\left[1-\frac{\alpha^2}{2k_1^2}-\frac{i\varepsilon\alpha}{k_1^2}\right], \quad (6.94)$$

and chose the negative sign in order to satisfy the condition $\mathrm{Im}\,k_{1z} > 0$. Analyticity ensures that this choice is also correct for values of k_r which lie close to the $\mathrm{Im}\,k_r$ axis in the first quadrant. The determination of the appropriate signs of $\sqrt{k_1^2 - k_r^2}$ in the other regions of interest in the complex k_r-plane is left as an exercise for the reader (cf. Prob. 6.8).

Using Eq. (6.91), we proceed with the evaluation of Eq. (6.90) and write

$$g_+ - g_- = \frac{N_+ D_- - N_- D_+}{D_+ D_-}, \tag{6.95}$$

where

$$N_\pm = \frac{i}{k_z}\left\{e^{ik_z|z-z_0|} - e^{ik_z(z+z_0)} + R_\pm e^{2ik_z h}\left[e^{-ik_z(z+z_0)} - e^{-ik_z|z-z_0|}\right]\right\}, \tag{6.96}$$

$$D_\pm = 1 + R_\pm e^{2ik_z h} . \tag{6.97}$$

The numerator of Eq. (6.95) is given by

$$\begin{aligned}
N_+ D_- - N_- D_+ &= N_+ - N_- + e^{2ik_z h}\left[N_+ R_- - N_- R_+\right] \\
&= \frac{i}{k_z}\left\{(R_+ - R_-)e^{2ik_z h}\left[e^{-ik_z(z+z_0)} - e^{-ik_z|z-z_0|}\right]\right\} \\
&\quad + \frac{ie^{2ik_z h}}{k_z}\left\{(R_- - R_+)\left[e^{ik_z|z-z_0|} - e^{ik_z(z+z_0)}\right]\right\} \\
&= \frac{i}{k_z}(R_+ - R_-)e^{2ik_z h}\left[e^{-ik_z(z+z_0)} - e^{-ik_z|z-z_0|} + e^{ik_z(z+z_0)} - e^{ik_z|z-z_0|}\right] \\
&= \frac{2i}{k_z}(R_+ - R_-)e^{2ik_z h}\left[\cos k_z(z+z_0) - \cos k_z(z-z_0)\right] \\
&= -\frac{4i}{k_z}(R_+ - R_-)e^{2ik_z h}\sin k_z z \sin k_z z_0 .
\end{aligned} \tag{6.98}$$

Here we have used the identity

$$\cos\alpha - \cos\beta = -2\sin\left(\frac{\alpha+\beta}{2}\right)\sin\left(\frac{\alpha-\beta}{2}\right) \tag{6.99}$$

in the last step of Eq. (6.98). Using Eq. (6.92), we can show that

$$R_+ - R_- = -\frac{4mk_z k_{1z}}{m^2k_z^2 - k_{1z}^2}, \quad R_+ + R_- = \frac{2\left(m^2k_z^2 + k_{1z}^2\right)}{m^2k_z^2 - k_{1z}^2}, \tag{6.100}$$

and therefore Eq. (6.98) becomes

$$N_+D_- - N_-D_+ = \frac{16imk_{1z}}{m^2k_z^2 - k_{1z}^2} e^{2ik_zh} \sin k_z z \sin k_z z_0 \ . \tag{6.101}$$

The denominator of Eq. (6.95) is given by

$$\begin{aligned} D_+D_- &= \left[1 + R_+e^{2ik_zh}\right]\left[1 + R_-e^{2ik_zh}\right] \\ &= 1 + R_+R_-e^{4ik_zh} + \left(R_+ + R_-\right)e^{2ik_zh} \ . \end{aligned} \tag{6.102}$$

But $R_+R_- = 1$, and using Eq. (6.100), we find that Eq. (6.102) becomes

$$\begin{aligned} D_+D_- &= 1 + e^{4ik_zh} + \frac{2\left(m^2k_z^2 + k_{1z}^2\right)}{m^2k_z^2 - k_{1z}^2}e^{2ik_zh} \\ &= e^{2ik_zh}\left[e^{-2ik_zh} + e^{2ik_zh} + \frac{2\left(m^2k_z^2 + k_{1z}^2\right)}{m^2k_z^2 - k_{1z}^2}\right] = 2e^{2ik_zh}\left[\cos 2k_zh + \frac{m^2k_z^2 + k_{1z}^2}{m^2k_z^2 - k_{1z}^2}\right] \\ &= 2e^{2ik_zh}\left[1 - 2\sin^2 k_zh + \frac{m^2k_z^2 + k_{1z}^2}{m^2k_z^2 - k_{1z}^2}\right] \\ &= 4e^{2ik_zh}\left[\frac{m^2k_z^2\left(1 - \sin^2 k_zh\right) + k_{1z}^2 \sin^2 k_zh}{m^2k_z^2 - k_{1z}^2}\right] \\ &= 4e^{2ik_zh}\frac{k_{1z}^2\left[\sin^2 k_zh + \dfrac{m^2k_z^2}{k_{1z}^2}\cos^2 k_zh\right]}{m^2k_z^2 - k_{1z}^2} \ . \end{aligned} \tag{6.103}$$

Here we have obtained the fourth step from the third step in Eq. (6.103) by using the identity

$$\cos 2\vartheta = 1 - 2\sin^2\vartheta \ . \tag{6.104}$$

Finally, combining Eqs. (6.101) and (6.103), we see that Eq. (6.95) becomes

$$g_+ - g_- = \frac{4im \sin k_z z \sin k_z z_0}{k_{1z}\left[\sin^2 k_z h + \dfrac{m^2 k_z^2}{k_{1z}^2}\cos^2 k_z h\right]} \ , \tag{6.105}$$

and therefore the branch line integral in Eq. (6.90) is given by

$$-\frac{1}{2}\int_{C_3} g(k_r) H_0^{(1)}(k_r r) k_r \, dk_r = -2im \int_{k_1}^{0} \frac{k_r \sin k_z z \sin k_z z_0 \, H_0^{(1)}(k_r r)}{k_{1z}\left[\sin^2 k_z h + \dfrac{m^2 k_z^2}{k_{1z}^2}\cos^2 k_z h\right]} dk_r$$

$$-2im \int_{0}^{i\infty} \frac{k_r \sin k_z z \sin k_z z_0 \, H_0^{(1)}(k_r r)}{k_{1z}\left[\sin^2 k_z h + \dfrac{m^2 k_z^2}{k_{1z}^2}\cos^2 k_z h\right]} dk_r . \tag{6.106}$$

But changing variables in Eq. (6.106),

$$k_z = \sqrt{k^2 - k_r^2} \Rightarrow dk_z = -\frac{k_r}{k_z} dk_r \ , \tag{6.107}$$

we find that the branch line integral is

$$-\frac{1}{2}\int_{C_3} g(k_r) H_0^{(1)}(k_r r) k_r \, dk_r$$

$$= \frac{2i\rho_1}{\rho} \int_{C_B} \frac{k_z}{k_{1z}} \frac{\sin k_z z_0 \sin k_z z}{\left[\sin^2 k_z h + \dfrac{\rho_1^2 k_z^2}{\rho^2 k_{1z}^2}\cos^2 k_z h\right]} H_0^{(1)}(k_r r) dk_z , \tag{6.108}$$

where the path of integration C_B in the complex k_z-plane is shown in Fig. 5.19. It is clear that Eq. (6.108) is identical to the continuum contribution in Eq. (5.175), for $0 \le z, z_0 \le h$, obtained using the method of normal modes.

In summary, we have seen that the Hankel transform and modal representations of the sound field in the Pekeris waveguide are entirely equivalent to one another. Specifically, we showed that the residue series and branch line integral obtained from the Hankel transform approach are exactly equal to the discrete and continuum parts, respectively, of the modal field:

$$p = \int_0^\infty \frac{i\left\{e^{ik_z|z-z_0|} - e^{ik_z(z+z_0)} + R_B e^{2ik_z h}\left[e^{-ik_z(z+z_0)} - e^{-ik_z|z-z_0|}\right]\right\}}{k_z\left[1 + R_B e^{2ik_z h}\right]}$$

$$\times J_0(k_r r) k_r \, dk_r$$

$$= \pi i \sum_{n=1}^{n_{\max}} \frac{k_n N(k_n) H_0^{(1)}(k_n r)}{\partial D(k_r)/\partial k_r\big|_{k_r = k_n}} - \frac{1}{2}\int_{C_3} g(k_r) H_0^{(1)}(k_r r) k_r \, dk_r$$

$$= \frac{i\pi}{\rho} \sum_{n=1}^{n_{\max}} A_n^2 \sin k_{zn} z_0 \sin k_{zn} z \, H_0^{(1)}(k_n r)$$

$$+ \frac{2i\rho_1}{\rho} \int_{C_B} \frac{k_z}{k_{1z}} \frac{\sin k_z z_0 \sin k_z z}{\left[\sin^2 k_z h + \dfrac{\rho_1^2 k_z^2}{\rho^2 k_{1z}^2} \cos^2 k_z h\right]} H_0^{(1)}(k_r r) \, dk_z \;, \quad (6.109)$$

where A_n is defined in Eq. (5.144).

PROBLEMS

6.1 Consider the pillbox function

$$f(k_r) = \begin{cases} 1, & 0 \le k_r \le K, \\ 0, & K < k_r \;. \end{cases} \quad (6.110)$$

(a) Show that the Hankel transform of $f(k_r)$ is given by

$$F(r)=\frac{KJ_1(Kr)}{r} \ , \tag{6.111}$$

where J_1 is the first order Bessel function (cf. Sec. 2.4).

Hint: Use the recursion relation [Abramowitz and Stegun, 1964]

$$J_{\nu-1}(x)=\frac{\nu}{x}J_\nu(x)+J'_\nu(x) \ . \tag{6.112}$$

(b) Compare the behavior of $F(r)$ for $Kr \ll 1$ and $Kr \gg 1$ with the behavior of the Fourier transform $S(\omega)$ of the boxcar function

$$s(t)=\begin{cases}1, & 0\le|t|\le T,\\ 0, & T<|t| \ ,\end{cases} \tag{6.113}$$

for $\omega T \ll 1$ and $\omega T \gg 1$. Discuss the similarities and differences between $F(r)$ and $S(\omega)$.

6.2 Show that the inverse Hankel transform [cf. Eq. (6.5)],

$$g(k_r)=\int_0^\infty p(r)J_0(k_r r)r\,dr \ , \tag{6.114}$$

can be generated exactly using the *Fourier-Bessel series*,

$$g(k_r)=\frac{2}{K^2}\sum_{n=1}^{\infty}\frac{p(\lambda_n/K)J_0(k_r\lambda_n/K)}{J_1^2(\lambda_n)}, \quad 0\le k_r\le K \ , \tag{6.115}$$

as long as $g(k_r)=0$ for $k_r>K$. Here J_1 is the first order Bessel function and λ_n are the zeros of J_0.

Hint: Use the orthonormality relation [Jackson, 1975]

$$\int_0^K J_\nu(k_r\lambda_n/K)J_\nu(k_r\lambda_m/K)k_r\,dk_r=\frac{K^2}{2}J_{\nu+1}^2(\lambda_n)\delta_{nm} \ , \tag{6.116}$$

where $J_\nu(\lambda_n)=0$, and δ_{nm} is the Kronecker delta [cf. Eq. (5.11)].

6.3 For the zeroth order ocean acoustic waveguide, show that the image method solution in Eqs. (6.24) and (6.25) obeys the principle of reciprocity [cf. Eq. (4.19)].

6.4 Show that in the limiting case of a rigid bottom, the alternative geometrical acoustics decomposition for the Pekeris waveguide in Eq. (6.61) reduces to the image method result in Eq. (6.24).

6.5 In the Hankel transform representation for the Pekeris waveguide in Eq. (6.64), verify that branch points of the integrand occur only at the locations $k_r = \pm k_1 = \pm \omega / c_1$.

6.6 In Eq. (6.74), prove that the integrals along the semi-circular paths C_1 and C_2 vanish as the radius of the semi-circle tends to ∞.

Hint: Use the asymptotic form for the Hankel function [cf. Eq. (4.119)], rewrite the integrals in terms of polar coordinates $\left(k_r = Ke^{i\theta}\right)$, and examine their behavior as $K \to \infty$.

6.7 In the case of the zeroth order ocean acoustic waveguide, only the residue series term in Eq. (6.76) arises from the complex contour integration analysis of the Hankel transform representation. Show that the residue series is equivalent to the normal mode sum in Eq. (5.25). In the process, demonstrate that the modal cycle distance L_n [cf. Eq. (5.56)] and the mode normalization constant A_n [cf. Eq. (5.24)] are related to one another by

$$L_n = \frac{4\rho k_n}{k_{zn} A_n^2} , \tag{6.117}$$

a result which also holds true for the Pekeris waveguide [cf. Eqs. (6.81), (6.82), (6.87), and (5.144)].

6.8 Determine the appropriate signs of the square root $\sqrt{k_1^2 - k_r^2}$ in the regions of the complex k_r-plane required to evaluate the branch line integral in Eq. (6.90).

6.9 Suppose that both the sound source and receiver in the Pekeris waveguide were located in the ocean bottom $(z, z_0 > h)$.

(a) Determine the Hankel transform representation for the field analogous to the result in Eq. (6.64), which was derived for the case $0 \leq z, z_0 \leq h$.

(b) Discuss the modal interpretation of the solution in part (a) in terms of discrete and continuous spectra (cf. Prob. 5.10).

6.10 Consider a point source in a homogeneous fluid layer with a pressure-release top and a homogeneous, lower velocity fluid bottom, as shown in Fig. 5.9. Using complex contour integration methods, show that the Hankel transform representation of the field for $0 \leq z, z_0 \leq h$ is equivalent to the modal continuum result in Eq. (5.94).

Chapter 7. Approximate Methods for Inhomogeneous Media: Ray and WKB Theory

How far that little candle throws his beams!
William Shakespeare, 1564-1616

Our life is frittered away by detail.
...Simplify, simplify.
Henry David Thoreau, 1817-1862

7.1 Introduction

In this chapter, we focus on the problem of wave propagation in a medium for which the index of refraction is a continuously varying function of depth. This discussion will complete a theoretical picture which has concentrated so far on applications to homogeneous media separated by discrete interfaces. We are also motivated by practical considerations, in that both the water column as well as layers within the seabed are excellent examples of media with continuously variable sound velocities.

In pursuing the issue of wave propagation in inhomogeneous media with continuous velocity variations, there are several approaches available to us. We can apply the general normal mode and Hankel transform methods developed in Chaps. 5 and 6 to velocity profiles for which the depth-dependent Eqs. (5.8) and (6.3) have closed form solutions in terms of special functions. Examples of some canonical profiles and the associated wave functions are described by Bucker [1970]. Complicated profiles can be accommodated by subdividing them into segments within each one of which a canonical profile is a suitable approximation. An alternative approach in complex environments is to compute the relevant wave quantities numerically, as is described by Porter and Reiss [1985] in a modal framework. On the other hand, we can approximate the continuously varying medium by a stack of homogeneous layers whose thicknesses approach zero [Brekhovskikh, 1980]. We then have the benefit of dealing with plane wave solutions, which are very familiar to us. Most of these approaches, in principle, provide exact, full wave solutions to the wave equation and are therefore of considerable interest to the ocean acoustician. However, we prefer to focus our attention on approximate methods which are appropriate for media in which the index of refraction is a slowly varying function of position. These ray and WKB theories provide us with powerful, new theoretical tools and add enormous physical insight into the nature of wave propagation in inhomogeneous media. Furthermore, in many instances, these theories can be applied with great success to sound propagation through the ocean and seabed because of the slowly varying nature of the environment.

First, we briefly describe some key characteristics of the refracting ocean environment. Then we derive the two-dimensional plane wave ray/WKB solution in an unbounded, horizontally stratified medium with no sources and discuss its features and underlying approximations. In three dimensions, we develop ray theory for an arbitrary unbounded medium. For the horizontally stratified case with a point source, we obtain expressions for the ray range, phase, travel time, and intensity, and apply them to the example of a linear gradient profile. We then consider the case of a medium with a velocity profile having a single minimum, for which we derive the ray series solution for a point source. We show that the ray result can also be obtained from an asymptotic evaluation of the Hankel transform representation of the field in the WKB approximation. We demonstrate that some of the key features of long-range SOFAR propagation can be obtained from the ray solution for the special case of the symmetric bilinear profile. In particular, we calculate the time-dependent behavior of a delta function pulse. We also derive the normal mode solution in the WKB approximation for the single minimum case and conclude by summarizing some of the deficiencies of classical ray theory.

7.2 The Refracting Ocean Environment

The sound velocity in the ocean can be described by the semi-empirical equation [Clay and Medwin, 1977],

$$\begin{aligned} c &= 1449.2 + 4.6T - 0.055T^2 + 0.00029T^3 \\ &\quad + (1.34 - 0.010T)(S-35) + 0.016z \ , \\ c &= \text{Sound Velocity (m/s)}, \quad S = \text{Salinity (ppt)}, \\ T &= \text{Temperature } (^\circ\text{C}), \quad z = \text{Depth (m)} \ , \end{aligned} \tag{7.1}$$

which is well suited to most applications. Additional references on the velocity of sound in sea water include the work of Wilson [1960], Leroy [1969], Del Grosso [1974], Chen and Millero [1977], Mackenzie [1981], and Spiesberger and Metzger [1991]. In deep water (thousands of meters depth), the variation in sound velocity is dominated by temperature in the upper ocean ($z < 1000$ m) and hydrostatic pressure in the deep ocean ($z > 1000$ m), with salinity playing a secondary role. A typical North Atlantic profile, illustrating this behavior, is shown in Fig. 7.1, which is based on the work of Urick [1975] and Tolstoy and Clay [1987]. In the upper tens of meters, the velocity may vary on a daily basis and is influenced by solar heating and wind, which generates surface gravity waves, bubbles, and various mixing processes. Below this surface layer lies the *thermocline*, which is an extensive region of rapidly decreasing temperature that causes a rapid decrease in sound velocity. The upper portion consists of a seasonal thermocline, which may be formed by solar heating during the summer. The lower portion is the main thermocline, which is essentially invariant under seasonal influences. At a depth of about 1000 m, the temperature approaches a constant value of 4° C, and the hydrostatic pressure becomes the primary influence on the variation in sound velocity. As a result, at great depths, the z-dependent term in Eq. (7.1) dominates, and the velocity gradient is simply $dc/dz \approx 0.016\ \mathrm{s}^{-1}$. The velocity minimum in Fig. 7.1 is perhaps the single most important acoustic characteristic of the ocean environment because, through the process of continuous total internal reflection, it creates a waveguide in which sound can be channeled to great distances. A judicious choice of source and receiver depths, for example on the *sound channel axis* (cf. Fig. 7.1), enables sound to propagate efficiently for thousands of kilometers. The discovery [Ewing and Worzel, 1948; Brekhovskikh, 1980; Brekhovskikh and Lysanov, 1991] of this *SOFAR* (SOound Fixing And Ranging) *channel* revolutionized ocean acoustics and led to a wealth of research following World War II. We will examine a simplified version of this underwater sound channel in Sec. 7.6. In shallow water (hundreds of meters depth and below), particularly in wintertime, the water column may have an isothermal character which creates a virtually homogeneous, isovelocity medium. The surface and bottom boundaries are the primary guiding mechanisms in shallow water, whereas the velocity profile is usually the dominant channeling mechanism in deep water. As a result, the effects of the surface and bottom are sometimes ignored altogether in deep water problems.

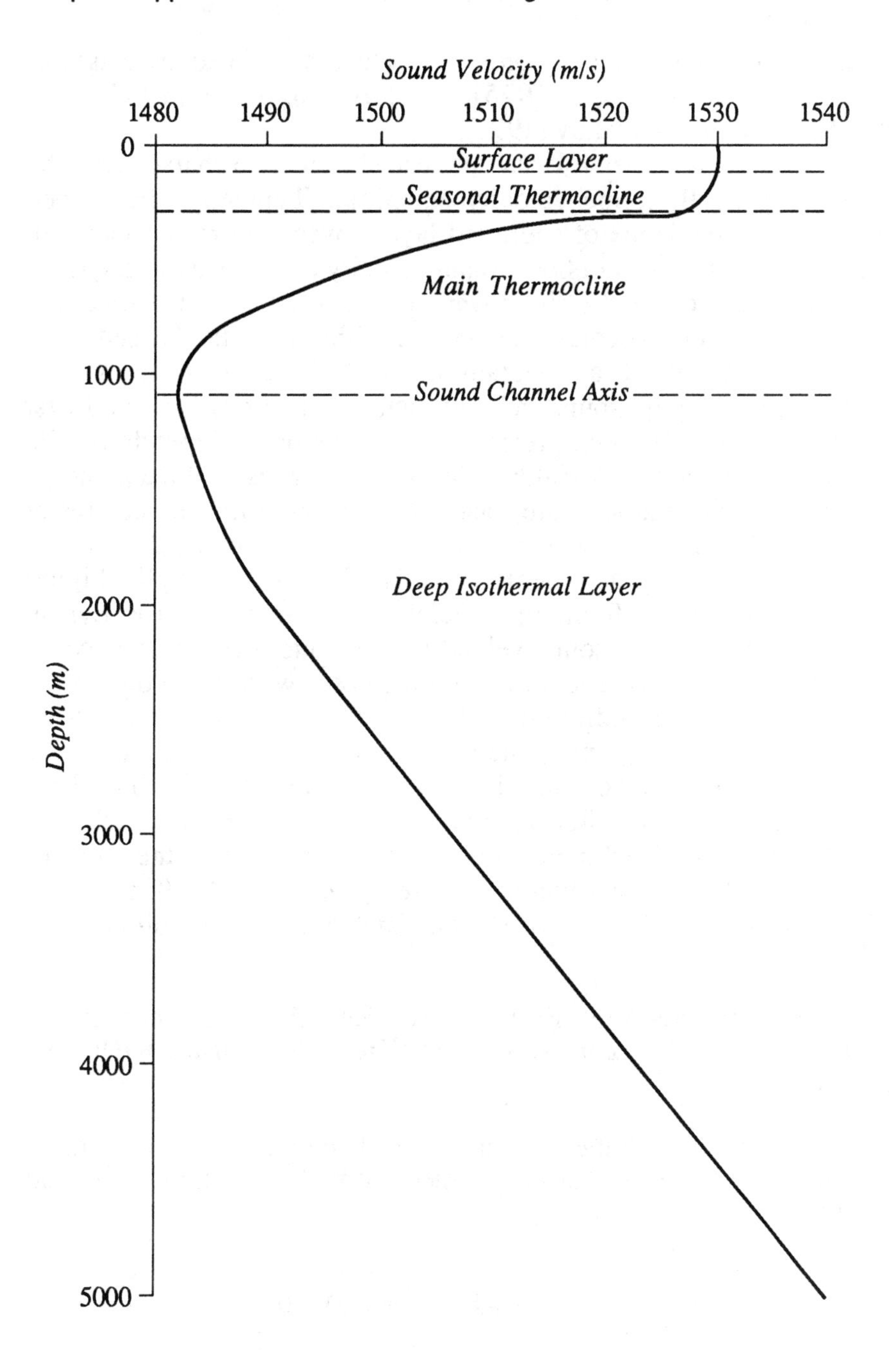

Figure 7.1 A typical North Atlantic sound velocity profile.

Additional canonical velocity profiles and their associated ray diagrams can be found in Urick [1975], Tolstoy and Clay [1987], and Brekhovskikh and Lysanov [1991].

Unlike the water column, the seabed cannot be characterized by a single, general, sound velocity equation. Typically, the seabed consists of a sequence of sediment layers overlying rock basement, and the discontinuities in both sound velocity and density at the layer interfaces, as well as the shear wave characteristics of the material, must be taken into account. However, within a particular sediment layer, the model of a constant density, fluid medium with a continuously varying sound velocity (and attenuation) is a good first approximation. The compressional wave velocity depends on the nature of the bottom material, its physical properties, and the geologic setting. Velocity gradients are usually two orders of magnitude greater than those found in the water column, with values of $1-2\ \mathrm{s}^{-1}$ being typical of silty-clay sediments (cf. Fig. 3.6); substantially higher values are sometimes found in surficial sand sediments [Frisk, Lynch, and Rajan, 1989]. The sound velocity in the bottom generally exceeds the velocity in the water column and increases with depth due to the compaction of the sediment material. However, localized, low-velocity regions may exist, such as the one which occurs near the water-bottom interface in silty-clay sediments (cf. Fig. 3.6). Considerably more detail on a variety of geoacoustic models, including shear properties of sediments and rocks, is available in the work of Hamilton [1980], Kuperman and Jensen [1980], Stoll [1985, 1989], Akal and Berkson [1986], and Hovem, Richardson, and Stoll [1991].

7.3 Two-Dimensional Plane Wave Ray/WKB Theory in an Unbounded, Horizontally Stratified Medium with No Sources

We begin with the two-dimensional Helmholtz equation for an unbounded, horizontally stratified medium with constant density and no sources [cf. Eq. (2.18)]:

$$\left[\frac{\partial^2}{\partial x^2}+\frac{\partial^2}{\partial z^2}+k^2(z)\right]p(x,z)=0 \ , \qquad (7.2)$$

where $k(z)=\omega/c(z)$. We assume a separable solution of the form $p(x,z)=v(x)u(z)$, which we substitute into Eq. (7.2) to obtain

$$\frac{1}{v(x)}\frac{d^2v(x)}{dx^2}=-\frac{1}{u(z)}\frac{d^2u(z)}{dz^2}-k^2(z) \quad . \tag{7.3}$$

Equation (7.3) must hold true for arbitrary values of x and z, and therefore the right- and left-hand sides of Eq. (7.3) must separately be equal to a separation constant which we call $-k_x^2$:

$$\frac{1}{v(x)}\frac{d^2v(x)}{dx^2}=-k_x^2=-\frac{1}{u(z)}\frac{d^2u(z)}{dz^2}-k^2(z) \quad . \tag{7.4}$$

This procedure thus yields two ordinary differential equations,

$$\frac{d^2v(x)}{dx^2}+k_x^2v(x)=0 \quad , \tag{7.5}$$

$$\frac{d^2u(z)}{dz^2}+k_z^2(z)u(z)=0 \quad , \tag{7.6}$$

with $k_z^2(z)=k^2(z)-k_x^2$. Assuming propagation in the direction of increasing x only, we can write the solution of Eq. (7.5) as

$$v(x)=Be^{ik_x(x-x_i)} \quad , \tag{7.7}$$

where B and x_i are arbitrary constants. For a slowly varying medium, we then postulate a solution of Eq. (7.6) having the plane wave-like form,

$$u(z)=A(z)e^{iS(z)} \quad , \tag{7.8}$$

where $A(z)$ and $S(z)$ are real functions. Substituting Eq. (7.8) into Eq. (7.6), we obtain

$$\frac{d^2A(z)}{dz^2}+2i\frac{dA(z)}{dz}\frac{dS(z)}{dz}+iA(z)\frac{d^2S(z)}{dz^2}+\left[k_z^2(z)-\left(\frac{dS(z)}{dz}\right)^2\right]A(z)=0. \tag{7.9}$$

Setting the real and imaginary parts of Eq. (7.9) separately equal to 0, we find that

$$\frac{d^2A(z)}{dz^2}+\left[k_z^2(z)-\left(\frac{dS(z)}{dz}\right)^2\right]A(z)=0 \ , \tag{7.10}$$

$$A(z)\frac{d^2S(z)}{dz^2}+2\frac{dA(z)}{dz}\frac{dS(z)}{dz}=0 \ . \tag{7.11}$$

In Eq. (7.10), which can be rewritten as

$$\frac{1}{k_z^2(z)A(z)}\frac{d^2A(z)}{dz^2}+1-\frac{1}{k_z^2(z)}\left(\frac{dS(z)}{dz}\right)^2=0 \ , \tag{7.12}$$

we assume that

$$\left|\frac{1}{k_z^2(z)A(z)}\frac{d^2A(z)}{dz^2}\right| << 1 \ . \tag{7.13}$$

Equation (7.12) therefore becomes

$$\left(\frac{dS(z)}{dz}\right)^2=k_z^2(z)\Rightarrow\frac{dS(z)}{dz}=\pm k_z(z) \ , \tag{7.14}$$

which has the solution

$$S(z)=\pm\int_{z_i}^{z}k_z(z)\,dz \ , \tag{7.15}$$

where z_i is an arbitrary constant. Substituting Eq. (7.14) into Eq. (7.11), we obtain

$$A(z)\frac{dk_z(z)}{dz}+2\frac{dA(z)}{dz}k_z(z)=0 \ , \tag{7.16}$$

which can be rewritten as

$$\frac{1}{A(z)}\frac{dA(z)}{dz}=-\frac{1}{2k_z(z)}\frac{dk_z(z)}{dz} \ , \tag{7.17}$$

and has the solution

$$A(z) = \frac{D}{\sqrt{k_z(z)}} \ , \tag{7.18}$$

where D is an arbitrary constant. Substituting Eqs. (7.15) and (7.18) into Eq. (7.8), we see that the depth-dependent solution is given by

$$u(z) = \frac{D}{\sqrt{k_z(z)}} \exp\left[\pm i \int_{z_i}^{z} k_z(z)\, dz \right] \ . \tag{7.19}$$

Finally, recalling that $p(x,z) = v(x)u(z)$, and using Eqs. (7.7) and (7.19), we obtain the full solution of Eq. (7.2),

$$p(x,z) = \frac{C}{\sqrt{k_z(z)}} \exp\left[ik_x(x - x_i) \right] \exp\left[\pm i \int_{z_i}^{z} k_z(z)\, dz \right] \ , \tag{7.20}$$

where we have combined the arbitrary multiplicative constants into a single constant $C = BD$. Equation (7.19) is the solution of the depth-dependent Eq. (7.6) in the *ray* or *WKB approximation*, while Eq. (7.20) is called the *two-dimensional plane wave ray* or *WKB solution* of the Helmholtz equation for a horizontally stratified medium. The term WKB acknowledges the contributions of Wentzel, Kramers, and Brillouin to the technique in a quantum mechanical context. However, Tolstoy [1973] indicates that the method was developed earlier by Jeffreys and perhaps even Rayleigh, and WKBJR may be a more appropriate acronym. Let us now examine this solution and the approximations used to obtain it in more detail.

First, we observe that the phase of the ray/WKB solution strongly resembles that of a plane wave. Indeed, the phase in Eq. (7.20), in the limit of a homogeneous medium $k(z) = k$, simply becomes

$$k_x(x - x_i) \pm \int_{z_i}^{z} k_z(z) dz = k_x(x - x_i) \pm k_z(z - z_i) \ , \tag{7.21}$$

and corresponds to the phase accumulated by an up- or down-going plane wave with wave vector

$$\mathbf{k} = (k_x, k_z) = (k \sin\theta,\ k\cos\theta) \ , \tag{7.22}$$

and propagation angle θ, as it travels from point (x_i, z_i) to point (x,z) (cf. Sec. 3.3). We can therefore associate the ray/WKB solution in Eq. (7.20) with an up- or down-going wave that has a locally planar wavefront and a wave vector, perpendicular to the wavefront, given by

$$\mathbf{k} = (k_x, k_z) = [k(z)\sin\theta(z),\ k(z)\cos\theta(z)] \ . \tag{7.23}$$

This wave vector describes a *ray* traveling through an inhomogeneous medium with an angle of propagation $\theta(z)$ that is a function of depth, as shown in Fig. 7.2. The vertical phase accumulated by the ray depends on the detailed nature of the continuous sound velocity variations. We see that the ray bends toward the region of lower sound velocity because the part of the wavefront in the region of lower velocity moves more slowly than the portion of the wavefront in the region of higher velocity. We reiterate that, in an inhomogeneous medium, these rays are associated with approximate solutions of the Helmholtz equation. In a homogeneous medium with no sources, on the other hand, the rays are connected with exact plane wave solutions of the Helmholtz equation.

Second, we notice that the separation constant $-k_x^2$ in Eq. (7.4) implies that the horizontal wavenumber,

$$k_x = k(z)\sin\theta(z) = Constant \ , \tag{7.24}$$

is a *ray invariant*. In fact, Eq. (7.24) is simply a statement of Snell's law [cf. Eq. (3.28)]. We may choose the values of $k(z)$ and $\theta(z)$ in Eq. (7.24) to be those at depth z_i, so that

$$\left.\begin{array}{l} k(z_i) = k_i = \omega/c(z_i) = \omega/c_i\,, \\ \theta(z_i) = \theta_i \end{array}\right\} \Rightarrow k_x = k_i \sin\theta_i \ , \tag{7.25}$$

and Eq. (7.20) becomes

$$p(x,z) = \frac{C}{\sqrt{k_z(z)}} \exp[ik_i \sin\theta_i (x - x_i)] \exp\left[\pm i \int_{z_i}^{z} k_z(z)\,dz\right] \ . \tag{7.26}$$

We note that the quantity $k_x/\omega = \sin\theta_i/c_i$ is called the *ray parameter* or *horizontal slowness*, and is another ray invariant used frequently in ray calculations.

Third, Tolstoy [1973] and Tolstoy and Clay [1987] indicate that the ray/WKB criterion in Eq. (7.13) is equivalent to

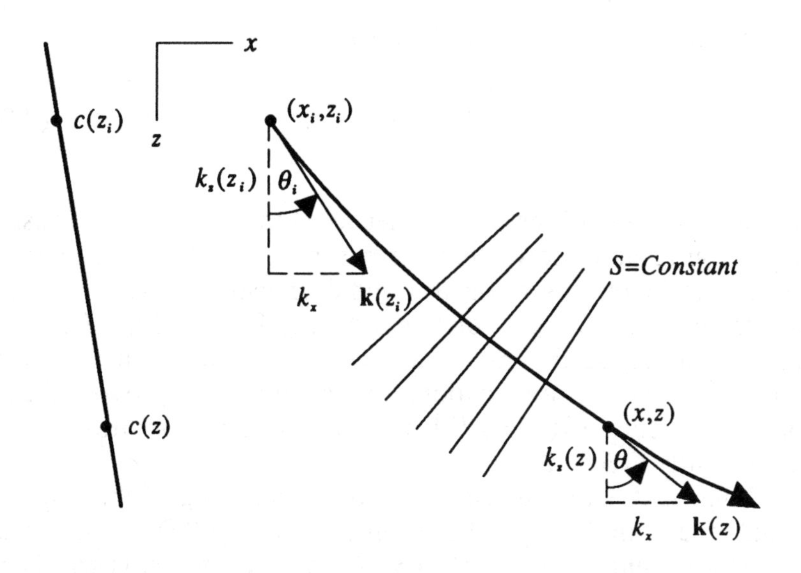

Figure 7.2 Two-dimensional ray trajectory for a monotonically increasing sound velocity profile $c(z)$.

$$\left|\frac{1}{k_z^2(z)}\frac{d^2}{dz^2}\ln k_z^2(z)-\frac{1}{k_z^2(z)}\left[\frac{d}{dz}\ln k_z(z)\right]^2\right|<<4 \quad , \tag{7.27}$$

which, for slowly varying $k_z(z)$, [i.e., locally $k_z(z)$ is linear] becomes (cf. Prob. 7.1)

$$\frac{1}{k_z(z)}\left|\frac{d}{dz}\ln k_z(z)\right|<<\frac{2}{\sqrt{3}}\approx 1 \quad . \tag{7.28}$$

Using $k_z = k(z)\cos\theta(z)$, we can show that Eq. (7.28) is

$$\left|\frac{1}{k^2(z)\cos\theta(z)}\frac{dk(z)}{dz}-\frac{\sin\theta(z)}{k(z)\cos^2\theta(z)}\frac{d\theta(z)}{dz}\right|<<1 \quad . \tag{7.29}$$

For rays propagating close to the vertical direction [i.e., $\theta(z)\approx 0$], Eq. (7.29) becomes

$$\left|\frac{1}{k^2(z)}\frac{dk(z)}{dz}\right|<<1 \quad , \tag{7.30}$$

which, using $k(z)=\omega/c(z)$, can be rewritten as

$$\frac{1}{\omega}\left|\frac{dc(z)}{dz}\right|<<1 \ . \tag{7.31}$$

Equation (7.31) quantifies the manner in which the medium must be slowly varying and shows that the ray and WKB approximations improve with increasing frequency and decreasing velocity gradient. However, as discussed by Tolstoy and Clay [1987], Eq. (7.31) is a necessary, but not sufficient, condition for the validity of ray theory for arbitrary angles of propagation. In fact, at a *turning point*, the ray becomes horizontal $[\theta(z)=\pi/2]$, and the general condition in Eq. (7.29) is clearly violated.

Finally, it is of interest to clarify the meaning of the $1/\sqrt{k_z(z)}$ amplitude factor in the ray/WKB solution of Eq. (7.26). First, we see that it becomes singular when $\theta(z)=\pi/2$, a result which is consistent with the breakdown of ray theory at a turning point. But, in addition, the presence of this term guarantees that the ray and WKB theories are *no reflection theories*. This behavior can be confirmed by calculating the time-averaged energy flux density of the field in Eq. (7.26) in the z direction and showing that, to leading order, it is a constant [cf. Prob. 7.3 and Brekhovskikh, 1980]. Thus, the ray/WKB solution does not take into account the continuous reflection process that occurs as a wave travels through a continuously varying, inhomogeneous medium. This feature further substantiates the failure of ray/WKB theory at turning points, which arise due to a process of local, total internal reflection.

7.4 Three-Dimensional Ray Theory for an Arbitrary Unbounded Medium with No Sources

The ray and WKB theories are identical in unbounded, horizontally stratified media with no sources. However, the WKB method is restricted to stratified media, whereas ray theory can be developed for an arbitrary unbounded medium with three-dimensional variations in sound velocity, as we shall see in this section. The two techniques also differ from one another in their determination of the acoustic field in the presence of sources. In the WKB method, the plane wave ray/WKB solution in Eq. (7.20) is incorporated into a full wave development, such as the Hankel transform approach, for stratified media. An example of this technique will be described in

Sec. 7.6.2. In ray theory, the intuitive notion of a diverging and converging ray bundle, combined with the conservation of energy, can be used to calculate the field intensity in the general three-dimensional case. We will demonstrate this approach for a point source in a horizontally stratified medium in Sec. 7.5.4.

The ray-theoretic development for three dimensions proceeds in a straightforward manner analogous to that described for two dimensions in Sec. 7.3. We begin with the three-dimensional Helmholtz equation for a constant density medium with no sources,

$$\left[\nabla^2 + k^2(\mathbf{r})\right]p(\mathbf{r}) = 0 \quad , \tag{7.32}$$

where $k(\mathbf{r}) = \omega/c(\mathbf{r})$, and $\mathbf{r} = (x,y,z)$ in Cartesian coordinates. We then assume a solution of the form

$$p(\mathbf{r}) = A(\mathbf{r})e^{iS(\mathbf{r})} \quad , \tag{7.33}$$

where $A(\mathbf{r})$ and $S(\mathbf{r})$ are real functions. Substituting Eq. (7.33) into Eq. (7.32), we obtain [cf. Eq. (7.9)]

$$\nabla^2 A(\mathbf{r}) + 2i\nabla A(\mathbf{r}) \bullet \nabla S(\mathbf{r}) + iA(\mathbf{r})\nabla^2 S(\mathbf{r}) + \left\{k^2(\mathbf{r}) - [\nabla S(\mathbf{r})]^2\right\}A(\mathbf{r}) = 0. \tag{7.34}$$

Setting the real and imaginary parts of Eq. (7.34) separately equal to 0, we find the analogues of Eqs. (7.10) and (7.11):

$$\nabla^2 A(\mathbf{r}) + \left\{k^2(\mathbf{r}) - [\nabla S(\mathbf{r})]^2\right\}A(\mathbf{r}) = 0 \quad , \tag{7.35}$$

$$A(\mathbf{r})\nabla^2 S(\mathbf{r}) + 2\nabla A(\mathbf{r}) \bullet \nabla S(\mathbf{r}) = 0 \quad , \tag{7.36}$$

where Eq. (7.36) is called the *transport equation* [Brekhovskikh and Lysanov, 1991]. If we rewrite Eq. (7.35) as [cf. Eq. (7.12)]

$$\frac{1}{k^2(\mathbf{r})A(\mathbf{r})}\nabla^2 A(\mathbf{r}) + 1 - \frac{1}{k^2(\mathbf{r})}[\nabla S(\mathbf{r})]^2 = 0 \quad , \tag{7.37}$$

and assume that [cf. Eq. (7.13)]

$$\left|\frac{1}{k^2(\mathbf{r})A(\mathbf{r})}\nabla^2 A(\mathbf{r})\right| << 1 \ , \tag{7.38}$$

we then obtain

$$[\nabla S(\mathbf{r})]^2 = k^2(\mathbf{r}) \ , \tag{7.39}$$

which is called the *eikonal equation* [cf Eq. (7.14)]. Analogous to the two-dimensional case, Eq. (7.39) describes rays normal to the wavefronts, which are surfaces of constant phase $S(\mathbf{r})$. Specifically, in Cartesian coordinates, Eq. (7.39) becomes

$$\left[\frac{\partial S(x,y,z)}{\partial x}\right]^2 + \left[\frac{\partial S(x,y,z)}{\partial y}\right]^2 + \left[\frac{\partial S(x,y,z)}{\partial z}\right]^2 = k^2(x,y,z) \ , \tag{7.40}$$

and the rays have direction cosines $k^{-1}\,\partial S/\partial x$, $k^{-1}\,\partial S/\partial y$, and $k^{-1}\,\partial S/\partial z$. It can be shown that the rays satisfy *Fermat's principle* [Officer, 1958; Tolstoy and Clay, 1987], that is, they travel along paths which correspond to extrema (normally minima) in travel time. In principle, these three-dimensional ray trajectories can be determined using the eikonal equation for arbitrary sound velocity profiles $c(\mathbf{r})$. In practice, the ray equations may not be solvable in closed form, and numerical methods may be required in order to execute this *ray tracing* process. In the subsequent development, we will therefore restrict our discussion to the horizontally stratified case.

7.5 Three-Dimensional Ray Theory for a Point Source in an Unbounded, Horizontally Stratified Medium

We consider a point source with cylindrical coordinates $\mathbf{r}_0 = (0,0,z_0)$ in a constant density, horizontally stratified medium, so that $k(\mathbf{r}) = \omega/c(\mathbf{r}) = \omega/c(z)$. Then the sound field is cylindrically symmetric about the z axis (cf. Sec. 5.2, Fig. 5.1, and Fig. 7.3), and, for the region exterior to the source in the ray approximation, is described by the eikonal equation [cf. Eq. (7.39)],

$$\left[\frac{\partial S(r,z)}{\partial r}\right]^2 + \left[\frac{\partial S(r,z)}{\partial z}\right]^2 = k^2(z) \ , \tag{7.41}$$

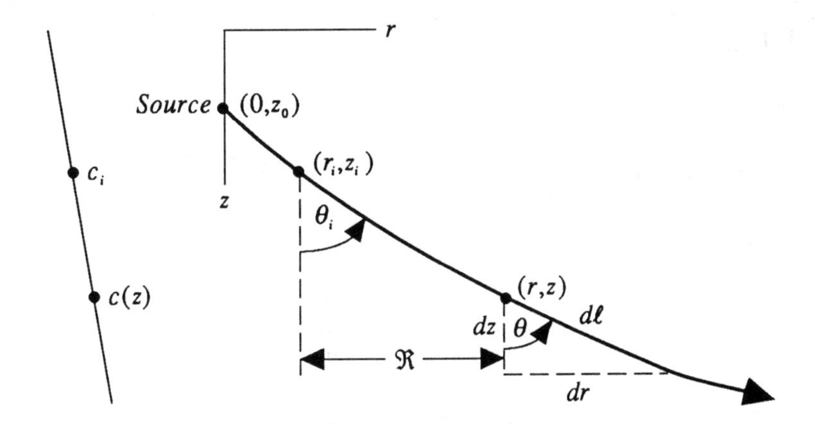

Figure 7.3 Three-dimensional ray trajectory for a point source in a horizontally stratified medium with a monotonically increasing sound velocity profile $c(z)$.

where we have used Eq. (5.3). Equation (7.41), combined with various geometrical considerations and the conservation of energy, will enable us to derive several key quantities characterizing the ray acoustic field. These include the ray range, phase, travel time, and intensity. First, however, we will use Eq. (7.41) in an alternative derivation of Snell's law.

7.5.1 An Alternative Derivation of Snell's Law

The infinitesimal path length $d\ell$ along a ray trajectory in cylindrical coordinates is given by (cf. Fig. 7.3)

$$d\ell = \sqrt{dr^2 + dz^2} \ , \tag{7.42}$$

with

$$\sin\theta = \frac{dr}{d\ell}, \quad \cos\theta = \frac{dz}{d\ell} \ . \tag{7.43}$$

Equation (7.43) enables us to write

$$k^2(z) = k^2(z)(\sin^2\theta + \cos^2\theta) = k^2(z)\left[\left(\frac{dr}{d\ell}\right)^2 + \left(\frac{dz}{d\ell}\right)^2\right], \tag{7.44}$$

while Eq. (7.44) allows us to rewrite the eikonal equation [cf. Eq. (7.41)] as

$$\left[\frac{\partial S(r,z)}{\partial r}\right]^2+\left[\frac{\partial S(r,z)}{\partial z}\right]^2=k^2(z)=k^2(z)\left[\left(\frac{dr}{d\ell}\right)^2+\left(\frac{dz}{d\ell}\right)^2\right] \ . \tag{7.45}$$

Equation (7.45) then implies that

$$\frac{\partial S(r,z)}{\partial r}=k(z)\frac{dr}{d\ell} \ , \tag{7.46}$$

$$\frac{\partial S(r,z)}{\partial z}=k(z)\frac{dz}{d\ell} \ . \tag{7.47}$$

Differentiating Eq. (7.46) with respect to ℓ, we obtain

$$\frac{d}{d\ell}\left[\frac{\partial S(r,z)}{\partial r}\right]=\frac{d}{d\ell}\left[k(z)\frac{dr}{d\ell}\right] \ , \tag{7.48}$$

which can be rewritten as

$$\frac{d}{d\ell}\left[k(z)\frac{dr}{d\ell}\right]=\frac{\partial}{\partial r}\left[\frac{dS(r,z)}{d\ell}\right] \ . \tag{7.49}$$

But, by definition, the total derivative of S with respect to ℓ is also given by

$$\frac{dS(r,z)}{d\ell}=\frac{\partial S(r,z)}{\partial r}\frac{dr}{d\ell}+\frac{\partial S(r,z)}{\partial z}\frac{dz}{d\ell} \ . \tag{7.50}$$

Substituting Eqs. (7.46) and (7.47) into Eq. (7.50), we obtain

$$\frac{dS(r,z)}{d\ell}=k(z)\left(\frac{dr}{d\ell}\right)^2+k(z)\left(\frac{dz}{d\ell}\right)^2=k(z) \ . \tag{7.51}$$

Substituting Eqs. (7.51) and (7.43) into Eq. (7.49), we find that

$$\frac{d}{d\ell}\left[k(z)\frac{dr}{d\ell}\right]=\frac{d}{d\ell}[k(z)\sin\theta]=\frac{\partial}{\partial r}[k(z)]=0\ , \tag{7.52}$$

which implies Snell's law [cf. Eq. (7.24)]:

$$k(z)\sin\theta=Constant\ . \tag{7.53}$$

Thus, we see that Snell's law, which determines the ray trajectories in a horizontally stratified medium, is a direct consequence of the eikonal equation.

7.5.2 Ray Range

We wish to determine the horizontal range $\Re = r - r_i$ traversed by a ray when it travels from the point (r_i, z_i) to the point (r, z), as shown in Fig. 7.3. Suppose that the ray has an initial angle θ_i and sound velocity c_i, and a final angle θ and velocity $c(z)$. We use Eqs. (7.43) and (7.53) and write Snell's law as

$$\frac{1}{c(z)}\sin\theta=\frac{1}{c(z)}\frac{dr}{d\ell}=\frac{1}{c_i}\sin\theta_i\ . \tag{7.54}$$

But using Eq. (7.42), we can rewrite Eq. (7.54) as

$$\frac{1}{c(z)}\frac{dr}{d\ell}=\frac{1}{c(z)}\frac{dr}{\sqrt{dr^2+dz^2}}=\frac{1}{c(z)}\frac{dr/dz}{\sqrt{(dr/dz)^2+1}}=\frac{\sin\theta_i}{c_i}\ . \tag{7.55}$$

Solving for dr/dz in Eq. (7.55), we find that

$$\frac{dr}{dz}=\frac{1}{\sqrt{\left[\frac{c_i}{c(z)\sin\theta_i}\right]^2-1}}\ , \tag{7.56}$$

and therefore the horizontal range is given by

$$\Re = \int_{z_i}^{z} \frac{dz}{\sqrt{\left[\frac{c_i}{c(z)\sin\theta_i}\right]^2 - 1}} \quad . \tag{7.57}$$

Linear Gradient Example

Let us consider the example of a linear gradient profile:

$$c(z) = c_0 + gz \quad , \tag{7.58}$$

where c_0 and g are constants. This profile is of interest because it is a good approximation to the variation in sound velocity for large regions of the water column and the seabed (cf. Figs. 7.1 and 3.6). Furthermore, complicated nonlinear profiles can be treated by subdividing them into linear segments. Finally, this profile is attractive because it yields all of the desired ray quantities in closed form.

We begin by rewriting Eq. (7.57),

$$\Re = \int_{z_i}^{z} \frac{c(z)\,dz}{\sqrt{\left(\frac{c_i}{\sin\theta_i}\right)^2 - c^2(z)}} \quad , \tag{7.59}$$

and changing variables,

$$c(z) = c_0 + gz \Rightarrow dc(z) = g\,dz \quad , \tag{7.60}$$

so that Eq. (7.59) becomes

$$\Re = \frac{1}{g}\int_{c_i}^{c(z)} \frac{c(z)\,dc(z)}{\sqrt{\left(\frac{c_i}{\sin\theta_i}\right)^2 - c^2(z)}} = -\frac{1}{g}\sqrt{\left(\frac{c_i}{\sin\theta_i}\right)^2 - c^2(z)}\,\Bigg|_{c_i}^{c(z)} \quad . \tag{7.61}$$

Evaluating the result in Eq. (7.61), we obtain

$$\Re = \frac{1}{g}\left[\sqrt{\left(\frac{c_i}{\sin\theta_i}\right)^2 - c_i^2} - \sqrt{\left(\frac{c_i}{\sin\theta_i}\right)^2 - c^2(z)}\right]$$

$$= \frac{c_i}{g\sin\theta_i}\left[\sqrt{1-\sin^2\theta_i} - \sqrt{1-\left(\frac{c(z)\sin\theta_i}{c_i}\right)^2}\right] . \tag{7.62}$$

Using Snell's law in Eq. (7.62) we find that

$$\Re = \frac{c_i}{g\sin\theta_i}(\cos\theta_i - \cos\theta) , \tag{7.63}$$

which we recognize as the parametric equation for the arc of a circle with radius $c_i/(g\sin\theta_i)$.

7.5.3 Ray Phase and Travel Time

We now wish to determine the phase S and travel time T for a ray as it propagates from the point (r_i, z_i) to the point (r, z) (cf. Fig. 7.3). The relationship between S and T becomes clear when we integrate Eq. (7.51) with respect to ℓ,

$$S = \int k(z)\,d\ell = \omega\int\frac{d\ell}{c(z)} = \omega\int dt = \omega T , \tag{7.64}$$

since the travel time dt accumulated along the infinitesimal path length $d\ell$ is $dt = d\ell/c(z)$. Using Eqs. (7.64) and (7.43), we proceed with a determination of S:

$$S = \omega\int\frac{d\ell}{c(z)} = \omega\int\frac{dz}{c(z)\cos\theta} = \omega\int\frac{(\cos^2\theta + \sin^2\theta)}{c(z)\cos\theta}dz$$

$$= \omega\int\frac{\cos\theta}{c(z)}dz + \omega\int\frac{\sin^2\theta}{c(z)\cos\theta}dz$$

$$= \omega\int\frac{\sqrt{1-\sin^2\theta}}{c(z)}dz + \omega\int\frac{\sin\theta}{c(z)}dr . \tag{7.65}$$

Equation (7.65) can also be written as

$$S = \int k(z)\cos\theta\, dz + \int k(z)\sin\theta\, dr = \int k_z(z)\, dz + \int k_r\, dr\,, \quad (7.66)$$

where the vertical wavenumber $k_z(z) = k(z)\cos\theta$ and the horizontal wavenumber $k_r = k(z)\sin\theta = Constant$. Equation (7.66) simply expresses the decomposition of the total phase into its vertical and horizontal components. Let us evaluate the integrals in Eq. (7.65) for a ray traveling between the points (r_i, z_i) and (r, z) with initial angle θ_i and sound velocity c_i. Using Snell's law, we obtain

$$\begin{aligned} S &= \omega \int_{z_i}^{z} \frac{\sqrt{1-\sin^2\theta}}{c(z)}\, dz + \omega \int_{r_i}^{r} \frac{\sin\theta}{c(z)}\, dr \\ &= \omega \int_{z_i}^{z} \sqrt{\frac{1}{c^2(z)} - \frac{\sin^2\theta_i}{c_i^2}}\, dz + \omega \int_{r_i}^{r} \frac{\sin\theta_i}{c_i}\, dr \\ &= \omega \int_{z_i}^{z} \sqrt{\frac{1}{c^2(z)} - \frac{\sin^2\theta_i}{c_i^2}}\, dz + \left(\frac{\omega}{c_i}\sin\theta_i\right)\Re \\ &= \int_{z_i}^{z} \sqrt{k^2(z) - k_i^2 \sin^2\theta_i}\, dz + (k_i \sin\theta_i)\Re\,. \end{aligned} \quad (7.67)$$

Here we recall that $\Re$ is the horizontal range [cf. Eq. (7.57)] and $k_i = \omega/c_i$ [cf. Eq. (7.25)]. The travel time T is then given by [cf. Eqs. (7.64) and (7.67)]

$$T = \int_{z_i}^{z} \sqrt{\frac{1}{c^2(z)} - \frac{\sin^2\theta_i}{c_i^2}}\, dz + \left(\frac{\sin\theta_i}{c_i}\right)\Re\,. \quad (7.68)$$

Linear Gradient Example

In order to calculate the ray phase S for the linear gradient example in Eq. (7.58), we use Eq. (7.65) and Snell's law to obtain

$$S = \omega \int_{z_i}^{z} \frac{dz}{c(z)\cos\theta} = \omega \int_{z_i}^{z} \frac{dz}{c(z)\sqrt{1 - c^2(z)\sin^2\theta_i / c_i^2}}\,. \quad (7.69)$$

Changing variables as in Eq. (7.60), we find that Eq. (7.69) becomes

$$S = \frac{\omega}{g} \int_{c_i}^{c(z)} \frac{dc(z)}{c(z)\sqrt{1 - c^2(z)\sin^2\theta_i / c_i^2}} \quad . \tag{7.70}$$

But from Gradshteyn and Ryzhik [1965], we have the result

$$\int \frac{dx}{x\sqrt{b + ax^2}} = \frac{1}{2\sqrt{b}} \ln\left[\frac{\sqrt{b} - \sqrt{b + ax^2}}{\sqrt{b} + \sqrt{b + ax^2}}\right], \quad a < 0, \quad b > 0, \tag{7.71}$$

so that Eq. (7.70) is given by

$$S = \frac{\omega}{2g} \ln\left[\frac{1 - \sqrt{1 - c^2(z)\sin^2\theta_i / c_i^2}}{1 + \sqrt{1 - c^2(z)\sin^2\theta_i / c_i^2}}\right]_{c_i}^{c(z)}$$

$$= \frac{\omega}{2g} \ln\left\{\left[\frac{1 - \sqrt{1 - c^2(z)\sin^2\theta_i / c_i^2}}{1 + \sqrt{1 - c^2(z)\sin^2\theta_i / c_i^2}}\right]\left[\frac{1 + \sqrt{1 - c_i^2\sin^2\theta_i / c_i^2}}{1 - \sqrt{1 - c_i^2\sin^2\theta_i / c_i^2}}\right]\right\} . \tag{7.72}$$

Using Snell's law in Eq. (7.72), we find that the ray phase is

$$S = \frac{\omega}{2g} \ln\left[\frac{(1 - \cos\theta)(1 + \cos\theta_i)}{(1 + \cos\theta)(1 - \cos\theta_i)}\right] , \tag{7.73}$$

while the travel time is

$$T = \frac{1}{2g} \ln\left[\frac{(1 - \cos\theta)(1 + \cos\theta_i)}{(1 + \cos\theta)(1 - \cos\theta_i)}\right] . \tag{7.74}$$

7.5.4 Ray Acoustic Intensity

The ray acoustic intensity is calculated by first determining the change in cross-sectional area of an infinitesimal ray bundle as it propagates between two points. Conservation of energy, which simply states that the total energy in the ray bundle must remain constant, is then used to determine the variation in intensity, which is a measure of the energy per unit area (cf. Sec. 2.7).

We consider an infinitesimal ray bundle with an initial angle θ_i and width $d\theta_i$, leaving the source located at the cylindrical coordinate position $(0, z_0)$, as shown in Fig. 7.4. Then, in spherical coordinates (r_0, θ, ϕ) (cf. Fig. 4.8), the total cross-sectional area dA_0 of this bundle at a unit distance $r_0 = 1$ from the source is given by [cf. Eqs. (4.65) and (4.67)]

$$dA_0 = 2\pi \sin\theta_i \, d\theta_i \quad . \tag{7.75}$$

Here we have integrated over the azimuthal angle ϕ from 0 to 2π. Suppose that this ray bundle propagates to the location (r,z), at which point it has the cross-sectional area dA. Then it is clear that the area dA and the area dA', which is associated with the cylindrical coordinate system in Fig. 7.4, are related to one another through the equation

$$\cos(\pi/2 - \theta) = \sin\theta = dA/dA' \quad . \tag{7.76}$$

But the cylindrical area element dA' is given by

$$dA' = 2\pi \Re \, dz \quad , \tag{7.77}$$

and therefore Eq. (7.76) becomes

$$dA = \sin\theta \, dA' = 2\pi \Re \sin\theta \, dz = 2\pi \Re \cos\theta \, dr \quad , \tag{7.78}$$

where we have used Eq. (7.43). If the intensity at unit distance from the source is I_0, and the intensity at horizontal range $\Re$ is I, then conservation of energy implies that

$$I_0 \, dA_0 = I \, dA \quad . \tag{7.79}$$

Substituting Eqs. (7.75) and (7.78) into Eq. (7.79), we obtain

$$\frac{I}{I_0} = \frac{dA_0}{dA} = \frac{\sin\theta_i \, d\theta_i}{\Re \cos\theta \, dr} = \frac{\sin\theta_i \, d\theta_i}{\Re \cos\theta \, d\Re} \quad , \tag{7.80}$$

since $d\Re = dr$. But $\Re = \Re(\theta_i, z)$, and therefore we have

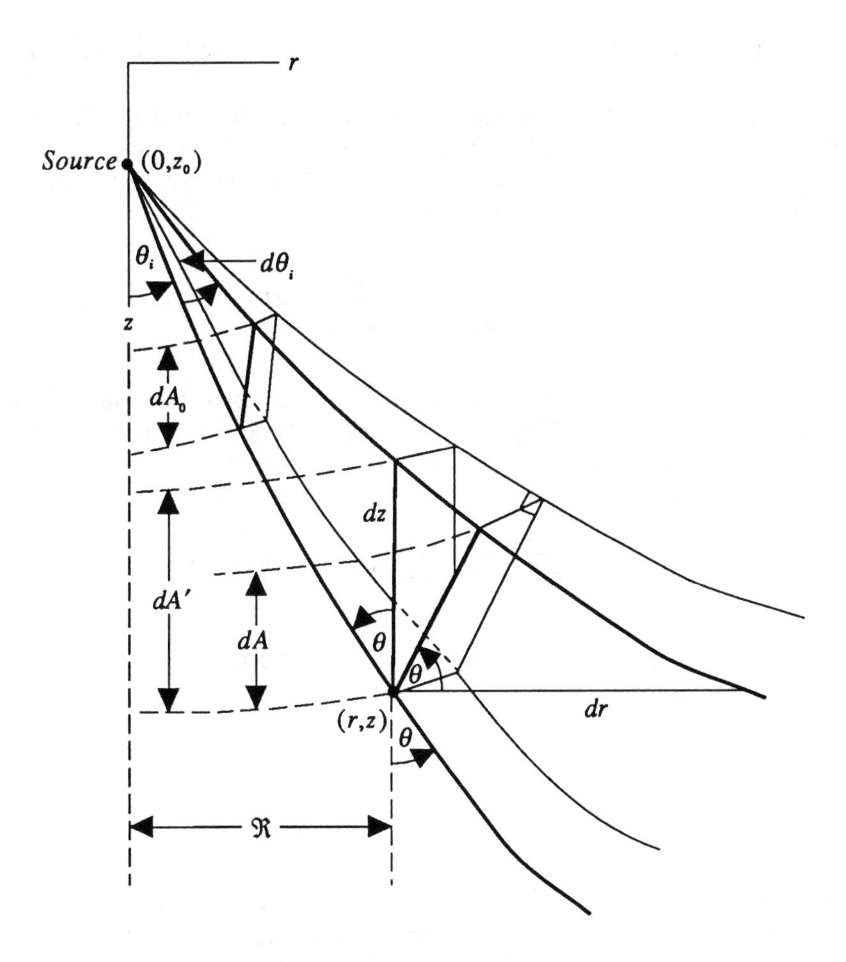

Figure 7.4 Geometry for calculation of the ray acoustic intensity.

$$d\Re = \frac{\partial \Re}{\partial \theta_i} d\theta_i + \frac{\partial \Re}{\partial z} dz = \frac{\partial \Re}{\partial \theta_i} d\theta_i \tag{7.81}$$

at a fixed depth z. Substituting Eq. (7.81) into Eq. (7.80), we see that the ray acoustic intensity is given by

$$\frac{I}{I_0} = \frac{\sin\theta_i}{\Re \cos\theta \left(\partial \Re / \partial \theta_i\right)} \quad . \tag{7.82}$$

The $1/\Re$ factor in Eq. (7.82) introduces cylindrical spreading into the ray field, while the remaining factors express the decrease or increase in intensity due to the spreading or contraction of the ray bundle along the path.

The relationship between the ray amplitude A in Eq. (7.33) and the ray intensity I is a subtle one. We recall that the time-averaged energy flux density $\mathbf{I}$ is given by Eq. (2.58). When we substitute Eq. (7.33) into Eq. (2.58), we find that

$$\mathbf{I} = \frac{A^2}{2\omega\rho_0}\nabla S = \frac{A^2}{2\omega\rho_0}\mathbf{k}(z) \ , \tag{7.83}$$

where we have used the eikonal equation and the property that A is real [cf. Eqs. (7.39) and (7.41)]. The ray intensity and ray amplitude are therefore related to one another through the equation:

$$I = |\mathbf{I}| = \frac{A^2 k(z)}{2\omega\rho_0} = \frac{A^2}{2\rho_0 c(z)} \Rightarrow A = \sqrt{2\rho_0 c(z) I} \ , \tag{7.84}$$

where we have used $k(z) = \omega/c(z)$. If the amplitude at unit distance from the source is A_0, then we have

$$\frac{A}{A_0} = \left[\frac{c(z)I}{c_i I_0}\right]^{1/2} \Rightarrow A = \left[\frac{c(z)I}{c_i I_0}\right]^{1/2} A_0 \ , \tag{7.85}$$

where c_i is the sound velocity at the source, and we recall that the density is constant. Typically, we assume that the source has unit amplitude $A_0 = 1$ at unit distance, and Eq. (7.85) becomes

$$A = \left[\frac{c(z)I}{c_i I_0}\right]^{1/2} = \left[\frac{c(z)\sin\theta_i}{c_i \Re \cos\theta (\partial\Re/\partial\theta_i)}\right]^{1/2} \ , \tag{7.86}$$

which, using Snell's law, can be rewritten as

$$A = \left[\frac{\sin\theta}{\Re\cos\theta(\partial\Re/\partial\theta_i)}\right]^{1/2} \ . \tag{7.87}$$

It is interesting to note that Eqs. (7.86) and (7.87) have exactly the same form as the amplitude of a particular geometrical acoustic path [cf. Eq. (6.60)] obtained from an asymptotic analysis of the Hankel transform representation of the field for the Pekeris waveguide (cf. Sec. 6.3.1). This connection between the ray acoustic field and the high-frequency approximation to the Hankel transform result will be explored further in Sec. 7.6.

Linear Gradient Example

For the linear gradient case, we have [cf. Eq. (7.63)]

$$\begin{aligned}\Re &= \frac{c_i}{g\sin\theta_i}\left[\cos\theta_i - \sqrt{1-\sin^2\theta}\right] \\ &= \frac{c_i}{g\sin\theta_i}\left[\cos\theta_i - \sqrt{1-\frac{c^2(z)}{c_i^2}\sin^2\theta_i}\right] ,\end{aligned} \tag{7.88}$$

and therefore we obtain

$$\begin{aligned}\frac{\partial\Re}{\partial\theta_i} &= -\frac{c_i\cos\theta_i}{g\sin^2\theta_i}\left[\cos\theta_i - \sqrt{1-\frac{c^2(z)}{c_i^2}\sin^2\theta_i}\right] \\ &+ \frac{c_i}{g\sin\theta_i}\left[-\sin\theta_i + \frac{c^2(z)}{c_i^2}\frac{\sin\theta_i\cos\theta_i}{\sqrt{1-c^2(z)\sin^2\theta_i/c_i^2}}\right] .\end{aligned} \tag{7.89}$$

It is then straightforward to show that

$$\begin{aligned}\cos\theta\frac{\partial\Re}{\partial\theta_i} &= \sqrt{1-\frac{c^2(z)}{c_i^2}\sin^2\theta_i}\,\frac{\partial\Re}{\partial\theta_i} \\ &= \frac{c_i}{g\sin^2\theta_i}\left[\cos\theta_i - \sqrt{1-\frac{c^2(z)}{c_i^2}\sin^2\theta_i}\right] = \frac{\Re}{\sin\theta_i} ,\end{aligned} \tag{7.90}$$

and Eq. (7.82) becomes

$$\frac{I}{I_0} = \frac{\sin^2\theta_i}{\Re^2} . \tag{7.91}$$

7.6 A Point Source in a Medium with a Velocity Profile Having a Single Minimum

A sound velocity profile for which there is a single minimum (cf. Fig. 7.5) holds considerable interest for the ocean acoustician because it is a good approximation to the canonical profile shown in Fig. 7.1. Furthermore, it is a fine example with which to illustrate the relationship between the ray series solution and the asymptotic approximation to the Hankel transform result. In addition, the special case of a symmetric, bilinear profile can be used to determine some of the key features of long-range SOFAR propagation. Finally, the single minimum profile lends itself to the development of a normal mode solution in the WKB approximation.

For simplicity, the development in this section assumes that there are no boundaries, that $c(z)$ tends monotonically toward ∞ for $z > 0$ $(z < 0)$ and $z \to \infty$ $(z \to -\infty)$, and that the velocity gradient dc/dz is discontinuous at the channel axis (i.e., the depth of the velocity minimum). The latter assumption is made in order to avoid pathological cases of the type to be discussed in Sec. 7.7.1.

7.6.1 The Ray Series Solution

The development of a ray series solution requires the identification of the specific *eigenrays* joining source and receiver, the computation of the amplitudes and phases of these rays, and the coherent summation over all of the ray contributions in order to obtain the total acoustic field. For an arbitrary sound velocity profile, this procedure may be executed by first tracing a number of rays leaving the source on a dense grid of initial angles. The ray tracing can be accomplished by subdividing the velocity profile into segments within each one of which the ray equations have closed form solutions [Cornyn, 1973; Roberts, 1974]. For a particular ray, the desired ray quantities are computed by combining the contributions associated with the individual segments. Then the rays which encompass the receiver position are found, and the ray quantities at the receiver location are computed through an interpolation technique. Alternatively, the ray equations may be solved entirely numerically [Jones, Riley, and Georges, 1986; Newhall et al., 1990].

On the other hand, for simpler profiles such as the single velocity minimum case, we may adopt a more intuitive approach which resembles that used to determine Green's functions by the method of images (cf. Sec. 4.6). This methodology, when combined with our previous experience in developing geometrical acoustics solutions for

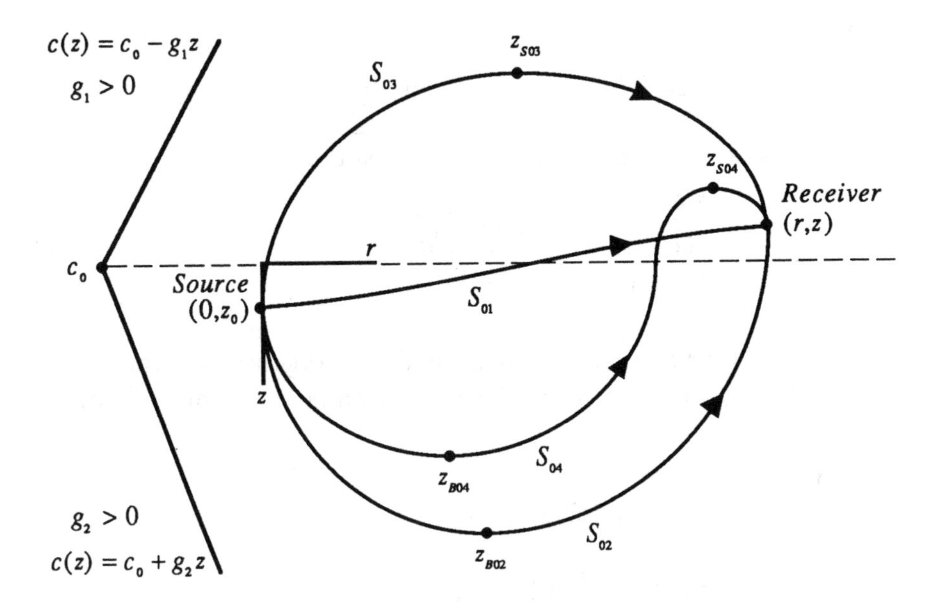

Figure 7.5 The four fundamental ray paths for a point source in a medium with a velocity profile having a single minimum. An asymmetric bilinear profile is shown as an example.

homogeneous waveguides with reflecting boundaries (cf. Sec. 6.3.1), provides us with a powerful way of computing ray acoustic fields.

For the single velocity minimum example with $z_0 \geq z$, we first identify the simplest distinct ray paths which join source and receiver. These are shown in Fig. 7.5 and consist of: a direct ray with no turning points; an initially downward propagating ray with a single lower turning point z_{B02}; an initially upward propagating ray with a single upper turning point z_{S03}; and an initially downward propagating ray with both lower and upper turning points, z_{B04} and z_{S04}. The ray phases for these four arrivals are [cf. Eqs. (7.66) and (7.67)]

$$S_{01} = \int_z^{z_0} k_{z01}(z)\,dz + k_{01}r,$$

$$S_{02} = \left[\int_{z_0}^{z_{B02}} + \int_z^{z_{B02}}\right] k_{z02}(z)\,dz + k_{02}r,$$

$$S_{03} = \left[\int_{z_{S03}}^{z_0} + \int_{z_{S03}}^{z}\right] k_{z03}(z)\,dz + k_{03}r,$$

$$S_{04} = \left[\int_z^{z_0} + 2\int_{z_{S04}}^{z} + 2\int_{z_0}^{z_{B04}}\right] k_{z04}(z)\,dz + k_{04}r \quad , \tag{7.92}$$

where

$$k_{0j} = k(z)\sin\theta_{0j}(z) = k(z_0)\sin\theta_{0ji},$$
$$k_{z0j}(z) = k(z)\cos\theta_{0j}(z) = k(z)\sqrt{1-\sin^2\theta_{0j}(z)}$$
$$= k(z)\sqrt{1-\frac{c^2(z)\sin^2\theta_{0ji}}{c^2(z_0)}} \quad . \tag{7.93}$$

Here $\theta_{0j}(z)$ is the angle of propagation for the eigenray of type j. The initial angles θ_{0ji} $(j=1,\dots 4)$ of these rays can be determined from the range equations,

$$r = k_{01}\int_z^{z_0} dz/k_{z01}(z),$$
$$r = k_{02}\left[\int_{z_0}^{z_{B02}} + \int_z^{z_{B02}}\right] dz/k_{z02}(z),$$
$$r = k_{03}\left[\int_{z_{S03}}^{z_0} + \int_{z_{S03}}^{z}\right] dz/k_{z03}(z),$$
$$r = k_{04}\left[\int_z^{z_0} + 2\int_{z_{S04}}^{z} + 2\int_{z_0}^{z_{B04}}\right] dz/k_{z04}(z) \quad , \tag{7.94}$$

where we have used Eq. (7.57), rewritten as

$$r = \int_{z_0}^{z} \frac{dz}{\sqrt{\left[\dfrac{c_i}{c(z)\sin\theta_i}\right]^2 - 1}} = \int_{z_0}^{z} \frac{dz}{\sqrt{\left[\dfrac{k(z)}{k_r}\right]^2 - 1}} = k_r\int_{z_0}^{z}\frac{dz}{k_z(z)} \quad . \tag{7.95}$$

The amplitudes A_{0j} $(j=1,\dots 4)$ of the four primary arrivals can be obtained using Eq. (7.86):

$$A_{0j} = \left[\frac{c(z)\sin\theta_{0ji}}{c_i r\cos\theta\left(\partial r/\partial\theta_{0ji}\right)}\right]^{1/2} \quad . \tag{7.96}$$

The four fundamental ray paths described above and depicted in Fig. 7.5 bear a striking resemblance to the four basic field constituents for a point source in a homogeneous fluid layer with impenetrable (cf. Fig. 4.11) and penetrable (cf. Fig. 6.2) boundaries. In the present continuously refracting case, the upper and lower turning points assume the roles which the top and bottom reflecting boundaries played in the homogeneous layer examples. We now exploit this analogy further in order to develop the full ray series solution for the case at hand. Specifically, we recall that the higher order terms in the solutions for the homogeneous layer cases corresponded to multiple reflections between the top and the bottom of the waveguide of the four basic field components. These multiples manifested themselves as additional vertical phase contributions for $n>0$ and $(j=1,\ldots 4)$ (cf. Sec. 6.3.1):

$$2nk_{znj}h = 2n\int_0^h k_{znj}\,dz \quad . \tag{7.97}$$

By analogy, we hypothesize that the higher order terms in the ray series solution for the continuously refracting case will contain additional vertical phase contributions of the form,

$$2n\int_{z_{Snj}}^{z_{Bnj}} k_{znj}(z)\,dz \quad , \tag{7.98}$$

for $n>0$ and $(j=1,\ldots 4)$. As a result, Eq. (7.92) becomes for $n\geq 0$:

$$S_{n1} = \left[\int_z^{z_0} + 2n\int_{z_{Sn1}}^{z_{Bn1}}\right] k_{zn1}(z)\,dz + k_{n1}r,$$

$$S_{n2} = \left[\int_{z_0}^{z_{Bn2}} + \int_z^{z_{Bn2}} + 2n\int_{z_{Sn2}}^{z_{Bn2}}\right] k_{zn2}(z)\,dz + k_{n2}r,$$

$$S_{n3} = \left[\int_{z_{Sn3}}^{z_0} + \int_{z_{Sn3}}^{z} + 2n\int_{z_{Sn3}}^{z_{Bn3}}\right] k_{zn3}(z)\,dz + k_{n3}r,$$

$$S_{n4} = \left[\int_z^{z_0} + 2\int_{z_{Sn4}}^{z} + 2\int_{z_0}^{z_{Bn4}} + 2n\int_{z_{Sn4}}^{z_{Bn4}}\right] k_{zn4}(z)\,dz + k_{n4}r \quad , \tag{7.99}$$

where the initial angle θ_{nji} of the nth ray of type j in Eq. (7.99) must satisfy [cf. Eq. (7.94)]

$$r = k_{n1}\left[\int_{z}^{z_0} + 2n\int_{z_{Sn1}}^{z_{Bn1}}\right]dz/k_{zn1}(z),$$

$$r = k_{n2}\left[\int_{z_0}^{z_{Bn2}} + \int_{z}^{z_{Bn2}} + 2n\int_{z_{Sn2}}^{z_{Bn2}}\right]dz/k_{zn2}(z),$$

$$r = k_{n3}\left[\int_{z_{Sn3}}^{z_0} + \int_{z_{Sn3}}^{z} + 2n\int_{z_{Sn3}}^{z_{Bn3}}\right]dz/k_{zn3}(z),$$

$$r = k_{n4}\left[\int_{z}^{z_0} + 2\int_{z_{Sn4}}^{z} + 2\int_{z_0}^{z_{Bn4}} + 2n\int_{z_{Sn4}}^{z_{Bn4}}\right]dz/k_{zn4}(z)\ , \quad (7.100)$$

and [cf. Eq. (7.93)]

$$k_{nj} = k(z)\sin\theta_{nj}(z) = k(z_0)\sin\theta_{nji},$$

$$k_{znj}(z) = k(z)\cos\theta_{nj}(z) = k(z)\sqrt{1-\sin^2\theta_{nj}(z)}$$

$$= k(z)\sqrt{1-\frac{c^2(z)\sin^2\theta_{nji}}{c^2(z_0)}}\ . \quad (7.101)$$

Here $\theta_{nj}(z)$ is the angle of propagation and z_{Snj} and z_{Bnj} are the upper and lower turning points, respectively, of the nth ray of type j.

The full ray series solution for the single velocity minimum case is therefore

$$p(r) = \sum_{n=0}^{\infty}\sum_{j=1}^{4} A_{nj}e^{iS_{nj}}\ , \quad (7.102)$$

where the θ_{nji} are obtained from Eqs. (7.100) and (7.101), the S_{nj} from Eq. (7.99), and the A_{nj} from the expression [cf. Eq. (7.96)]

$$A_{nj} = \left[\frac{c(z)\sin\theta_{nji}}{c_i r\cos\theta\left(\partial r/\partial\theta_{nji}\right)}\right]^{1/2}\ . \quad (7.103)$$

We note that each of the S_{nj} in Eq. (7.99) describes the total phase accumulated by an eigenray of type j which has executed n full cycles, i.e., the ray has passed through n upper and n lower turning points. The phase ΔS_{njr} accumulated in the lateral direction between successive upper or lower turning points is then given by [cf. Eq. (6.54)]

$$\begin{aligned}\Delta S_{njr} &= 2nk_{nj}^2 \int_{z_{Snj}}^{z_{Bnj}} dz/k_{znj}(z) - 2(n-1)k_{nj}^2 \int_{z_{Snj}}^{z_{Bnj}} dz/k_{znj}(z) \\ &= 2k_{nj}^2 \int_{z_{Snj}}^{z_{Bnj}} dz/k_{znj}(z) \ . \qquad (7.104)\end{aligned}$$

where we have substituted Eq. (7.100) into Eq. (7.99). In Sec. 7.6.4, we will compare the modal cycle distance for the same problem with the result in Eq. (7.104).

7.6.2 Asymptotic Evaluation of the Hankel Transform Representation in the WKB Approximation

The field due to a point source in a medium with a single minimum velocity profile can also be obtained by using the Hankel transform method combined with the WKB approximation. In this approach, WKB solutions are used in the endpoint method to determine the depth-dependent Green's function. We will show that an asymptotic evaluation of the Hankel transform of the Green's function then leads to the ray series solution derived in the previous section.

We recall (cf. Secs. 6.2 and 6.3) that the depth-dependent Green's function $g(k_r;z,z_0)$ can be constructed from two linearly independent solutions of the homogeneous version of Eq. (6.3), which, for the present case, is identical to Eq. (7.6). These functions, $p_S(k_r;z)$ and $p_B(k_r;z)$, satisfy homogeneous boundary conditions at the upper and lower endpoints, respectively, of the interval within which we wish to determine $g(k_r;z,z_0)$. Using the WKB result in Eq. (7.19), we can write these solutions as

$$p_S(k_r;z) = \frac{1}{\sqrt{k_z(z)}}\left\{A\exp\left[-i\int_{z_i}^{z} k_z(z)\,dz\right] + B\exp\left[i\int_{z_i}^{z} k_z(z)\,dz\right]\right\}, \qquad (7.105)$$

$$p_B(k_r;z) = \frac{1}{\sqrt{k_z(z)}}\left\{D\exp\left[-i\int_{z_i}^{z} k_z(z)\,dz\right] + E\exp\left[i\int_{z_i}^{z} k_z(z)\,dz\right]\right\}, \qquad (7.106)$$

where A, B, D, and E are arbitrary constants to be determined by the boundary conditions. Since $c(z)$ tends monotonically toward ∞ for $z>0$ $(z<0)$ and $z \to \infty$ $(z \to -\infty)$, the boundary conditions require that total reflection occur at the upper and lower turning points, z_S and z_B. Because of the increasing nature of the sound velocity as the waves in Eqs. (7.105) and (7.106) approach the turning points, we approximate them as rigid boundaries with reflection coefficients $R_S = 1$ and $R_B = 1$. Recalling the definition of the reflection coefficient as the ratio of the reflected and the incident fields evaluated on the boundary (cf. Sec. 3.3), we then have:

$$R_S = \left. \frac{B \exp\left[i \int_{z_i}^{z} k_z(z)\, dz \right]}{A \exp\left[-i \int_{z_i}^{z} k_z(z)\, dz \right]} \right|_{z=z_S} = \frac{B \exp\left[i \int_{z_i}^{z_S} k_z(z)\, dz \right]}{A \exp\left[-i \int_{z_i}^{z_S} k_z(z)\, dz \right]} = 1 \ , \quad (7.107)$$

$$R_B = \left. \frac{D \exp\left[-i \int_{z_i}^{z} k_z(z)\, dz \right]}{E \exp\left[i \int_{z_i}^{z} k_z(z)\, dz \right]} \right|_{z=z_B} = \frac{D \exp\left[-i \int_{z_i}^{z_B} k_z(z)\, dz \right]}{E \exp\left[i \int_{z_i}^{z_B} k_z(z)\, dz \right]} = 1 \ . \quad (7.108)$$

Equation (7.107) requires that $z_i = z_S$ and $B/A = 1$, while Eq. (7.108) requires that $z_i = z_B$ and $D/E = 1$. The solutions in Eqs. (7.105) and (7.106) therefore become

$$p_S(k_r;z) = \frac{A}{\sqrt{k_z(z)}} \left\{ \exp\left[-i \int_{z_S}^{z} k_z(z)\, dz \right] + \exp\left[i \int_{z_S}^{z} k_z(z)\, dz \right] \right\} \ , \quad (7.109)$$

$$p_B(k_r;z) = \frac{D}{\sqrt{k_z(z)}} \left\{ \exp\left[-i \int_{z_B}^{z} k_z(z)\, dz \right] + \exp\left[i \int_{z_B}^{z} k_z(z)\, dz \right] \right\} \ . \quad (7.110)$$

Taking the derivatives of Eqs. (7.109) and (7.110) with respect to z, we obtain

$$p_S'(k_r;z) = -iA\sqrt{k_z(z)} \left\{ \exp\left[-i \int_{z_S}^{z} k_z(z)\, dz \right] - \exp\left[i \int_{z_S}^{z} k_z(z)\, dz \right] \right\}$$
$$- \frac{1}{2} \frac{1}{k_z(z)} \frac{dk_z(z)}{dz} p_S(k_r;z) \ , \quad (7.111)$$

$$p_B'(k_r;z) = -iD\sqrt{k_z(z)}\left\{\exp\left[-i\int_{z_B}^{z} k_z(z)\,dz\right] - \exp\left[i\int_{z_B}^{z} k_z(z)\,dz\right]\right\}$$
$$-\frac{1}{2}\frac{1}{k_z(z)}\frac{dk_z(z)}{dz}p_B(k_r;z) \ . \qquad (7.112)$$

We then use Eqs. (7.109), (7.110), (7.111), and (7.112) in a straightforward manner to calculate the Wronskian $W(z_0)$,

$$W(z_0) = p_S(z_0)p_B'(z_0) - p_S'(z_0)p_B(z_0)$$
$$= 2iAD\exp\left[-i\int_{z_S}^{z_B} k_z(z)\,dz\right]\left\{1 - \exp\left[2i\int_{z_S}^{z_B} k_z(z)\,dz\right]\right\} \ , \qquad (7.113)$$

and to show (cf. Prob. 7.8) that the depth-dependent Green's function $g(k_r;z,z_0)$ [cf. Eq. (6.7)],

$$g(k_r) = -\frac{2}{W(z_0)}p_S(k_r;z)p_B(k_r;z_0), \quad -\infty < z \le z_0 < \infty \ , \qquad (7.113a)$$

is given by

$$g(k_r) = i\frac{\left\{\begin{matrix}\exp\left[i\int_z^{z_0} k_z\,dz\right] + \exp\left[i\left(\int_{z_0}^{z_B} + \int_z^{z_B}\right)k_z\,dz\right] \\ + \exp\left[i\left(\int_{z_S}^{z_0} + \int_{z_S}^{z}\right)k_z\,dz\right] + \exp\left[i\left(\int_z^{z_0} + 2\int_{z_S}^{z} + 2\int_{z_0}^{z_B}\right)k_z\,dz\right]\end{matrix}\right\}}{\sqrt{k_z k_{zi}}\left\{1 - \exp\left[2i\int_{z_S}^{z_B} k_z\,dz\right]\right\}} ,$$

(7.114)

where $k_z = k_z(z)$ and $k_{zi} = k_z(z_0)$. The acoustic field is obtained by taking the Hankel transform of Eq. (7.114) [cf. Eq. (6.4)]:

$$p(r) = \int_0^\infty g(k_r)J_0(k_r r)k_r\,dk_r \ . \qquad (7.115)$$

The asymptotic evaluation of Eq. (7.115) proceeds in a manner analogous to the geometrical acoustics decomposition of the field due to a point source in a homogeneous fluid layer bounded by arbitrary horizontally stratified media (cf. Sec. 6.3.1). First, we use the expansion [cf. Eq. (6.20)]

$$\frac{1}{1-\exp\left[2i\int_{z_S}^{z_B} k_z\,dz\right]}=\sum_{n=0}^{\infty}\exp\left[2in\int_{z_S}^{z_B} k_z\,dz\right] , \qquad (7.116)$$

where we assume a small amount of attenuation, so that

$$\left|\exp\left[2i\int_{z_S}^{z_B} k_z\,dz\right]\right|<1 , \qquad (7.117)$$

thus assuring convergence in Eq. (7.116). We substitute Eq. (7.116) into Eq. (7.114), and Eq. (7.115) then becomes [cf. Eq. (6.31)]

$$\begin{aligned}
p=&\sum_{n=0}^{\infty}\int_0^{\infty}\frac{i}{\sqrt{k_z k_{zi}}}\exp\left[i\left(\int_z^{z_0}+2n\int_{z_S}^{z_B}\right)k_z\,dz\right]J_0(k_r r)k_r\,dk_r && I\\
&+\sum_{n=0}^{\infty}\int_0^{\infty}\frac{i}{\sqrt{k_z k_{zi}}}\exp\left[i\left(\int_{z_0}^{z_B}+\int_z^{z_B}+2n\int_{z_S}^{z_B}\right)k_z\,dz\right]J_0(k_r r)k_r\,dk_r && II\\
&+\sum_{n=0}^{\infty}\int_0^{\infty}\frac{i}{\sqrt{k_z k_{zi}}}\exp\left[i\left(\int_{z_S}^{z_0}+\int_{z_S}^{z}+2n\int_{z_S}^{z_B}\right)k_z\,dz\right]J_0(k_r r)k_r\,dk_r && III\\
&+\sum_{n=0}^{\infty}\int_0^{\infty}\frac{i}{\sqrt{k_z k_{zi}}}\exp\left[i\left(\int_z^{z_0}+2\int_{z_S}^{z}+2\int_{z_0}^{z_B}+2n\int_{z_S}^{z_B}\right)k_z\,dz\right]J_0(k_r r)k_r\,dk_r . && IV
\end{aligned}$$

$$(7.118)$$

Analogous to our interpretation of Eq. (6.31), we interpret the $n=0$ term in each of the four expressions in Eq. (7.118) as a plane wave ray decomposition of a particular wave species emanating from the source plane $z=z_0$. *Term I* consists of a spectrum of rays leaving the source plane and propagating directly to the receiver. *Term II* is composed of rays which emanate from the source plane and pass through a lower turning point z_B, while *Term III* is a synthesis of rays which pass through an upper turning point z_S. *Term IV* is a spectrum of rays which emanate from the source plane and pass through a lower and then an upper turning point. The higher order terms ($n>0$) correspond to multiple excursions between lower and upper turning points of these fundamental plane wave ray types.

Our asymptotic analysis begins with an approximate evaluation of Eq. (7.118) for $k_r r >> 1$ [cf. Eq. (6.32)]:

$$p \sim \sum_{n=0}^{\infty} a \int_{-\infty}^{\infty} \frac{\sqrt{K_r}}{\kappa_z} \exp\left\{ ik_0 \left[\left(\int_z^{z_0} + 2n \int_{z_S}^{z_B} \right) K_z \, dz + K_r r \right] \right\} dK_r \qquad I$$

$$+ \sum_{n=0}^{\infty} a \int_{-\infty}^{\infty} \frac{\sqrt{K_r}}{\kappa_z} \exp\left\{ ik_0 \left[\left(\int_{z_0}^{z_B} + \int_z^{z_B} + 2n \int_{z_S}^{z_B} \right) K_z \, dz + K_r r \right] \right\} dK_r \qquad II$$

$$+ \sum_{n=0}^{\infty} a \int_{-\infty}^{\infty} \frac{\sqrt{K_r}}{\kappa_z} \exp\left\{ ik_0 \left[\left(\int_{z_S}^{z_0} + \int_{z_S}^{z} + 2n \int_{z_S}^{z_B} \right) K_z \, dz + K_r r \right] \right\} dK_r \qquad III$$

$$+ \sum_{n=0}^{\infty} a \int_{-\infty}^{\infty} \frac{\sqrt{K_r}}{\kappa_z} \exp\left\{ ik_0 \left[\left(\int_z^{z_0} + 2\int_{z_S}^{z} + 2\int_{z_0}^{z_B} + 2n \int_{z_S}^{z_B} \right) K_z \, dz + K_r r \right] \right\} dK_r, \qquad IV \qquad (7.119)$$

where

$$a = \frac{\sqrt{k_0} e^{i\pi/4}}{\sqrt{2\pi r}}, \quad \kappa_z = \sqrt{K_z K_{zi}} \ , \qquad (7.120)$$

and [cf. Eq. (6.33)]

$$K_r = \frac{k_r}{k_0} = \frac{k \sin\theta}{k_0}, \quad K_z = \frac{k_z}{k_0} = \frac{k\cos\theta}{k_0}, \quad K_{zi} \doteq \frac{k_{zi}}{k_0} = \frac{k_i \cos\theta_i}{k_0}. \qquad (7.121)$$

Here $k_0 = \omega/c_0$ is the total wavenumber on the channel axis, k_i and $k = k(z)$ are the total wavenumbers at the source and receiver, respectively, and $\theta = \theta(z)$ is the propagation angle of a particular plane wave ray component. Each of the four types of integrals in Eq. (7.119) is in a canonical form for the application of asymptotic methods for large values of the parameter k_0 [cf. Eq. (4.123)]. Following the developments of Secs. 4.8.1 and 6.3.1, we determine the saddle points associated with the phase factor in each integral and evaluate the integral using the modified method of stationary phase. For *Term I*, the saddle points occur at values $K_r = K_{n1} = k \sin\theta_{n1}/k_0$ which satisfy the equation

$$\frac{d}{dK_r}\left[\left(\int_z^{z_0}+2n\int_{z_S}^{z_B}\right)K_z\,dz+K_r r\right]=0\ , \tag{7.122}$$

so that

$$\begin{aligned} r &= \left[\int_z^{z_0}+2n\int_{z_{Sn1}}^{z_{Bn1}}\right]K_{n1}dz/K_{zn1}(z) \\ &= k_{n1}\left[\int_z^{z_0}+2n\int_{z_{Sn1}}^{z_{Bn1}}\right]dz/k_{zn1}(z)\ . \end{aligned} \tag{7.123}$$

But Eq. (7.123) is precisely the first line of Eq. (7.100). Furthermore, it is clear that the saddle points associated with *Terms II-IV* simply reproduce the remaining lines of Eq. (7.100). Thus, the saddle points arising in an asymptotic evaluation of the Hankel transform representation are associated with the angles of propagation $\theta_{nj}(z)$ $(j=1,\ldots 4)$ of the eigenrays in the ray series solution.

Proceeding with an evaluation of the integrals in Eq. (7.119) using the modified method of stationary phase described in Sec. 4.8.1 [cf. Eqs. (4.129)-(4.133)], we find that

$$p(r)\sim\sum_{n=0}^{\infty}\sum_{j=1}^{4}e^{i\pi/2}\left[\frac{k_0\sin\theta_{nj}(z)}{k_i\cos\theta_{nj}(z)\cos\theta_{nji}}\right]^{1/2}\frac{e^{ik_0H(K_{nj})}}{\sqrt{rH''(K_{nj})}}, \tag{7.124}$$

which, using Snell's law,

$$\sin\theta_{nj}(z)=\frac{k_i\sin\theta_{nji}}{k(z)}\ , \tag{7.125}$$

can be rewritten as

$$p(r)\sim\sum_{n=0}^{\infty}\sum_{j=1}^{4}e^{i\pi/2}\left[\frac{k_0\sin\theta_{nji}}{k(z)\cos\theta_{nj}(z)\cos\theta_{nji}}\right]^{1/2}\frac{e^{ik_0H(K_{nj})}}{\sqrt{rH''(K_{nj})}}\ . \tag{7.126}$$

When $j=1$ in Eqs. (7.124) and (7.126), we have

$$k_0 H(K_{n1}) = \left[\int_z^{z_0} + 2n\int_{z_{Sn1}}^{z_{Bn1}}\right] k_{zn1}(z)\,dz + k_{n1}r = S_{n1} \quad , \quad (7.127)$$

where S_{n1} is the ray phase defined in Eq. (7.99). We also obtain

$$H''(K_{n1}) = \left.\frac{\partial^2 H(K_r)}{\partial^2 K_r}\right|_{K_r = K_{n1}} = -\left[\int_z^{z_0} + 2n\int_{z_{Sn1}}^{z_{Bn1}}\right]\frac{dz}{K_{zn1}^3} \ . \quad (7.128)$$

But from Eq. (7.123), we have

$$\frac{\partial r}{\partial K_{n1}} = \left[\int_z^{z_0} + 2n\int_{z_{Sn1}}^{z_{Bn1}}\right]\frac{dz}{K_{zn1}^3} = -H''(K_{n1}) \ , \quad (7.129)$$

and furthermore, we know that

$$\frac{\partial r}{\partial \theta_{n1i}} = \frac{\partial r}{\partial K_{n1}}\frac{\partial K_{n1}}{\partial \theta_{n1i}} = -H''(K_{n1})\frac{k(z)}{k_0}\frac{\partial \sin\theta_{n1}}{\partial \theta_{n1i}}$$

$$= -H''(K_{n1})\frac{k_i}{k_0}\frac{\partial \sin\theta_{n1i}}{\partial \theta_{n1i}} = -H''(K_{n1})\frac{k_i}{k_0}\cos\theta_{n1i} \ , \quad (7.130)$$

and therefore

$$H''(K_{n1}) = -\frac{k_0}{k_i}\frac{1}{\cos\theta_{n1i}}\frac{\partial r}{\partial \theta_{n1i}} \ . \quad (7.131)$$

The $j = 1$ term p_{n1} in Eq. (7.126) then becomes

$$p_{n1} \sim \sum_{n=0}^{\infty}\left[\frac{c(z)\sin\theta_{n1i}}{c_i r\cos\theta_{n1}\left(\partial r/\partial\theta_{n1i}\right)}\right]^{1/2} e^{iS_{n1}} \ . \quad (7.132)$$

The analysis of terms $j = 2,\ldots 4$ in Eq. (7.126) proceeds in a manner analogous to that used for $j = 1$, and we obtain the final result

$$p(r) \sim \sum_{n=0}^{\infty} \sum_{j=1}^{4} A_{nj} e^{iS_{nj}} \quad , \tag{7.133}$$

where S_{nj} and A_{nj} are defined in Eqs (7.99) and (7.103), respectively. Thus, the asymptotic evaluation of the Hankel transform representation exactly reproduces the ray series solution in Eq. (7.102). A key ingredient, common to the two approaches, is the high-frequency nature of the approximations used. This characteristic manifests itself as the large k_0 requirement in the asymptotic Hankel transform technique and the high-frequency condition in Eq. (7.31) for the ray method.

Finally, we recall that in previous calculations of point source fields in homogeneous media (cf. Secs. 4.8.1 and 6.3.1), the asymptotic evaluation of the Hankel transform representation gave rise to the geometrical acoustics approximation to the total field. Similarly, for inhomogeneous media, ray theory constitutes the solution of the wave equation in the geometrical acoustics limit. The underlying, unifying themes are the approximate, high-frequency evaluation of integral representations of the fields using saddle point methods and the association of the saddle points with geometrical acoustic paths. In fact, the geometrical acoustics approximation corresponds to the first term (or asymptotic form) in an asymptotic expansion of the field in inverse powers of k_0 (cf. p. 96). Specifically, the solution of the Helmholtz Eq. (7.32) can be written as [Ahluwalia and Keller, 1977]

$$p(\mathbf{r}) = e^{iS(\mathbf{r})} \sum_{n=0}^{\infty} \frac{A_n(\mathbf{r})}{(ik_0)^n} \quad , \tag{7.134}$$

where k_0 is a characteristic wavenumber (e.g., the total wavenumber on the channel axis). Suppose we substitute Eq. (7.134) into the Helmholtz equation for inhomogeneous media [cf. Eq. (7.32)], and group coefficients of equal powers of k_0. Setting these coefficients individually equal to zero, we find that the terms corresponding to $n = 0$ satisfy the transport and eikonal equations [cf. Eqs. (7.36) and (7.39)]. Thus, the first term in Eq. (7.134) is simply the ray solution in Eq. (7.33). The role played by higher order terms ($n > 0$) in Eq. (7.134) as corrections to the geometrical acoustics approximation is intimately connected to the properties of asymptotic expansions, which are discussed extensively by Erdelyi [1956] and Bleistein and Handelsman [1986]. In order to clarify this point, we rewrite Eq. (7.134) as

$$p(\mathbf{r}) = e^{iS(\mathbf{r})} \sum_{n=0}^{N} \frac{A_n(\mathbf{r})}{(ik_0)^n} + R_N(\mathbf{r}) \ , \tag{7.135}$$

where the remainder term $R_N(\mathbf{r})$ is defined as

$$R_N(\mathbf{r}) = e^{iS(\mathbf{r})} \sum_{n=N+1}^{\infty} \frac{A_n(\mathbf{r})}{(ik_0)^n} \ . \tag{7.136}$$

An asymptotic expansion with fixed N has the property

$$R_N(\mathbf{r}) \to 0 \ \text{ for } k_0 \to \infty \ , \tag{7.137}$$

which we contrast with the behavior of a convergent Taylor series [e.g., Eq. (3.40)], for which the remainder term tends to zero as $N \to \infty$ when the expansion variable is fixed. In some cases, for fixed k_0, an asymptotic expansion may actually diverge as N increases. Also, the error associated with retaining N terms in an asymptotic expansion can sometimes be related to the magnitude of term $(N+1)$. Amidst these subtleties, we can say with certainty that the geometrical acoustics solution is exact in the limit of infinite frequency $(k_0 \to \infty)$.

7.6.3 Long-Range SOFAR Propagation for a Symmetric Bilinear Profile

We will now examine ray propagation for a specific type of single minimum velocity profile, namely, the symmetric bilinear profile (cf. Fig. 7.5 with $g_1 = g_2 = g$):

$$c(z) = \begin{cases} c_0 - gz, & z \le 0, \\ c_0 + gz, & z \ge 0 \ . \end{cases} \tag{7.138}$$

This simplification, combined with the assumption that both the source and receiver lie on the channel axis $(z_0 = z = 0)$, will enable us to determine analytically some of the key properties of long-range SOFAR propagation. Of particular interest is the time-dependent behavior of a delta function pulse, which we will calculate in closed form.

For source and receiver lying on the channel axis in a symmetric bilinear profile, the lower and upper turning points in Sec. 7.6.1 are related to one another by

$$z_{Bnj} = -z_{Snj} \equiv z_{nj}, \quad j = 1, \ldots 4 \ . \tag{7.139}$$

Using Eq. (7.139), we find that Eq. (7.99) for the ray phase becomes

$$\begin{aligned}
S_{n1} &= 2n \int_{-z_{n1}}^{z_{n1}} k_{zn1}(z)\,dz + k_{n1} r, \\
S_{n2} &= \left[2\int_{0}^{z_{n2}} + 2n \int_{-z_{n2}}^{z_{n2}} \right] k_{zn2}(z)\,dz + k_{n2} r, \\
S_{n3} &= \left[2\int_{-z_{n3}}^{0} + 2n \int_{-z_{n3}}^{z_{n3}} \right] k_{zn3}(z)\,dz + k_{n3} r, \\
S_{n4} &= \left[2\int_{-z_{n4}}^{z_{n4}} + 2n \int_{-z_{n4}}^{z_{n4}} \right] k_{zn4}(z)\,dz + k_{n4} r \ .
\end{aligned} \tag{7.140}$$

Recognizing that

$$k_{znj}(z) = k_{znj}(-z) \ \Rightarrow \ \int_{-z_{nj}}^{0} k_{znj}(z)\,dz = \int_{0}^{z_{nj}} k_{znj}(z)\,dz, \tag{7.141}$$

we simplify Eq. (7.140) further to obtain

$$\begin{aligned}
S_{n1} &= 4n \int_{0}^{z_{n1}} k_{zn1}(z)\,dz + k_{n1} r, \\
S_{n2} &= 2(2n+1) \int_{0}^{z_{n2}} k_{zn2}(z)\,dz + k_{n2} r, \\
S_{n3} &= 2(2n+1) \int_{0}^{z_{n3}} k_{zn3}(z)\,dz + k_{n3} r, \\
S_{n4} &= 4(n+1) \int_{0}^{z_{n4}} k_{zn4}(z)\,dz + k_{n4} r \ .
\end{aligned} \tag{7.142}$$

The initial angle θ_{nji} of the nth ray of type j in Eq. (7.142) must satisfy [cf. Eq. (7.100)]

$$r = 4nk_{n1}\int_0^{z_{n1}} dz/k_{zn1}(z),$$

$$r = 2(2n+1)k_{n2}\int_0^{z_{n2}} dz/k_{zn2}(z),$$

$$r = 2(2n+1)k_{n3}\int_0^{z_{n3}} dz/k_{zn3}(z),$$

$$r = 4(n+1)k_{n4}\int_0^{z_{n4}} dz/k_{zn4}(z) \quad , \tag{7.143}$$

and [cf. Eq. (7.101)]

$$k_{nj} = k(z)\sin\theta_{nj}(z) = k_0 \sin\theta_{nji},$$

$$k_{znj}(z) = k(z)\cos\theta_{nj}(z) = k(z)\sqrt{1-\sin^2\theta_{nj}(z)}$$

$$= k(z)\sqrt{1-\frac{c^2(z)\sin^2\theta_{nji}}{c_0^2}} \quad , \tag{7.144}$$

where $k_0 = \omega/c_0$. The ray amplitude is given by [cf. Eq. (7.103)]

$$A_{nj} = \left[\frac{\sin\theta_{nji}}{r\cos\theta\left(\partial r/\partial\theta_{nji}\right)}\right]^{1/2} . \tag{7.145}$$

We now focus on the behavior of rays which execute many cycles, i.e., rays for which $n >> 1$, so that Eq. (7.143) becomes

$$r \approx 4nk_n\int_0^{z_n} dz/k_{zn}(z) \quad , \tag{7.146}$$

with $\theta_{nji} \approx \theta_{ni}$ $(j = 1,\ldots 4)$, $z_{nj} \approx z_n$, and

$$k_{nj} \approx k_n = k(z)\sin\theta_n(z) = k_0\sin\theta_{ni},$$

$$k_{znj}(z) \approx k_{zn}(z) = k(z)\cos\theta_n(z) = k(z)\sqrt{1-\sin^2\theta_n(z)}$$

$$= k(z)\sqrt{1-\frac{c^2(z)\sin^2\theta_{ni}}{c_0^2}} \ . \tag{7.147}$$

The ray series in Eq. (7.102) then has the form

$$p \approx \sum_{n=0}^{n_{\min}-1}\left[A_{n1}e^{iS_{n1}} + 2A_{n2}e^{iS_{n2}} + A_{n4}e^{iS_{n4}}\right]$$
$$+ \sum_{n=n_{\min}}^{\infty} A_n e^{iS_n}\left\{1 + 2\exp\left[2i\int_0^{z_n} k_{zn}(z)\,dz\right] + \exp\left[4i\int_0^{z_n} k_{zn}(z)\,dz\right]\right\}. \quad (7.148)$$

Here S_{nj} $(j = 1,\ldots 4)$ satisfies Eqs. (7.142)-(7.144) and A_{nj} satisfies Eq. (7.145) for $n < n_{\min}$, where $n_{\min}$ is the minimum value of n for which the criterion $n >> 1$ is satisfied. For $n \geq n_{\min}$, the approximate ray phase and amplitude, S_n and A_n, are given by

$$S_n \approx 4n\int_0^{z_n} k_{zn}(z)\,dz + k_n r \ , \quad (7.149)$$

$$A_n \approx \left[\frac{\sin\theta_{ni}}{r\cos\theta\left(\partial r/\partial\theta_{ni}\right)}\right]^{1/2} , \quad (7.150)$$

with θ_{ni} satisfying Eqs. (7.146) and (7.147). Thus, in Eq. (7.148), we use exact ray results for $n < n_{\min}$ and approximate ray results for $n \geq n_{\min}$.

Ray Range, Initial Angle, and Turning Depth

For a linear gradient and $n < n_{\min}$, it is clear from Eq. (7.63) that the initial angles in Eq. (7.143) satisfy

$$r = \frac{4nc_0}{g}\tan\alpha_{n1i} \Rightarrow \tan\alpha_{n1i} = \frac{gr}{4nc_0} , \quad (7.151)$$

$$r = \frac{2(2n+1)c_0}{g}\tan\alpha_{n2i} \Rightarrow \tan\alpha_{n2i} = \frac{gr}{2(2n+1)c_0} , \quad (7.152)$$

$$r = \frac{4(n+1)c_0}{g}\tan\alpha_{n4i} \Rightarrow \tan\alpha_{n4i} = \frac{gr}{4(n+1)c_0} , \quad (7.153)$$

where we have introduced the initial grazing angles $\alpha_{nji} = \pi/2 - \theta_{nji}$ for convenience. It is obvious that the ray angle $\alpha_{01i} = 0$ corresponds to the direct path along the channel axis joining source and receiver. The initial angles of rays for which $n \geq n_{\min}$ are described by [cf. Eqs. (7.63) and (7.146)]

$$r \approx \frac{4nc_0}{g} \tan \alpha_{ni} \;\Rightarrow\; \tan \alpha_{ni} \approx \frac{gr}{4nc_0} \;, \tag{7.154}$$

where $\alpha_{ni} = \pi/2 - \theta_{ni}$. Thus, with the exception of the direct path, the initial grazing angle decreases as the number n of ray cycles increases.

From Snell's law, the turning depth z_t associated with a ray having initial angle α_i and initial velocity c_0 in a linear gradient medium satisfies the equation

$$k_0 \cos \alpha_i = k(z_t) = \omega/(c_0 + gz_t) \;, \tag{7.155}$$

which can be rewritten as

$$z_t = \frac{c_0}{g}\left[\frac{1}{\cos \alpha_i} - 1\right] . \tag{7.156}$$

The turning depths z_{n1}, z_{n2}, z_{n4}, and z_n associated with the rays in Eqs. (7.151)-(7.154) are therefore given by

$$z_{n1} = \frac{c_0}{g}\left[\frac{1}{\cos \alpha_{n1i}} - 1\right], \; z_{n2} = \frac{c_0}{g}\left[\frac{1}{\cos \alpha_{n2i}} - 1\right],$$

$$z_{n4} = \frac{c_0}{g}\left[\frac{1}{\cos \alpha_{n4i}} - 1\right] , \tag{7.157}$$

$$z_n = \frac{c_0}{g}\left[\frac{1}{\cos \alpha_{ni}} - 1\right] . \tag{7.158}$$

Thus, we see that the ray turning depth increases with increasing initial grazing angle.

Ray Phase and Travel Time

For $n < n_{\min}$, we use Eqs. (7.142) and (7.73) to obtain

$$S_{n1} = \frac{2n\omega}{g} \ln\left[\frac{1+\sin\alpha_{n1i}}{1-\sin\alpha_{n1i}}\right], \quad S_{n2} = \frac{(2n+1)\omega}{g} \ln\left[\frac{1+\sin\alpha_{n2i}}{1-\sin\alpha_{n2i}}\right],$$

$$S_{n4} = \frac{2(n+1)\omega}{g} \ln\left[\frac{1+\sin\alpha_{n4i}}{1-\sin\alpha_{n4i}}\right], \tag{7.159}$$

where it is clear that $S_{01} = k_0 r$. Similarly, for $n \geq n_{\min}$, we have [cf. Eqs. (7.149) and (7.73)]

$$S_n \approx \frac{2n\omega}{g} \ln\left[\frac{1+\sin\alpha_{ni}}{1-\sin\alpha_{ni}}\right]. \tag{7.160}$$

Using the first two terms in the expansion [Abramowitz and Stegun, 1964],

$$\ln\left[\frac{1+z}{1-z}\right] = 2z\left(1+\frac{z^2}{3}+\frac{z^4}{5}+\ldots\right) \quad \text{for } |z|<1 , \tag{7.161}$$

we obtain approximate expressions for Eqs. (7.159) and (7.160), valid for small grazing angles ($\sin\alpha << 1$):

$$S_{n1} \approx \frac{4n\omega}{g} \sin\alpha_{n1i}\left(1+\frac{\sin^2\alpha_{n1i}}{3}\right),$$

$$S_{n2} \approx \frac{2(2n+1)\omega}{g} \sin\alpha_{n2i}\left(1+\frac{\sin^2\alpha_{n2i}}{3}\right),$$

$$S_{n4} \approx \frac{4(n+1)\omega}{g} \sin\alpha_{n4i}\left(1+\frac{\sin^2\alpha_{n4i}}{3}\right), \tag{7.162}$$

$$S_n \approx \frac{4n\omega}{g}\sin\alpha_{ni}\left(1+\frac{\sin^2\alpha_{ni}}{3}\right) . \tag{7.163}$$

From Eqs. (7.162) and (7.163), we see that the ray travel time T is simply

$$T_{n1} \approx \frac{4n}{g}\sin\alpha_{n1i}\left(1+\frac{\sin^2\alpha_{n1i}}{3}\right),$$

$$T_{n2} \approx \frac{2(2n+1)}{g}\sin\alpha_{n2i}\left(1+\frac{\sin^2\alpha_{n2i}}{3}\right),$$

$$T_{n4} \approx \frac{4(n+1)}{g}\sin\alpha_{n4i}\left(1+\frac{\sin^2\alpha_{n4i}}{3}\right) , \tag{7.164}$$

$$T_n \approx \frac{4n}{g}\sin\alpha_{ni}\left(1+\frac{\sin^2\alpha_{ni}}{3}\right) , \tag{7.165}$$

where $S = \omega T$ and $T_{01} = r/c_0$. Using Eqs. (7.151)-(7.154), we can rewrite Eqs. (7.164) and (7.165) as

$$T_{n1} \approx T_{01}\cos\alpha_{n1i}(1+\ldots),\ T_{n2} \approx T_{01}\cos\alpha_{n2i}(1+\ldots),$$

$$T_{n4} \approx T_{01}\cos\alpha_{n4i}(1+\ldots) , \tag{7.166}$$

$$T_n \approx T_{01}\cos\alpha_{ni}(1+\ldots) . \tag{7.167}$$

Equations (7.166) and (7.167) indicate that the ray travel time increases with decreasing initial grazing angle. This result is consistent with the intuitive notion that, despite their longer path lengths, high-angle rays arrive at the receiver before low-angle rays because the steeper rays probe higher velocity regions of the sound velocity structure.

Finally, we can substitute the first two terms of the expansion,

$$\sin z = z\left(1-\frac{z^2}{3!}+\frac{z^4}{5!}-\ldots\right) , \tag{7.168}$$

into Eqs. (7.164) and (7.165) to obtain approximate results for $\alpha_i << 1$:

$$T_{n1} \approx \frac{4n\alpha_{n1i}}{g}\left(1+\frac{\alpha_{n1i}^2}{6}\right), \quad T_{n2} \approx \frac{2(2n+1)\alpha_{n2i}}{g}\left(1+\frac{\alpha_{n2i}^2}{6}\right),$$

$$T_{n4} \approx \frac{4(n+1)\alpha_{n4i}}{g}\left(1+\frac{\alpha_{n4i}^2}{6}\right), \tag{7.169}$$

$$T_n \approx \frac{4n\alpha_{ni}}{g}\left(1+\frac{\alpha_{ni}^2}{6}\right). \tag{7.170}$$

Equation (7.170) can be approximately evaluated still further, since we know from Eq. (7.154) that

$$\alpha_{ni} \approx \tan^{-1}\frac{gr}{4nc_0}, \tag{7.171}$$

and for $n >> 1$, we can assume that

$$\frac{gr}{4nc_0} << 1 . \tag{7.172}$$

Using the first two terms of the expansion [Abramowitz and Stegun, 1964]

$$\tan^{-1} z = z\left(1-\frac{z^2}{3}+\frac{z^4}{5}-\ldots\right) \quad \text{for } |z| \le 1,\ z^2 \ne -1 , \tag{7.173}$$

in Eq. (7.171), we see that

$$\alpha_{ni} \approx \frac{gr}{4nc_0}\left[1-\frac{1}{3}\left(\frac{gr}{4nc_0}\right)^2\right], \tag{7.174}$$

which we substitute into Eq. (7.170) to obtain

$$T_n \approx T_{01}\left[1-\frac{1}{6}\left(\frac{gr}{4nc_0}\right)^2\right] . \tag{7.175}$$

Equation (7.175) is consistent with the results of Brekhovskikh and Lysanov [1991], who examined the related case of ray propagation through a medium with a linear gradient profile bounded by a pressure-release surface. It is evident from Eqs. (7.166), (7.167), and (7.175) that the shortest travel time T_{02} is associated with the deepest diving ray having a single turning point, while the longest time T_{01} is associated with the direct path between source and receiver.

Finally, when we take the derivative of T_n in Eq. (7.175) with respect to n,

$$\frac{dT_n}{dn} \approx \frac{T_{01}}{3n^3}\left(\frac{gr}{4c_0}\right)^2 , \tag{7.176}$$

we see that the time difference between adjacent arrivals approaches zero at the rate $1/n^3$ as $n \to \infty$. Thus, the ray arrivals become temporally closer to one another as n increases and the corresponding grazing angles decrease. This property will help elucidate some of the features of pulse propagation, to be discussed in a later section.

Ray Amplitude

For $n < n_{\min}$, we use Eq. (7.145) to obtain

$$A_{n1} = \frac{\cos\alpha_{n1i}}{r}, \quad A_{n2} = \frac{\cos\alpha_{n2i}}{r}, \quad A_{n4} = \frac{\cos\alpha_{n4i}}{r} , \tag{7.177}$$

which is simply the linear gradient result in Eq. (7.88) evaluated at the appropriate initial angle. Similarly, for $n \geq n_{\min}$, we find that

$$A_n \approx \frac{\cos\alpha_{ni}}{r} . \tag{7.178}$$

Equations (7.177) and (7.178) show that the ray amplitude increases with decreasing grazing angle, reaching its maximum value, $A_{01} = 1/r$, for the direct ray.

Time-Dependent Behavior of a Delta Function Pulse

Having determined the appropriate ray quantities, we are now in a position to evaluate the ray series in Eq. (7.148) and the field associated with a delta function pulse. First, we notice that the infinite sum in Eq. (7.148) can be written as

$$\sum_{n=n_{\min}}^{\infty} A_n e^{iS_n}\left\{1+2\exp\left[2i\int_0^{z_n} k_{zn}(z)\,dz\right]+\exp\left[4i\int_0^{z_n} k_{zn}(z)\,dz\right]\right\}$$

$$= \sum_{n=n_{\min}}^{\infty} A_n e^{iS_n}\left\{1+\exp\left[2i\int_0^{z_n} k_{zn}(z)\,dz\right]\right\}^2 \approx 4\sum_{n=n_{\min}}^{\infty} A_n e^{iS_n} \ . \qquad (7.179)$$

Here we have used the approximation

$$\exp\left[2i\int_0^{z_n} k_{zn}(z)\,dz\right] \approx 1 \ , \qquad (7.180)$$

which is justified for $n >> 1$, since [cf. Eq. (7.158)]

$$\lim_{n\to\infty}\begin{cases} z_n = 0, \\ k_{zn} = k(z)\sin\alpha_n(z) = 0 \ , \end{cases} \qquad (7.181)$$

where $\alpha_n(z)$ is the grazing angle of the nth ray at depth z. Furthermore, we know from Eq. (7.176) that

$$\lim_{n\to\infty} \Delta S_n = S_{n+1} - S_n = 0 \ , \qquad (7.182)$$

and we can therefore approximate the sum in Eq. (7.179) by an integral:

$$4\sum_{n=n_{\min}}^{\infty} A_n e^{iS_n} \approx 4\int_{n_{\min}}^{\infty} A_n e^{iS_n}\,dn \ . \qquad (7.183)$$

Using Eq. (7.154), we change variables in Eq. (7.183),

$$\alpha_{ni} \approx \tan^{-1}\left(\frac{gr}{4nc_0}\right) \Rightarrow d\alpha_{ni} \approx -\frac{4c_0}{gr}\sin^2\alpha_{ni}\,dn \ , \qquad (7.184)$$

and also make the approximation [cf. Eq. (7.163)]

$$S_n \approx \frac{4nc_0k_0}{g}\sin\alpha_{ni} \approx k_0 r\cos\alpha_{ni} \ . \tag{7.185}$$

The integral in Eq. (7.183) then becomes

$$4\int_{n_{\min}}^{\infty} A_n e^{iS_n}\,dn \approx \frac{g}{c_0}\int_0^{\alpha_{n_{\min}i}} \frac{\cos\alpha_{ni}}{\sin^2\alpha_{ni}}\exp[ik_0 r\cos\alpha_{ni}]\,d\alpha_{ni} \ , \tag{7.186}$$

where [cf. Eqs. (7.153) and (7.154)]

$$\alpha_{n_{\min}i} = \tan^{-1}\left(\frac{gr}{4n_{\min}c_0}\right) = \alpha_{n_{\min}-1,4i} \ , \tag{7.187}$$

and we have used Eq. (7.178). Changing variables in Eq. (7.186),

$$\tau = \frac{r}{c_0}\cos\alpha_{ni} = T_{01}\cos\alpha_{ni} \Rightarrow d\tau = -T_{01}\sqrt{1-\tau^2/T_{01}^2}\,d\alpha_{ni} \ , \tag{7.188}$$

we obtain

$$\frac{g}{c_0}\int_0^{\alpha_{n_{\min}i}} \frac{\cos\alpha_{ni}}{\sin^2\alpha_{ni}}\exp[ik_0 r\cos\alpha_{ni}]\,d\alpha_{ni} = \frac{g}{r}\int_{T_{n_{\min}}}^{T_{01}} \frac{(\tau/T_{01})e^{i\omega\tau}}{\left(1-\tau^2/T_{01}^2\right)^{3/2}}\,d\tau \ , \tag{7.189}$$

where $T_{n_{\min}} = T_{n_{\min}-1,4}$ [cf. Eqs. (7.169) and (7.170)]. Using Eqs. (7.179), (7.183), and (7.189), we can rewrite Eq. (7.148) as

$$p \approx \sum_{n=0}^{n_{\min}-1}\left[A_{n1}e^{iS_{n1}} + 2A_{n2}e^{iS_{n2}} + A_{n4}e^{iS_{n4}}\right] + \frac{g}{r}\int_{T_{n_{\min}}}^{T_{01}} \frac{(\tau/T_{01})e^{i\omega\tau}}{\left(1-\tau^2/T_{01}^2\right)^{3/2}}\,d\tau \ . \tag{7.190}$$

Proceeding with the time-dependent calculation, we recall that the acoustic field due to a point source with temporal behavior $S(t)$

satisfies the inhomogeneous, time-dependent wave equation [cf. Eq. (4.1)]:

$$\nabla^2 P(\mathbf{r},t) - \frac{1}{c^2(z)} \frac{\partial^2 P(\mathbf{r},t)}{\partial t^2} = -4\pi\delta(\mathbf{r}-\mathbf{r}_0)S(t) \quad . \tag{7.191}$$

It is straightforward to show that [cf. Eq. (1.6)]

$$P(\mathbf{r},t) = \frac{1}{\sqrt{2\pi}} \int_{-\infty}^{\infty} p(\mathbf{r},\omega)s(\omega)e^{-i\omega t} d\omega \quad , \tag{7.192}$$

where $p(\mathbf{r},\omega)$ satisfies the inhomogeneous Helmholtz equation [cf. Eq. (4.3)],

$$\left[\nabla^2 + k^2(z)\right] p(\mathbf{r},\omega) = -4\pi\delta(\mathbf{r}-\mathbf{r}_0) \quad , \tag{7.193}$$

and $s(\omega)$ is given by

$$s(\omega) = \frac{1}{\sqrt{2\pi}} \int_{-\infty}^{\infty} S(t)e^{i\omega t} dt \quad . \tag{7.194}$$

For a delta function pulse initiated at $t = 0$, we have $S(t) = \delta(t)$, and therefore Eqs. (7.194) and (7.192) become [cf. Eq. (4.7)]

$$s(\omega) = \frac{1}{\sqrt{2\pi}} \int_{-\infty}^{\infty} \delta(t)e^{i\omega t} dt = \frac{1}{\sqrt{2\pi}} \quad , \tag{7.195}$$

$$P(\mathbf{r},t) = \frac{1}{2\pi} \int_{-\infty}^{\infty} p(\mathbf{r},\omega)e^{-i\omega t} d\omega \quad . \tag{7.196}$$

Substituting Eq. (7.190) into Eq. (7.196), we obtain

$$\begin{aligned} P \approx {} & \frac{1}{2\pi} \int_{-\infty}^{\infty} \left\{ \sum_{n=0}^{n_{\min}-1} \left[A_{n1}e^{i\omega T_{n1}} + 2A_{n2}e^{i\omega T_{n2}} + A_{n4}e^{i\omega T_{n4}} \right] \right\} e^{-i\omega t} d\omega \\ & + \frac{g}{2\pi r} \int_{-\infty}^{\infty} \left\{ \int_{T_{n_{\min}}}^{T_{01}} \frac{(\tau/T_{01})}{\left(1-\tau^2/T_{01}^2\right)^{3/2}} e^{i\omega\tau} d\tau \right\} e^{-i\omega t} d\omega \quad , \end{aligned} \tag{7.197}$$

where we have used the relation $S = \omega T$. Rewriting Eq. (7.197) as

$$P \approx \frac{1}{2\pi} \sum_{n=0}^{n_{\min}-1} \int_{-\infty}^{\infty} \left[A_{n1} e^{i\omega(T_{n1}-t)} + 2A_{n2} e^{i\omega(T_{n2}-t)} + A_{n4} e^{i\omega(T_{n4}-t)} \right] d\omega$$

$$+ \frac{g}{2\pi r} \int_{T_{n_{\min}}}^{T_{01}} \frac{(\tau/T_{01})}{\left(1-\tau^2/T_{01}^2\right)^{3/2}} \left[\int_{-\infty}^{\infty} e^{i\omega(\tau-t)} \, d\omega \right] d\tau \ , \qquad (7.198)$$

and using the integral representation for the delta function [cf. Eq. (4.8)],

$$\delta(t-T) = \delta(T-t) = \frac{1}{2\pi} \int_{-\infty}^{\infty} e^{i\omega(T-t)} \, d\omega \ , \qquad (7.199)$$

we find that

$$P \approx \sum_{n=0}^{n_{\min}-1} \left[A_{n1} \delta(t-T_{n1}) + 2A_{n2} \delta(t-T_{n2}) + A_{n4} \delta(t-T_{n4}) \right]$$

$$+ \frac{g}{r} \int_{T_{n_{\min}}}^{T_{01}} \frac{(\tau/T_{01})}{\left(1-\tau^2/T_{01}^2\right)^{3/2}} \delta(t-\tau) \, d\tau \ . \qquad (7.200)$$

Finally, using Eqs. (4.5), (4.7), and (7.177), we obtain

$$P \approx \sum_{n=0}^{n_{\min}-1} \left[\frac{\cos\alpha_{n1i}}{r} \delta(t-T_{n1}) + 2\frac{\cos\alpha_{n2i}}{r} \delta(t-T_{n2}) + \frac{\cos\alpha_{n4i}}{r} \delta(t-T_{n4}) \right]$$

$$+ \Pi(t) \frac{g}{r} \frac{(t/T_{01})}{\left(1-t^2/T_{01}^2\right)^{3/2}} \ , \qquad (7.201)$$

where

$$\Pi(t) = \begin{cases} 1, & T_{n_{\min}} < t < T_{01}, \\ 0, & \text{otherwise} \ . \end{cases} \qquad (7.202)$$

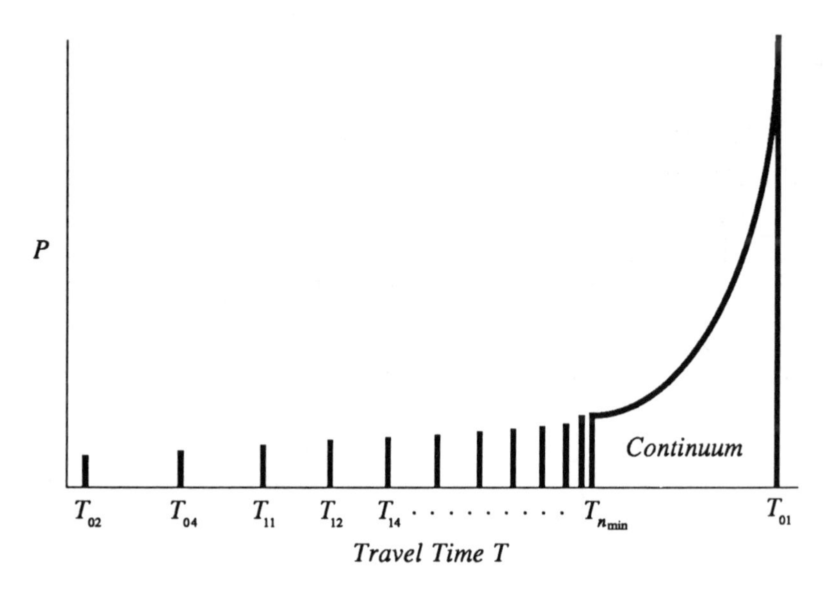

Figure 7.6 Temporal behavior of the acoustic field at a fixed source-receiver separation for a delta function pulse propagating through a symmetric, bilinear sound velocity profile.

Let us consider the temporal behavior of the field described by Eqs. (7.201) and (7.202) at a fixed range r from the source. We see that the original delta function pulse gives rise initially to a set of discrete arrivals associated with specific ray types having large grazing angles and executing a small number $n < n_{min}$ of cycles. As time progresses, the eigenray angles decrease while the corresponding number of ray cycles increases, and the temporal separation between the received pulses decreases until the signals meld into a continuum at $t = T_{n_{min}}$. The accumulation of these arrivals for $t > T_{n_{min}}$ also causes a pronounced increase in signal level, which peaks at the time $t = T_{01}$ of the last arrival traveling along the channel axis, and then abruptly drops to zero. This behavior is sketched in Fig. 7.6 and is characteristic of observed long-range SOFAR propagation [Ewing and Worzel, 1948; Urick, 1975; Brekhovskikh, 1980; Brekhovskikh and Lysanov, 1991]. In realistic ocean environments, the details of the arrival structure, including the onset of the continuum, depend upon the actual sound velocity profile as well as the influence of the surface and bottom boundaries. In addition, the classical shape of the continuum tends to emerge only at long ranges (hundreds of kilometers), since rays which cycle many times traverse large distances in the real ocean. We note that measured ray arrival times can be used to infer the sound velocity structure of the ocean using tomographic inversion methods, as described by Munk and Wunsch [1979], Cornuelle et al. [1985], Howe, Worcester, and Spindel [1987], and Desaubies, Tarantola, and Zinn-Justin [1990]. The relationship between sound velocity and temperature [e.g., Eq. (7.1)] can then be exploited to determine physical oceanographic behavior.

7.6.4 Normal Modes in the WKB Approximation

We now return to the general case of the single minimum velocity profile and recall that the Hankel transform representation of the sound field is given by [cf. Eqs. (7.114) and (7.115)]

$$p(r)=\int_0^\infty g(k_r)J_0(k_r r)k_r\,dk_r\ , \tag{7.203}$$

$$g(k_r)=i\frac{\left\{\begin{aligned}&\exp\left[i\int_z^{z_0}k_z\,dz\right]+\exp\left[i\left(\int_{z_0}^{z_B}+\int_z^{z_B}\right)k_z\,dz\right]\\&+\exp\left[i\left(\int_{z_S}^{z_0}+\int_{z_S}^{z}\right)k_z\,dz\right]+\exp\left[i\left(\int_z^{z_0}+2\int_{z_S}^{z}+2\int_{z_0}^{z_B}\right)k_z\,dz\right]\end{aligned}\right\}}{\sqrt{k_z k_{zi}}\left\{1-\exp\left[2i\int_{z_S}^{z_B}k_z\,dz\right]\right\}}. \tag{7.204}$$

The normal mode solution of Eqs. (7.203) and (7.204) proceeds in a manner analogous to the modal decomposition of the field for the Pekeris waveguide described in Sec. 6.3.2. As before, the mode sum is associated with pole-type singularities of the integrand [zeros of the denominator in Eq. (7.204)], whereas the ray series is developed from the expansion of the denominator (cf. Sec. 7.6.2). This intimate connection between the mathematical treatment of the denominator of $g(k_r)$ and the resulting wave types thus emerges as a central theme in the solution of wave propagation problems [Felsen, 1984].

We begin by determining the nature of any singularities of the integrand in Eqs. (7.203) and (7.204). Since $c(z)\to\infty$ as $z\to\pm\infty$, there are no spatial regions in which the field may become unbounded, and therefore the integrand has no branch points (cf. pp. 187-188). We then define $g(k_r)=N(k_r)/D(k_r)$, with

$$N(k_r)=\frac{i}{\sqrt{k_z k_{zi}}}\left\{\exp\left[i\int_z^{z_0} k_z\,dz\right]+\exp\left[i\left(\int_{z_0}^{z_B}+\int_z^{z_B}\right)k_z\,dz\right]\right.$$
$$\left.+\exp\left[i\left(\int_{z_S}^{z_0}+\int_{z_S}^{z}\right)k_z\,dz\right]+\exp\left[i\left(\int_z^{z_0}+2\int_{z_S}^{z}+2\int_{z_0}^{z_B}\right)k_z\,dz\right]\right\}, \quad (7.205)$$

$$D(k_r)=1-\exp\left[2i\int_{z_S}^{z_B} k_z\,dz\right] \;. \quad (7.206)$$

It is clear that the integrand has an infinite number of real poles $k_r=\pm k_n$ $(k_n>0)$ which satisfy the equation

$$D(k_r)=1-\exp\left[2i\int_{z_S}^{z_B} k_z(z)\,dz\right]=1-\exp\left[2i\int_{z_S}^{z_B}\sqrt{k^2(z)-k_r^2}\,dz\right]$$
$$=1-\exp\left[2i\int_{z_S}^{z_B}\sqrt{\omega^2/c^2(z)-k_r^2}\,dz\right]=0 \;. \quad (7.207)$$

Equation (7.207) is a natural extension to slowly varying media of the eigenvalue equation developed for a homogeneous fluid layer bounded by arbitrary horizontally stratified media [cf. Eq. (5.71)]. For the case at hand, the modes are trapped by the upper and lower turning points, z_S and z_B, rather than the bounding media. Equation (7.207) implies that

$$\exp\left[2i\int_{z_{Sn}}^{z_{Bn}} k_{zn}(z)\,dz\right]=1=\exp[2i(n-1)\pi],\quad n=1,2,3,\ldots \;, \quad (7.208)$$

and therefore

$$\int_{z_{Sn}}^{z_{Bn}} k_{zn}(z)\,dz=(n-1)\pi,\quad n=1,2,3,\ldots \;. \quad (7.209)$$

Having determined the eigenvalue equation governing the pole positions, we can now evaluate Eq. (7.203) using complex contour integration methods. We proceed with the steps described in Sec. 6.3.2, which are simplified due to the absence of any branch points or branch cuts. As a result, the entire solution can be expressed in terms of a residue series [cf. Eq. (6.76)]:

$$p(r) = \pi i \sum_{n=1}^{\infty} \frac{k_n N(k_n) H_0^{(1)}(k_n r)}{\partial D(k_r)/\partial k_r|_{k_r = k_n}} \quad . \tag{7.210}$$

Here $N(k_n)$ is given by [cf. Eq. (7.205)]

$$N = \frac{i}{\sqrt{k_{zn} k_{zin}}} \eta,$$

$$\eta = \exp\left[i \int_z^{z_0} k_{zn}\, dz\right] + \exp\left[i\left(\int_{z_0}^{z_{Bn}} + \int_z^{z_{Bn}}\right) k_{zn}\, dz\right]$$
$$+ \exp\left[i\left(\int_{z_{Sn}}^{z_0} + \int_{z_{Sn}}^{z}\right) k_{zn}\, dz\right] + \exp\left[i\left(\int_z^{z_0} + 2\int_{z_{Sn}}^{z} + 2\int_{z_0}^{z_{Bn}}\right) k_{zn}\, dz\right], \tag{7.211}$$

where z_{Sn} and z_{Bn} are the upper and lower turning points associated with the nth mode, and

$$k_{zn} = \sqrt{k^2(z) - k_n^2}, \ \ k_{zin} = \sqrt{k^2(z_0) - k_n^2} \quad . \tag{7.212}$$

The quantity η can be rewritten as

$$\eta = \exp\left[i \int_z^{z_0} k_{zn}\, dz\right] + \exp\left[i\left(\int_{z_{Sn}}^{z_0} + \int_{z_{Sn}}^{z}\right) k_{zn}\, dz\right]$$
$$+ \exp\left[2i \int_{z_{Sn}}^{z_{Bn}} k_{zn}\, dz\right]\left\{\exp\left[i\left(\int_{z_0}^{z_{Bn}} + \int_z^{z_{Bn}} - 2\int_{z_{Sn}}^{z_{Bn}}\right) k_{zn}\, dz\right]\right.$$
$$\left. + \exp\left[i\left(\int_z^{z_0} + 2\int_{z_{Sn}}^{z} + 2\int_{z_0}^{z_{Bn}} - 2\int_{z_{Sn}}^{z_{Bn}}\right) k_{zn}\, dz\right]\right\}$$
$$= \exp\left[i \int_z^{z_0} k_{zn}\, dz\right] + \exp\left[i\left(\int_{z_{Sn}}^{z_0} + \int_{z_{Sn}}^{z}\right) k_{zn}\, dz\right]$$
$$+ \exp\left[i\left(\int_{z_0}^{z_{Bn}} + \int_z^{z_{Bn}} - 2\int_{z_{Sn}}^{z_{Bn}}\right) k_{zn}\, dz\right]$$
$$+ \exp\left[i\left(\int_z^{z_0} + 2\int_{z_{Sn}}^{z} + 2\int_{z_0}^{z_{Bn}} - 2\int_{z_{Sn}}^{z_{Bn}}\right) k_{zn}\, dz\right] , \tag{7.213}$$

where we have used the eigenvalue equation [cf. Eq. (7.208)]. Combining the phase integrals in each of the last two terms of Eq. (7.213), we obtain

$$\eta = \exp\left[i\int_{z}^{z_0} k_{zn}\,dz\right] + \exp\left[i\left(\int_{z_{Sn}}^{z_0} + \int_{z_{Sn}}^{z}\right)k_{zn}\,dz\right]$$
$$+\exp\left[-i\left(\int_{z}^{z_0} + 2\int_{z_{Sn}}^{z}\right)k_{zn}\,dz\right] + \exp\left[-i\int_{z}^{z_0} k_{zn}\,dz\right]. \quad (7.214)$$

But the second term in Eq. (7.214) is given by

$$\exp\left[i\left(\int_{z_{Sn}}^{z_0} + \int_{z_{Sn}}^{z}\right)k_{zn}\,dz\right] = \exp\left[i\left(\int_{z}^{z_0} + 2\int_{z_{Sn}}^{z}\right)k_{zn}\,dz\right], \quad (7.215)$$

so that Eq. (7.214) becomes

$$\eta = \exp\left[i\int_{z}^{z_0} k_{zn}\,dz\right]\left\{1 + \exp\left[2i\int_{z_{Sn}}^{z} k_{zn}\,dz\right]\right\}$$
$$+\exp\left[-i\int_{z}^{z_0} k_{zn}\,dz\right]\left\{1 + \exp\left[-2i\int_{z_{Sn}}^{z} k_{zn}\,dz\right]\right\}$$
$$= \exp\left[i\left(\int_{z}^{z_0} + \int_{z_{Sn}}^{z}\right)k_{zn}\,dz\right]\left\{\exp\left[-i\int_{z_{Sn}}^{z} k_{zn}\,dz\right] + \exp\left[i\int_{z_{Sn}}^{z} k_{zn}\,dz\right]\right\}$$
$$+\exp\left[-i\left(\int_{z}^{z_0} + \int_{z_{Sn}}^{z}\right)k_{zn}\,dz\right]\left\{\exp\left[i\int_{z_{Sn}}^{z} k_{zn}\,dz\right] + \exp\left[-i\int_{z_{Sn}}^{z} k_{zn}\,dz\right]\right\}$$
$$= 4\cos\left[\left(\int_{z}^{z_0} + \int_{z_{Sn}}^{z}\right)k_{zn}\,dz\right]\cos\left[\int_{z_{Sn}}^{z} k_{zn}\,dz\right]. \quad (7.216)$$

Finally, using the relation

$$\left(\int_{z}^{z_0} + \int_{z_{Sn}}^{z}\right)k_{zn}\,dz = \int_{z_{Sn}}^{z_0} k_{zn}\,dz\ , \quad (7.217)$$

in Eq. (7.216), we obtain

$$\eta = 4\cos\left[\int_{z_{Sn}}^{z_0} k_{zn}\,dz\right]\cos\left[\int_{z_{Sn}}^{z} k_{zn}\,dz\right], \quad (7.218)$$

and therefore [cf. Eq. (7.211)]

$$N(k_n) = \frac{4i}{\sqrt{k_{zn}k_{zin}}} \cos\left[\int_{z_{Sn}}^{z_0} k_{zn}\, dz\right] \cos\left[\int_{z_{Sn}}^{z} k_{zn}\, dz\right] \quad . \qquad (7.219)$$

Turning now to the denominator of Eq. (7.210) [cf. Eq. (7.206)], we find that

$$\left.\frac{\partial D(k_r)}{\partial k_r}\right|_{k_r = k_n} = -2i \exp\left[2i \int_{z_S}^{z_B} k_z\, dz\right] \left.\int_{z_S}^{z_B} \frac{\partial k_z}{\partial k_r} dz\right|_{k_r = k_n}$$

$$= 2i \exp\left[2i \int_{z_S}^{z_B} k_z\, dz\right] \left.\int_{z_S}^{z_B} \frac{k_r}{k_z} dz\right|_{k_r = k_n}$$

$$= 2i \exp\left[2i \int_{z_{Sn}}^{z_{Bn}} k_{zn}\, dz\right] \int_{z_{Sn}}^{z_{Bn}} \frac{k_n}{k_{zn}} dz = 2i \int_{z_{Sn}}^{z_{Bn}} \frac{k_n}{k_{zn}} dz \quad , \qquad (7.220)$$

where we have again used the eigenvalue equation [cf. Eq. (7.208)]. But using Eq. (7.209), we can show that the modal cycle distance L_n, which equals the interference wavelength $\lambda_{n,n+1}$ between adjacent modes [cf. Eq. (5.170)], is given by

$$L_n = -\frac{2\pi}{dk_n/dn} = -2\pi \frac{dn}{dk_n} = 2 \int_{z_{Sn}}^{z_{Bn}} \frac{k_n}{k_{zn}} dz \quad , \qquad (7.221)$$

so that [cf. Eq. (6.81) for the Pekeris waveguide]

$$\left.\frac{\partial D(k_r)}{\partial k_r}\right|_{k_r = k_n} = iL_n \quad . \qquad (7.222)$$

Substituting Eqs. (7.219) and (7.222) into Eq. (7.210), we obtain the normal mode solution for the single minimum profile in the WKB approximation:

$$p(r) = 4i\pi \sum_{n=1}^{\infty} \frac{k_n}{L_n \sqrt{k_{zn}k_{zin}}} \cos\left[\int_{z_{Sn}}^{z_0} k_{zn}\, dz\right] \cos\left[\int_{z_{Sn}}^{z} k_{zn}\, dz\right] H_0^{(1)}(k_n r). \quad (7.223)$$

Equation (7.223) can also be derived using the method of eigenfunction expansion and the WKB approximation (cf. Prob. 7.9). Furthermore, it is straightforward to show that with a suitable limiting procedure, Eq. (7.223) reduces to the normal mode solution for a homogeneous fluid layer with rigid boundaries (cf. Prob. 7.10).

Finally, we recall that the ray series solution for the single minimum profile (cf. Sec. 7.6.1) produced a phase accumulation ΔS_{njr} in the lateral direction given by [cf. Eq. (7.104)]

$$\Delta S_{njr} = 2k_{nj}^2 \int_{z_{Snj}}^{z_{Bnj}} dz / k_{znj}(z) \ , \tag{7.224}$$

for the nth ray of type j. Equation (7.224) corresponds to a *ray cycle distance* Λ_{njr} between successive upper or lower turning points described by

$$\Lambda_{njr} = \Delta S_{njr} / k_{nj} = 2k_{nj} \int_{z_{Snj}}^{z_{Bnj}} dz / k_{znj}(z) \ . \tag{7.225}$$

Comparing Eq. (7.225) with Eq. (7.221), we see that the ray and modal cycle distances are identical. This result further corroborates the intimate relationship between rays and interfering modes, which has been mentioned previously for other cases on pp. 123-126, 155-156, and 183-184. Summarizing these examples, we conclude that the interference between adjacent, closely spaced modes gives rise to a ray having a cycle distance which consists, in general, of a geometrical acoustic (ray) part and a beam displacement. In situations where there are no penetrable reflecting boundaries, such as the single minimum profile, the beam displacement is zero and only the geometrical acoustic portion contributes.

7.7 Deficiencies of Classical Ray Theory

The beauty of classical ray theory lies in its ability to provide us with an intuitive, physical picture of the manner in which sound fields behave in inhomogeneous media. We gain this intuition first by using Snell's law to generate a ray diagram, which gives us a qualitative sense for the key propagation characteristics associated with a particular environment and source/receiver geometry. For example, we expect high sound levels at receiver positions where there is a high concentration of rays and low sound levels at locations where there is a low concentration of rays. Proceeding further, we can utilize the ray acoustic equations derived in this chapter to obtain quantitative estimates of acoustic field properties. However, the accuracy of these computations depends entirely on the extent to which a particular situation satisfies the underlying assumptions associated with ray theory. In cases where these assumptions are violated, the errors in the field calculations can be large, even infinite. To deal with these circumstances, classical ray theory is sometimes patched up by introducing wave-theoretic corrections as they are required. This approach, however, can lead to a complicated computational methodology which takes away from the elegance and simplicity of ray theory in its most basic form. Our purpose in this section is to identify the important deficiencies of classical ray theory and, in some cases, to indicate ways in which it can be improved. Extensive discussions of corrections to ray theory can be found in Sachs and Silbiger [1971], Sachs [1972], Brekhovskikh [1980], Boyles [1984], Tolstoy and Clay [1987], Brekhovskikh and Godin [1990], and Brekhovskikh and Lysanov [1991]. Many of these improvements are based on the fact that ray theory is the leading order term in an asymptotic expansion of the field in inverse powers of a characteristic wavenumber (cf. pp. 238-239). Corrections can therefore be obtained by calculating higher order terms in the expansion. Uniform asymptotic expansions [Bleistein and Handelsman, 1986] are also useful tools in developing improvements to pure ray theory. In addition, the concept of classical rays can be extended to encompass complex and diffracted rays, thereby providing both additional physical insight and more accurate computational capabilities [Ahluwalia and Keller, 1977; Choudhary and Felsen, 1978].

7.7.1 Intensity Predictions

Perhaps the most glaring deficiency of classical ray theory is its inability to predict acoustic intensity accurately under certain conditions. Specifically, ray theory may predict regions of zero sound pressure level, called *shadow zones*, whereas exact wave theory predicts a small, but finite, field in those cases. This behavior reflects the inability of ray theory to model the diffraction process which enables sound to leak into regions that appear inaccessible according to Snell's law. There are other features, called *perfect foci*, *caustics*, and *convergence zones*, in which ray theory predicts an infinite sound pressure level, whereas wave theory predicts a large, but finite, field. In the discussion that follows, we will present examples of each one of these phenomena.

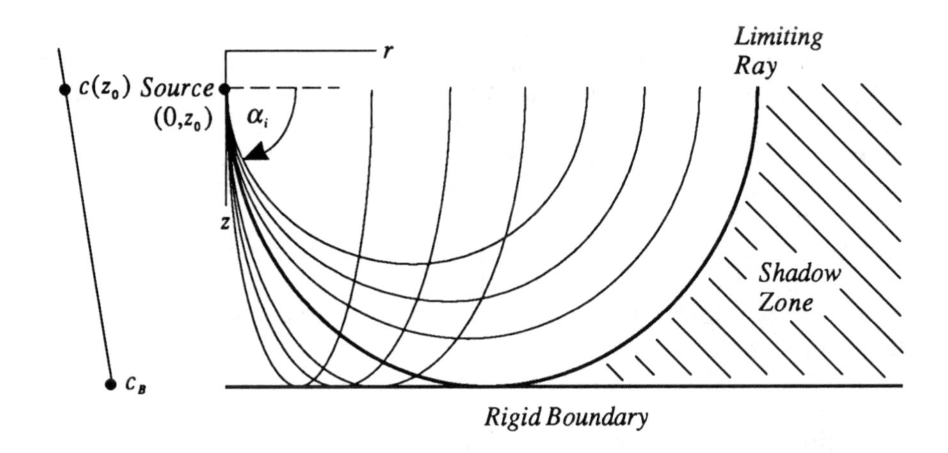

Figure 7.7 Formation of a shadow zone in a medium with a linear gradient profile bounded below by a rigid reflector. The trajectory of the limiting ray defines the shadow zone boundary.

Shadow Zones

Shadow zones arise in situations where ray theory, via Snell's law, does not allow sound to penetrate into certain regions. Because the ray intensity calculation is based on determining the concentration of rays in a particular region and no rays enter the shadow zone, the ray intensity is zero.

As an example, we consider a point source in a medium with a linear gradient profile bounded below by a rigid reflector, as shown in Fig. 7.7. This situation is an approximation to the realistic case of a source in the vicinity of the water-bottom interface in the deep ocean (cf. Fig. 7.1). Because of the upward-refracting nature of the profile, rays leaving the source at small grazing angles α_i have turning points which lie above the reflecting boundary. As the grazing angle increases, however, the turning point approaches the interface, until at the angle α_{ic}, given by

$$\alpha_{ic} = \cos^{-1}\left[\frac{c(z_0)}{c_B}\right] , \tag{7.226}$$

the turning point just touches the boundary. For steeper angles, $\alpha_i > \alpha_{ic}$, the rays are specularly reflected from the interface. In this manner, a shadow zone is created having a boundary defined by the trajectory of the *limiting ray* with initial angle α_{ic}.

Perfect Foci, Caustics, and Convergence Zones

In some circumstances, ray theory predicts infinite values of the acoustic field. These situations can be identified by examining the expression for the ray acoustic intensity [cf. Eq. (7.82)],

$$\frac{I}{I_0} = \frac{\sin\theta_i}{\Re\cos\theta(\partial\Re/\partial\theta_i)} \quad , \tag{7.227}$$

and determining the conditions under which the denominator of Eq. (7.227) vanishes. It is clear that the intensity becomes infinite when

$$\cos\theta = 0 \;\Rightarrow\; \theta = \pi/2 \quad , \tag{7.228}$$

i.e., at turning points. The failure of ray theory at turning points is not surprising, since we have already encountered similar behavior in the two-dimensional plane wave ray/WKB solution [cf. Eq. (7.26)]. But Eq. (7.227) indicates that the ray intensity also becomes infinite when

$$\partial\Re/\partial\theta_i = 0 \quad , \tag{7.229}$$

i.e., when an infinitesimal change in the initial angle of a ray produces no change in the horizontal range traversed by the ray. In other words, Eq. (7.229) describes positions for which two or more rays, differing in their initial angles by infinitesimal amounts, will have trajectories that cross one another. Intuitively, we expect infinite intensity at these locations, since the ray intensity calculation is based on determining the energy per unit cross-sectional area of an infinitesimal ray bundle. The detailed nature of the solutions of Eq. (7.229) depends on the specific profile under consideration and the source location.

Perfect Foci: Tolstoy and Clay [1987] describe a situation for which the solutions of Eq. (7.229) are perfect foci. These can occur when all of the derivatives of the sound velocity profile are continuous on the sound channel axis, for example, in the case

$$c(z) = c_0 \cosh bz \quad , \tag{7.230}$$

where c_0 and b are constants. Then, for a source on the sound channel axis (cf. Fig. 7.8), the horizontal range traversed by a ray is given by

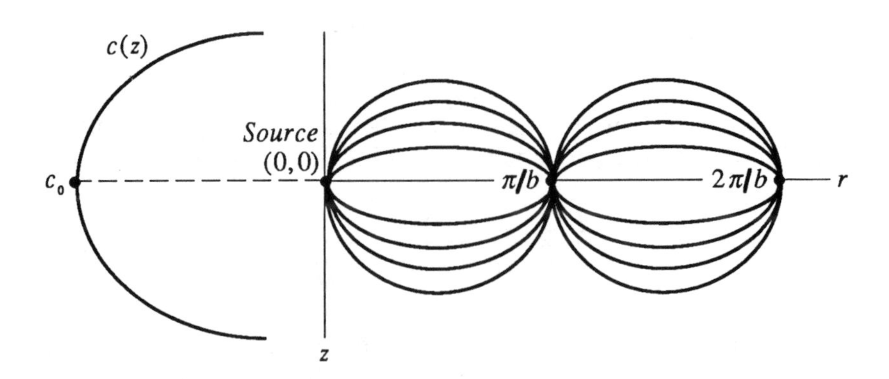

Figure 7.8 Formation of perfect foci in a medium with the sound velocity profile $c(z)=c_0 \cosh bz$.

$$\Re=\int_0^z \frac{\sin\theta_i}{\sqrt{\text{sech}^2 bz-\sin^2\theta_i}}dz=\frac{1}{b}\sin^{-1}\left[\frac{\sin\theta_i}{\sqrt{1-\sin^2\theta_i}}\sinh bz\right]$$

$$=\frac{1}{b}\sin^{-1}\left[\tan\theta_i \sinh bz\right] \ , \tag{7.231}$$

where we have used Eq. (7.57) and a table of integrals [Gradshteyn and Ryzhik, 1965]. Rewriting Eq. (7.231),

$$\sin b\Re = \tan\theta_i \sinh bz \ , \tag{7.232}$$

we find that Eq. (7.229) becomes

$$\frac{\partial\Re}{\partial\theta_i}=\frac{\sinh bz}{b\cos^2\theta_i \cos b\Re}=0 \ . \tag{7.233}$$

Equation (7.233) has the solution $z=0$, which, when substituted into Eq. (7.231), implies that

$$\Re=\frac{1}{b}\sin^{-1}[0]=\frac{n\pi}{b}, \ \ n=1,2,3,\ldots \ . \tag{7.234}$$

Thus, we obtain a sequence of equally spaced, perfect foci, with spacing $\Delta\Re=\pi/b$, as shown in Fig. 7.8.

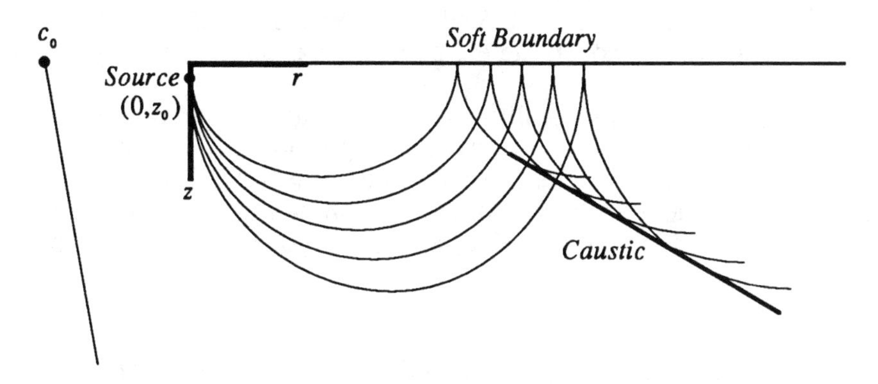

Figure 7.9 Formation of a caustic in a medium with a linear gradient profile bounded above by a soft reflector.

Caustics and Convergence Zones: We saw above that very specific conditions are required in order to generate perfect foci. Caustics are imperfect foci which correspond to a locus of intersecting rays and can therefore occur under a much broader range of conditions. Caustics may also define a shadow zone boundary. As an example, we consider a point source in a medium with a linear gradient profile bounded above by a soft reflector, as shown in Fig. 7.9. This situation is an approximation to the realistic case of a source in the vicinity of the ocean surface in a deep Arctic environment. Here the isothermal character of the water allows the hydrostatic pressure term in Eq. (7.1) to dominate, and a linear gradient profile is created [cf. Eq. (7.58)].

Assuming that the source is sufficiently close to the surface to make the approximation $z_0 \approx 0$, we trace only down-going rays which have passed through one turning point and undergone one surface reflection. It is clear that adjacent ray paths intersect before encountering the second turning point, thereby producing a caustic. Using Eq. (7.63), we can determine the horizontal range traversed by these rays:

$$\begin{aligned}\Re &= 3\frac{c_0}{g}\cot\theta_i - \frac{c_0}{g}\frac{\cos\theta}{\sin\theta_i} \\ &= 3\frac{c_0}{g}\cot\theta_i - \frac{c_0}{g\sin\theta_i}\sqrt{1-\left(\frac{c(z)\sin\theta_i}{c_0}\right)^2} \quad . \qquad (7.235)\end{aligned}$$

It is left as an exercise for the reader (cf. Prob. 7.11) to show that Eqs. (7.235) and (7.229) constitute the parametric equations of a hyperbola [Officer, 1958].

Under certain conditions in realistic ocean environments, regions of high acoustic intensity occur periodically in range due to repeated caustic formation. Specifically, for a shallow source and receiver in a medium having the profile shown in Fig. 7.1, it can be shown that the spacing of these convergence zones is about 60 km (cf. Prob. 7.12).

7.7.2 Reflection Coefficients at Discontinuous Interfaces and Turning Points

As discussed in Sec. 7.3, ray theory is a no reflection theory in the sense that it does not take into account the continuous reflection process which occurs as a ray propagates through a continuously varying medium. In addition, classical ray theory does not include the reflection effects which occur at discontinuous boundaries, such as the water-bottom interface. These deficiencies can be remedied, at least partially, by taking advantage of the local plane wave-like behavior of a ray and incorporating appropriate plane wave reflection coefficients as required. For example, when a ray strikes a boundary at an angle θ, its contribution to the classical ray series can be multiplied by the reflection coefficient $R(\theta)$ associated with that boundary. The reflection process which occurs at a turning point fits naturally into this framework, since pure ray theory treats a turning point as a rigid boundary, with $R=1$ (cf. Secs. 7.6.1 and 7.6.2). In fact, wave-theoretic calculations show that a ray undergoes a $-\pi/2$ phase shift when it passes through a turning point or touches a caustic [Brekhovskikh, 1980; Brekhovskikh and Godin, 1990], and therefore $R=\exp(-i\pi/2)$ is a more accurate reflection coefficient for those circumstances.

Summarizing then, we can improve the accuracy of the classical ray series [cf. Eq. (7.102)] by introducing plane wave reflection coefficients as needed:

$$p(r)=\sum_{n=0}^{\infty} A_n' e^{iS_n} \quad \text{with} \quad A_n' = R_S^j R_B^m e^{-il\pi/2} A_n \ . \tag{7.236}$$

Here, for the nth ray, A_n and S_n are the classical ray amplitude and phase, R_S and R_B are the surface and bottom reflection coefficients, j

and m are the total number of surface and bottom reflections, and l is the total number of turning point and caustic interactions.

PROBLEMS

7.1 Starting with Eq. (7.13), derive the ray/WKB criteria in Eqs. (7.27) and (7.28).

7.2 For typical values of the sound velocity gradient in the water column and the seabed, derive the frequency requirements which must be met in order for the ray/WKB criterion in Eq. (7.31) to be satisfied.

7.3 Determine the leading order behavior of the time-averaged energy flux density in the z direction for the ray/WKB solution in Eq. (7.26).

7.4 Consider the following sound velocity profile for which the ray equations have closed form solutions:

$$\frac{1}{c^2(z)} = \frac{1}{c_0^2} - \delta z \ , \tag{7.237}$$

where c_0 and δ are constants. This profile is of particular interest because it also generates exact, Airy function solutions to the Helmholtz equation [Brekhovskikh, 1980]. Show that the ray acoustic expressions for horizontal range $\Re$ and travel time T for the $1/c^2$-linear profile are given by [cf. Eqs. (7.57) and (7.68)]

$$\Re = \frac{2}{\delta}\left(\frac{\sin\theta_i}{c_i}\right)^2 \left(\cot\theta_i - \cot\theta\right) \ , \tag{7.238}$$

$$T = \frac{2}{3\delta}\left(\frac{\sin\theta_i}{c_i}\right)^3 \left[\cot^3\theta_i - \cot^3\theta + 3\left(\cot\theta_i - \cot\theta\right)\right] \ . \tag{7.239}$$

7.5 Consider the following sound velocity profile for which the ray equations have closed form solutions [Kawahara and Frisk, 1984]:

$$c^3(z) = c_0^3 + \beta z \ , \tag{7.240}$$

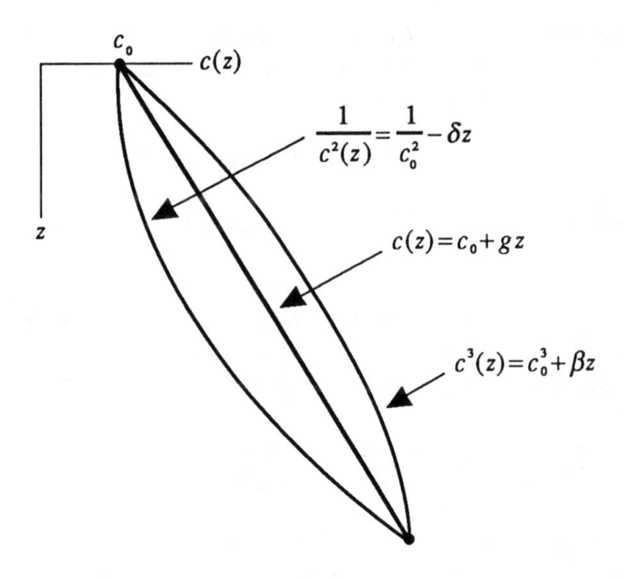

Figure 7.10 Comparison of three canonical sound velocity profiles with fixed endpoints and positive δ, g, and β.

where c_0 and β are constants. This profile is useful because it provides an excellent fit to experimentally measured velocities and velocity gradients in bottom sediments. It is compared with two other canonical profiles, namely the c-linear and $1/c^2$-linear profiles, in Fig. 7.10 [cf. Eqs. (7.58) and (7.237)]. Show that the ray acoustic expressions for horizontal range $\Re$ and travel time T for the c^3-linear profile are given by [cf. Eqs. (7.57) and (7.68)]

$$\Re = \frac{1}{\beta}\left(\frac{c_i}{\sin\theta_i}\right)^3 \left[\cos\theta_i\left(2+\sin^2\theta_i\right) - \cos\theta\left(2+\sin^2\theta\right)\right], \quad (7.241)$$

$$T = \frac{3}{\beta}\left(\frac{c_i}{\sin\theta_i}\right)^2 \left(\cos\theta_i - \cos\theta\right) \ . \quad (7.242)$$

Hint: You may need the following integral in the calculation:

$$\int \frac{x^3}{\sqrt{a^2-x^2}}\,dx = -\frac{1}{3}\sqrt{a^2-x^2}\left(x^2+2a^2\right) \ . \quad (7.243)$$

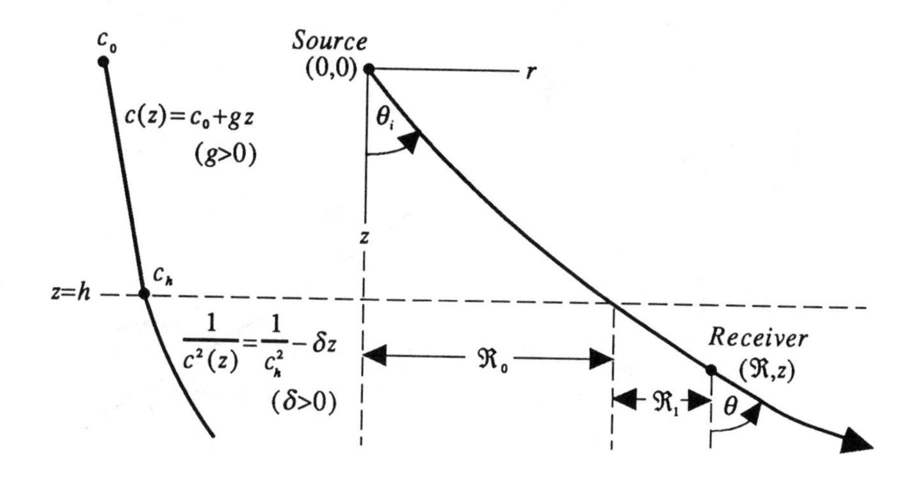

Figure 7.11 Sound velocity profile and ray trajectory for Prob. 7.6.

7.6 Consider the sound velocity profile shown in Fig. 7.11. Suppose that a ray leaves the source at an angle θ_i in the linear gradient region and arrives at the receiver at an angle θ in the $1/c^2$-linear region. Assume that c is continuous at $z = h$.

(a) What is the horizontal range $\Re = \Re_0 + \Re_1$ traversed by the ray in terms of θ_i, θ, h, and the parameters of the profiles?

(b) Consider a ray which leaves the source at an initial angle $\theta_i = \pi/4$ and reaches a turning point in the $1/c^2$-linear region. Derive the value of δ, in terms of c_0, g, and h, for which the horizontal ranges traversed in the two regions are equal $(\Re_0 = \Re_1)$.

(c) What is the minimum value of g, in terms of c_0 and h, for which the ray with an inital angle $\theta_i = \pi/4$ turns before entering the $1/c^2$-linear region?

7.7 Consider the sound velocity profile shown in Fig. 7.12. Suppose that a ray leaves the source at an angle θ_i in the linear gradient region and arrives at the receiver in the constant velocity region.

(a) What is the horizontal range $\Re = \Re_0 + \Re_1$ traversed by the ray in terms of θ_i, g, h, z, c_0, c_h, and c_1?

(b) Calculate the ray acoustic intensity at the receiver in terms of θ_i, c_0, c_h, c_1, $\Re_0$, and $\Re_1$ (do not leave derivatives of $\Re_0$ and $\Re_1$).

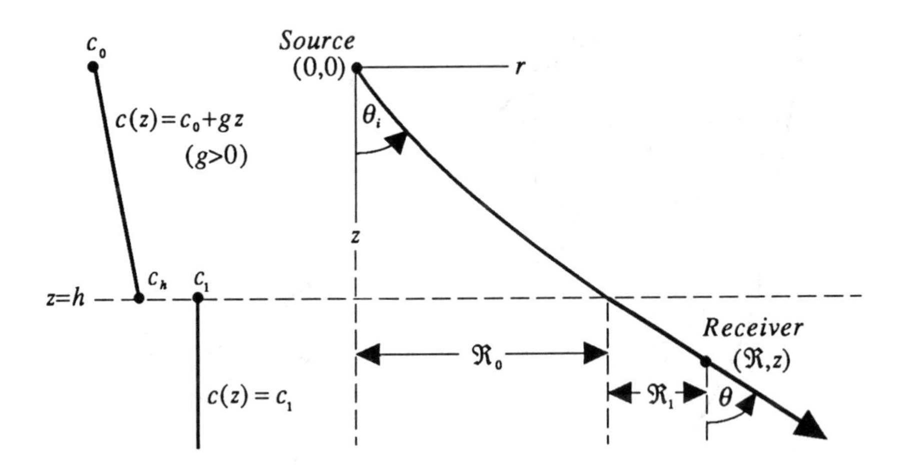

Figure 7.12 Sound velocity profile and ray trajectory for Prob. 7.7.

Assume a source of unit intensity at unit distance and neglect reflection at the discontinuous interface.

(c) Show that the result for the intensity in part (b) has the correct behavior in the limit of a homogeneous medium ($c_0 = c_h = c_1$).

7.8 For the case of the single minimum velocity profile, derive the expression for the depth-dependent Green's function in Eq. (7.114).

7.9 Using the method of eigenfunction expansion and the WKB approximation, derive the normal mode result for the single minimum profile in Eq. (7.223).

7.10 Using a suitable limiting procedure, show that the normal mode solution for the single minimum profile in Eq. (7.223) reduces to the mode solution for a homogeneous fluid layer with rigid boundaries (cf. Prob. 5.5).

7.11 Show that Eqs. (7.235) and (7.229), which describe the caustic shown in Fig. 7.9, constitute the parametric equations of a hyperbola.

7.12 For a shallow source and receiver in a medium having the velocity profile shown in Fig. 7.1, show that the convergence zone spacing is approximately 60 km.

Appendix A. Adiabatic Mode/WKB Theory for a Slowly Varying, Range-Dependent Environment

In the previous chapters, we focused on sound propagation in horizontally stratified media, for which the acoustic properties are a function of only one dimension. In general, however, the acoustic characteristics of the ocean and seabed vary in three dimensions. As a step toward accommodating this full multidimensional variability, we consider an intermediate, two dimensional case in which the acoustic properties vary in both range and depth. For simplicity, we assume that the variation in range is much slower than the variation in depth, which is a good approximation to the real ocean environment. This assumption enables us to synthesize the sound field from modal solutions in depth and WKB solutions in range. The early work on this adiabatic mode/WKB solution was performed by Pierce [1965], Milder [1969], and Weinberg and Burridge [1974].

Our development for the range-dependent case parallels the derivation for the range-independent case described in Sec. 5.2. We assume a point source with cylindrical coordinates $\mathbf{r}_0 = (0,0,z_0)$ in a fluid medium with density $\rho(r,z)$ and sound velocity $c(r,z)$ (cf. Fig. 5.1). Then the inhomogeneous, time-independent wave equation is given by [cf. Eq. (5.1)]

$$\rho(r,z)\nabla \bullet \left[\frac{1}{\rho(r,z)} \nabla p(\mathbf{r}) \right] + k^2(r,z)p(\mathbf{r}) = -4\pi \frac{\delta(r)}{r} \delta(\theta)\delta(z-z_0), \quad \text{(A.1)}$$

where $\mathbf{r} = (r,\theta,z)$ and the wavenumber $k(r,z) = \omega/c(r,z)$. Equation (A.1) can be rewritten as

$$\nabla^2 p(\mathbf{r}) + \rho(r,z)\nabla\left[\frac{1}{\rho(r,z)} \right] \bullet \nabla p(\mathbf{r}) + k^2(r,z)p(\mathbf{r})$$
$$= -4\pi \frac{\delta(r)}{r} \delta(\theta)\delta(z-z_0) \ , \quad \text{(A.2)}$$

where

$$\nabla = \mathbf{i}_r \frac{\partial}{\partial r} + \mathbf{i}_\theta \frac{1}{r}\frac{\partial}{\partial \theta} + \mathbf{i}_z \frac{\partial}{\partial z},$$

$$\nabla^2 = \frac{1}{r}\frac{\partial}{\partial r}\left(r\frac{\partial}{\partial r}\right) + \frac{1}{r^2}\frac{\partial^2}{\partial \theta^2} + \frac{\partial^2}{\partial z^2} , \qquad \text{(A.3)}$$

and $\mathbf{i}_r$, $\mathbf{i}_\theta$, and $\mathbf{i}_z$ are unit vectors in the r, θ, and z directions, respectively. Positioning the source at $\mathbf{r}_0 = (0,0,z_0)$ and assuming cylindrical symmetry in the acoustic properties imply cylindrical symmetry in the field about the z axis. Therefore, the θ derivative terms in Eq. (A.2) vanish, and we obtain

$$\frac{1}{r}\frac{\partial}{\partial r}\left[r\frac{\partial p(r,z)}{\partial r}\right] + \frac{\partial^2 p(r,z)}{\partial z^2} + \rho(r,z)\left\{\frac{\partial}{\partial z}\left[\frac{1}{\rho(r,z)}\right]\frac{\partial p(r,z)}{\partial z}\right.$$
$$\left. + \frac{\partial}{\partial r}\left[\frac{1}{\rho(r,z)}\right]\frac{\partial p(r,z)}{\partial r}\right\} + k^2(r,z)p(r,z) = -4\pi\frac{\delta(r)}{r}\delta(\theta)\delta(z-z_0). \text{(A.4)}$$

Integrating both sides of Eq. (A.4) with respect to θ from 0 to 2π, we find that

$$\frac{1}{r}\frac{\partial}{\partial r}\left[r\frac{\partial p(r,z)}{\partial r}\right] + \frac{\partial^2 p(r,z)}{\partial z^2} + \rho(r,z)\left\{\frac{\partial}{\partial z}\left[\frac{1}{\rho(r,z)}\right]\frac{\partial p(r,z)}{\partial z}\right.$$
$$\left. + \frac{\partial}{\partial r}\left[\frac{1}{\rho(r,z)}\right]\frac{\partial p(r,z)}{\partial r}\right\} + k^2(r,z)p(r,z) = -2\frac{\delta(r)}{r}\delta(z-z_0) . \qquad \text{(A.5)}$$

We then propose a partially separable solution of Eq. (A.5) in terms of the eigenfunctions $R_n(r)$ and $u_n(r,z)$, having amplitudes $a_n(z_0)$ [cf. Eq. (5.6)]:

$$p(r,z) = \sum_n a_n(z_0)u_n(r,z)R_n(r) . \qquad \text{(A.6)}$$

Here we assume that at range r, $u_n(r,z)$ satisfies a local eigenvalue equation of Sturm-Liouville type [cf. Eqs. (5.7) and (5.9)],

$$\frac{1}{\rho(r,z)}\frac{d^2u_n(r,z)}{dz^2}+\frac{d}{dz}\left[\frac{1}{\rho(r,z)}\right]\frac{du_n(r,z)}{dz}+\frac{k_{zn}^2}{\rho(r,z)}u_n(r,z)=0 \quad , \qquad (A.7)$$

where $k_{zn}^2 = k^2(r,z) - k_n^2(r)$, and k_n and k_{zn} are the local discrete values of the horizontal and vertical wavenumbers, respectively, associated with the eigenfunction $u_n(r,z)$. Furthermore, we assume that Eq. (A.7), combined with the prescribed boundary conditions at range r, constitute a proper Sturm-Liouville problem (cf. pp. 114-115). Substituting Eq. (A.6) into Eq. (A.5), we obtain

$$\sum_n a_n(z_0)\left\{\frac{u_n(r,z)}{r}\frac{d}{dr}\left[r\frac{dR_n(r)}{dr}\right]+\frac{R_n(r)}{r}\frac{\partial}{\partial r}\left[r\frac{\partial u_n(r,z)}{\partial r}\right]\right.$$

$$+2\frac{\partial u_n(r,z)}{\partial r}\frac{dR_n(r)}{dr}+R_n(r)\left[\frac{\partial^2 u_n(r,z)}{\partial z^2}+\rho(r,z)\frac{\partial}{\partial z}\left(\frac{1}{\rho(r,z)}\right)\frac{\partial u_n(r,z)}{\partial z}\right.$$

$$\left.+\rho(r,z)\frac{\partial}{\partial r}\left(\frac{1}{\rho(r,z)}\right)\frac{\partial u_n(r,z)}{\partial r}\right]+\rho(r,z)u_n(r,z)\frac{\partial}{\partial r}\left(\frac{1}{\rho(r,z)}\right)\frac{dR_n(r)}{dr}$$

$$\left.+k^2(r,z)u_n(r,z)R_n(r)\right\}=\sum_n\left\{-2\frac{\delta(r)}{r}\frac{1}{\rho(r,z_0)}u_n^*(r,z_0)u_n(r,z)\right\} \quad ,$$

$$(A.8)$$

where we have applied the closure relation [cf. Eq. (5.12)] to the right-hand side of Eq. (A.5):

$$\sum_n \frac{1}{\rho(r,z_0)}u_n^*(r,z_0)u_n(r,z)=\delta(z-z_0) \quad . \qquad (A.9)$$

The slowly varying nature of the medium in the radial direction implies that we can neglect terms containing

$$\frac{\partial u_n(r,z)}{\partial r}, \quad \frac{\partial}{\partial r}\left(\frac{1}{\rho(r,z)}\right) \quad , \qquad (A.10)$$

on the left-hand side of Eq. (A.8). In addition, we can apply the delta function identity,

$$\delta(x-x_0)f(x)=\delta(x-x_0)f(x_0) \quad , \qquad (A.11)$$

to the right-hand side of Eq. (A.8), which then becomes

$$\sum_n a_n(z_0)\left\{\frac{u_n(r,z)}{r}\frac{d}{dr}\left[r\frac{dR_n(r)}{dr}\right]\right.$$

$$\left.+R_n(r)\underbrace{\left[\frac{d^2u_n(r,z)}{dz^2}+\rho(r,z)\frac{d}{dz}\left(\frac{1}{\rho(r,z)}\right)\frac{du_n(r,z)}{dz}+k^2(r,z)u_n(r,z)\right]}_{Term\,I}\right\}$$

$$=\sum_n\left\{-2\frac{\delta(r)}{r}\frac{1}{\rho(0,z_0)}u_n^*(0,z_0)u_n(r,z)\right\} \quad . \tag{A.12}$$

From the eigenvalue equation [cf. Eq. (A.7)], it is clear that

$$Term\,I = k_n^2(r)u_n(r,z) \quad , \tag{A.13}$$

and therefore Eq. (A.12) becomes

$$\sum_n a_n(z_0)u_n(r,z)\left\{\frac{1}{r}\frac{d}{dr}\left[r\frac{dR_n(r)}{dr}\right]+k_n^2(r)R_n(r)\right\}$$

$$=\sum_n\left\{-2\frac{\delta(r)}{r}\frac{1}{\rho(0,z_0)}u_n^*(0,z_0)u_n(r,z)\right\} \quad . \tag{A.14}$$

Since Eq. (A.14) must hold true for arbitrary values of r, z, and z_0, we separately equate functions of r, z, and z_0 on the left- and right-hand sides of the equation and obtain

$$a_n(z_0)=u_n^*(0,z_0)/\rho(0,z_0) \quad , \tag{A.15}$$

and

$$\frac{1}{r}\frac{d}{dr}\left[r\frac{dR_n(r)}{dr}\right]+k_n^2(r)R_n(r)=-2\frac{\delta(r)}{r} \quad . \tag{A.16}$$

Making the transformation,

$$P_n(r)=\sqrt{r}R_n(r) \quad , \tag{A.17}$$

in Eq. (A.16), we obtain

$$\frac{1}{\sqrt{r}}\left\{\frac{d^2P_n(r)}{dr^2}+k_n^2(r)\left[1+\frac{1}{4k_n^2(r)r^2}\right]P_n(r)\right\}=-2\frac{\delta(r)}{r} \quad , \qquad (A.18)$$

which, for $k_n(r)r >> 1$, can be approximated by

$$\frac{d^2P_n(r)}{dr^2}+k_n^2(r)P_n(r)\sim 0 \quad . \qquad (A.19)$$

The slowly varying nature of the medium in the radial direction implies that we can solve Eq. (A.19) using the WKB approximation [cf. Eq. (7.19)]:

$$P_n(r)\sim\frac{A}{\sqrt{k_n(r)}}\exp\left[i\int_{r_i}^{r}k_n(r)\,dr\right] \quad , \qquad (A.20)$$

where A and r_i are arbitrary constants, and we have used the radiation condition to eliminate the solution which propagates radially inward. Using Eq. (A.17), it is clear that the solution of Eq. (A.16) for $k_n(r)r >> 1$ is

$$R_n(r)\sim\frac{A}{\sqrt{k_n(r)r}}\exp\left[i\int_{r_i}^{r}k_n(r)\,dr\right] \quad . \qquad (A.21)$$

The arbitrary constants in Eq. (A.21) can be determined by requiring that, in the limit of a range-independent medium $\left[k_n(r)=k_n=Const.\right]$, $R_n(r)$ must reduce to the asymptotic form for the two-dimensional free-space Green's function in Eq. (5.19):

$$R_n(r)=i\pi H_0^{(1)}(k_nr)\sim\sqrt{\frac{2\pi}{k_nr}}e^{i\pi/4}e^{ik_nr} \quad . \qquad (A.22)$$

Comparing Eqs. (A.21) and (A.22), we find that

$$A=\sqrt{2\pi}e^{i\pi/4}, \quad r_i=0 \quad , \qquad (A.23)$$

so that Eq. (A.21) becomes

$$R_n(r) \sim \sqrt{\frac{2\pi}{k_n(r)r}} e^{i\pi/4} \exp\left[i \int_0^r k_n(r)\, dr \right] . \tag{A.24}$$

Combining Eqs. (A.6), (A.15), and (A.24), we obtain the complete adiabatic mode/WKB solution of Eq. (A.5) [cf. Eq. (5.22)]:

$$p(r,z) \sim \frac{\sqrt{2\pi} e^{i\pi/4}}{\rho(0,z_0)} \sum_{n=1}^{\infty} u_n^*(0,z_0) u_n(r,z) \frac{\exp\left[i \int_0^r k_n(r)\, dr \right]}{\sqrt{k_n(r)r}} . \tag{A.25}$$

This approximate result for a slowly varying, range-dependent environment has several interesting features. It indicates that the range-varying vertical eigenfunctions can be synthesized from the local eigenfunctions at the source and the receiver. These eigenfunctions thus adapt individually to the local source/receiver environments and do not exchange energy among themselves through any mode coupling process. The range-dependent phase accumulation associated with the WKB phase integral in Eq. (A.25) also reflects the adaptation of the modal field to the local environment. In addition, Eq. (A.25) allows modes which are excited at the source to be cut off and subsequently re-excited as they propagate along the waveguide, depending upon the local acoustic properties. However, the detailed process of transition into or out of the modal continuum is not properly modeled (cf. p. 163). In conclusion, we expect that the adiabatic mode/WKB solution will be most useful in situations where a fixed number of modes propagate without cutoff along a waveguide whose properties do not change abruptly with range.

Appendix B. Sound Propagation in a Wedge with Impenetrable Boundaries

The determination of the sound field in a wedge with impenetrable boundaries is of interest to us for several reasons. First, this environment is a good first approximation to realistic ocean conditions found in continental slope and shelf regions. Second, a combination of Fourier transform, eigenfunction expansion, and endpoint methods enables us to obtain an exact, closed form solution to this problem for arbitrary wedge angles (cf. Prob. 4.6). Third, we can use this exact solution as a benchmark for evaluating approximate solutions for laterally varying environments [Graves et al., 1975]. The early work on this problem was performed by Bradley and Hudimac [1970].

We consider a point source with cylindrical coordinates $\mathbf{r}_0 = (r_0, \theta_0, 0)$ situated in a wedge of infinite extent in the z direction, as shown in Fig. B.1 (note that this cylindrical coordinate system has a different orientation from that used in previous examples). The interior portion of the wedge is a homogeneous fluid medium with constant density ρ and sound velocity c. The wedge boundaries are assumed to be soft at $\theta = 0$ and hard at $\theta = \beta$. Then the inhomogeneous, time-independent wave equation is given by [cf. Eq. (5.1)]

$$\nabla^2 p(\mathbf{r}) + k^2 p(\mathbf{r}) = -4\pi \frac{\delta(r - r_0)}{r} \delta(\theta - \theta_0)\delta(z) \ , \qquad \text{(B.1)}$$

where $\mathbf{r} = (r, \theta, z)$ and the wavenumber $k = \omega/c$. Using Eq. (A.3), we can rewrite Eq. (B.1) as

$$\frac{1}{r}\frac{\partial}{\partial r}\left[r \frac{\partial p(r,\theta,z)}{\partial r}\right] + \frac{1}{r^2}\frac{\partial^2 p(r,\theta,z)}{\partial \theta^2} + \frac{\partial^2 p(r,\theta,z)}{\partial z^2}$$

$$+ k^2 p(r,\theta,z) = -4\pi \frac{\delta(r - r_0)}{r} \delta(\theta - \theta_0)\delta(z) \ . \qquad \text{(B.2)}$$

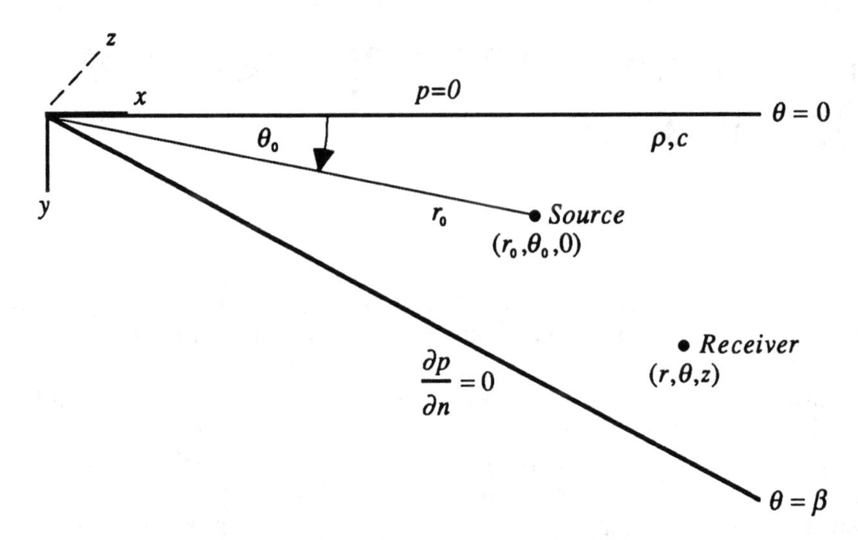

Figure B.1 Geometry for calculation of the acoustic field due to a point source in a homogeneous fluid wedge with impenetrable boundaries.

Applying the one-dimensional inverse Fourier transform operator,

$$I.F.T.\{\bullet\} = \frac{1}{\sqrt{2\pi}} \int_{-\infty}^{\infty} \{\bullet\} e^{-ik_z z}\, dz \ , \tag{B.3}$$

to both sides of Eq. (B.2), we obtain

$$\frac{1}{r}\frac{\partial}{\partial r}\left[r \frac{\partial g(r,\theta;k_z)}{\partial r} \right] + \frac{1}{r^2}\frac{\partial^2 g(r,\theta;k_z)}{\partial \theta^2} + k_r^2 g(r,\theta;k_z)$$
$$= -2\sqrt{2\pi}\frac{\delta(r-r_0)}{r}\delta(\theta-\theta_0) \ , \tag{B.4}$$

where the radial wavenumber $k_r = \sqrt{k^2 - k_z^2}$, and $p(r,\theta;z)$ and $g(r,\theta;k_z)$ are conjugate Fourier transform pairs (cf. pp. 85-86)

$$p(r,\theta;z) = \frac{1}{\sqrt{2\pi}} \int_{-\infty}^{\infty} g(r,\theta;k_z) e^{ik_z z}\, dk_z \ , \tag{B.5}$$

$$g(r,\theta;k_z) = \frac{1}{\sqrt{2\pi}} \int_{-\infty}^{\infty} p(r,\theta;z) e^{-ik_z z}\, dz \ . \tag{B.6}$$

In order to solve Eq. (B.4), we assume an eigenfunction expansion of the form

$$g(r,\theta;k_z)=\sum_n a_n(\theta_0)f_n(\theta)R_n(r) \ , \tag{B.7}$$

which, when substituted into Eq. (B.4), yields

$$\sum_n a_n(\theta_0)\left\{\frac{f_n(\theta)}{r}\frac{d}{dr}\left[r\frac{dR_n(r)}{dr}\right]+\frac{R_n(r)}{r^2}\frac{d^2f_n(\theta)}{d\theta^2}+k_r^2f_n(\theta)R_n(r)\right\}$$
$$=\sum_n\left\{-2\sqrt{2\pi}\frac{\delta(r-r_0)}{r}f_n(\theta_0)f_n(\theta)\right\} \ . \tag{B.8}$$

Since Eq. (B.8) must hold true for arbitrary values of r, θ, and θ_0, we separately equate functions of r, θ, and θ_0 on the left- and right-hand sides of the equation and obtain

$$a_n(\theta_0)=f_n(\theta_0) \ , \tag{B.9}$$

$$\frac{f_n(\theta)}{r}\frac{d}{dr}\left[r\frac{dR_n(r)}{dr}\right]+\frac{R_n(r)}{r^2}\frac{d^2f_n(\theta)}{d\theta^2}+k_r^2f_n(\theta)R_n(r)$$
$$=-2\sqrt{2\pi}\frac{\delta(r-r_0)}{r}f_n(\theta) \ . \tag{B.10}$$

Dividing both sides of Eq. (B.10) by $f_n(\theta)R_n(r)/r^2$, we obtain

$$\frac{r}{R_n(r)}\frac{d}{dr}\left[r\frac{dR_n(r)}{dr}\right]+\frac{1}{f_n(\theta)}\frac{d^2f_n(\theta)}{d\theta^2}+k_r^2r^2=-2\sqrt{2\pi}\frac{r\delta(r-r_0)}{R_n(r)} \ , \tag{B.11}$$

which can be rewritten as

$$\frac{r}{R_n(r)}\frac{d}{dr}\left[r\frac{dR_n(r)}{dr}\right]+k_r^2r^2+2\sqrt{2\pi}\frac{r\delta(r-r_0)}{R_n(r)}=-\frac{1}{f_n(\theta)}\frac{d^2f_n(\theta)}{d\theta^2} \ . \tag{B.12}$$

Here we have grouped all functions of r on the left-hand side of the equation and all functions of θ on the right-hand side. Equation

(B.12) must hold true for arbitrary values of r and θ, and therefore the right- and left-hand sides of Eq. (B.12) must separately be equal to a separation constant which we call α^2:

$$\frac{d^2 f_n(\theta)}{d\theta^2} + \alpha^2 f_n(\theta) = 0 \ , \tag{B.13}$$

$$\frac{1}{r}\frac{d}{dr}\left[r\frac{dR_n(r)}{dr}\right] + \left(k_r^2 - \frac{\alpha^2}{r^2}\right)R_n(r) = -2\sqrt{2\pi}\,\frac{\delta(r-r_0)}{r} \ . \tag{B.14}$$

Equation (B.13) has the solution

$$f_n(\theta) = A_n \sin\alpha\theta + B_n \cos\alpha\theta \ , \tag{B.15}$$

where A_n and B_n are arbitrary constants. Imposing the soft boundary condition at $\theta = 0$ in Eq. (B.15),

$$p(r,\theta,z)\big|_{\theta=0} = 0 \ \Rightarrow \ f_n(\theta)\big|_{\theta=0} = 0 \ , \tag{B.16}$$

we find that $B_n = 0$, while the hard boundary condition at $\theta = \beta$,

$$\left.\frac{\partial p(r,\theta,z)}{\partial n}\right|_{\theta=\beta} = \left.\frac{\partial p(r,\theta,z)}{\partial \theta}\right|_{\theta=\beta} = 0 \ \Rightarrow \ \left.\frac{df_n(\theta)}{d\theta}\right|_{\theta=\beta} = 0 \ , \tag{B.17}$$

yields

$$f_n(\theta) = A_n \sin\alpha_n\theta, \quad \alpha_n = \frac{(n-1/2)\pi}{\beta}, \quad n = 1,2,3,\ldots \ . \tag{B.18}$$

It is clear that Eq. (B.13), combined with the boundary conditions in Eqs. (B.16) and (B.17), constitute a proper Sturm-Liouville problem (cf. pp. 113-115). Normalizing the eigenfunctions,

$$1 = \int_0^\beta A_n^2 \sin^2\alpha_n\theta\, d\theta = \frac{1}{2}\int_0^\beta A_n^2 (1 - \cos 2\alpha_n\theta)\, d\theta = \frac{A_n^2\beta}{2} \ , \tag{B.19}$$

we obtain

$$f_n(\theta) = \sqrt{\frac{2}{\beta}} \sin \alpha_n \theta, \quad \alpha_n = \frac{(n - 1/2)\pi}{\beta}, \quad n = 1,2,3,\ldots \ . \tag{B.20}$$

We can solve the radial Eq. (B.14) using the endpoint method. First we rewrite Eq. (B.14) as

$$\frac{d^2 R_n(r)}{dr^2} + \frac{1}{r}\frac{dR_n(r)}{dr} + \left(k_r^2 - \frac{\alpha_n^2}{r^2}\right) R_n(r) = -2\sqrt{2\pi}\frac{\delta(r - r_0)}{r} \ , \tag{B.21}$$

and divide both sides of Eq. (B.21) by k_r^2, to obtain

$$\frac{1}{k_r^2}\frac{d^2 R_n(r)}{dr^2} + \frac{1}{k_r^2 r}\frac{dR_n(r)}{dr} + \left(1 - \frac{\alpha_n^2}{k_r^2 r^2}\right) R_n(r) = -2\sqrt{2\pi}\frac{\delta(r - r_0)}{k_r^2 r} . \tag{B.22}$$

Changing variables, $x = k_r r$, we find that Eq. (B.22) becomes

$$\frac{d^2 R_n(x)}{dx^2} + \frac{1}{x}\frac{dR_n(x)}{dx} + \left(1 - \frac{\alpha_n^2}{x^2}\right) R_n(x) = -2\sqrt{2\pi}\frac{\delta(x - x_0)}{x} \ , \tag{B.23}$$

where $x_0 = k_r r_0$, and we have used the delta function identity ($a = Const.$):

$$\delta[a(x - x_0)] = \frac{1}{a}\delta(x - x_0) \ . \tag{B.24}$$

We can rewrite Eq. (B.23) in standard form [cf. Eq. (4.15)] for application of the endpoint method:

$$\frac{x}{2\sqrt{2\pi}}\left[\frac{d^2 R_n(x)}{dx^2} + \frac{1}{x}\frac{dR_n(x)}{dx} + \left(1 - \frac{\alpha_n^2}{x^2}\right) R_n(x)\right] = -\delta(x - x_0) \ . \tag{B.25}$$

Equation (B.25) has the solution [cf. Eqs. (4.83)-(4.85)]

$$R_n(x, x_0) = \begin{cases} -\dfrac{1}{C} R_n^{(0)}(x) R_n^{(\infty)}(x_0), & 0 \le x \le x_0, \\ -\dfrac{1}{C} R_n^{(0)}(x_0) R_n^{(\infty)}(x), & x_0 \le x < \infty \end{cases} \ , \tag{B.26}$$

where

$$C = \frac{x_0}{2\sqrt{2\pi}} W\left[R_n^{(0)}(x_0),\ R_n^{(\infty)}(x_0)\right]$$

$$= \frac{x_0}{2\sqrt{2\pi}}\left[R_n^{(0)}(x_0)R_n^{(\infty)\prime}(x_0) - R_n^{(0)\prime}(x_0)R_n^{(\infty)}(x_0)\right] . \quad \text{(B.27)}$$

Here $R_n^{(0)}(x)$ and $R_n^{(\infty)}(x)$ satisfy the homogeneous version of Eq. (B.25),

$$\frac{d^2R_n(x)}{dx^2} + \frac{1}{x}\frac{dR_n(x)}{dx} + \left(1 - \frac{\alpha_n^2}{x^2}\right)R_n(x) = 0 , \quad \text{(B.28)}$$

and appropriate boundary conditions at $x = 0$ and $x = \infty$, respectively. But Eq. (B.28) is simply Bessel's equation of order α_n [cf. Eq. (2.23)]. We require that $R_n^{(0)}(x)$ be finite (specifically, zero) at $x = 0$ and that $R_n^{(\infty)}(x)$ satisfy the Sommerfeld radiation condition at $x = \infty$, and thus obtain

$$R_n^{(0)}(x) = J_{\alpha_n}(x),\ R_n^{(\infty)}(x) = H_{\alpha_n}^{(1)}(x),\ C = \frac{i}{\sqrt{2}\pi^{3/2}} , \quad \text{(B.29)}$$

where J_{α_n} is a Bessel function and $H_{\alpha_n}^{(1)}$ is a Hankel function of the first kind, and we have used Eq. (5.205). Equation (B.26) then becomes

$$R_n(r) = i\sqrt{2}\pi^{3/2}J_{\alpha_n}(k_r r_<)H_{\alpha_n}^{(1)}(k_r r_>),\ 0 \le r < \infty , \quad \text{(B.30)}$$

where $r_<$ ($r_>$) is the lesser (greater) of r and r_0. Combining Eqs. (B.5), (B.7), (B.9), (B.20), and (B.30), we obtain the complete solution for a point source in a homogeneous fluid wedge with impenetrable boundaries:

$$p = \frac{2i\pi}{\beta}\sum_{n=1}^{\infty}\sin\alpha_n\theta_0 \sin\alpha_n\theta \int_{-\infty}^{\infty} J_{\alpha_n}(k_r r_<)H_{\alpha_n}^{(1)}(k_r r_>)e^{ik_z z}\, dk_z . \quad \text{(B.31)}$$

References

Abramowitz, Milton, and Irene A. Stegun, *Handbook of Mathematical Functions* (National Bureau of Standards Applied Mathematics Series 55, Washington, DC, 1964).

Achenbach, J. D., *Wave Propagation in Elastic Solids* (North-Holland, Amsterdam, 1975).

Ahluwalia, Daljit S., and Joseph B. Keller, "Exact and Asymptotic Representations of the Sound Field in a Stratified Ocean," in *Wave Propagation and Underwater Acoustics*, edited by Joseph B. Keller and John S. Papadakis (Springer-Verlag, Berlin, 1977).

Akal, Tuncay, and John M. Berkson, Eds., *Ocean Seismo-Acoustics: Low Frequency Underwater Acoustics* (Plenum, New York, 1986).

Aki, Keiiti, and Paul G. Richards, *Quantitative Seismology Theory and Methods* (Freeman, San Francisco, 1980).

Ammicht, Egbert, and David C. Stickler, "Uniform Asymptotic Evaluation of the Continuous Spectrum Contribution for a Stratified Ocean," J. Acoust. Soc. Am. **76**, 186-191 (1984).

Arfken, George, *Mathematical Methods for Physicists* (Academic, Orlando, 1985).

Bartlett, John, and Justin Kaplan, General Ed., *Familiar Quotations* (Little, Brown and Company, Boston, 1992).

Bleistein, Norman, and Richard A. Handelsman, *Asymptotic Expansions of Integrals* (Dover, New York, 1986).

Boyles, C. Allan, *Acoustic Waveguides: Applications to Oceanic Science* (Wiley, New York, 1984).

Bradley, David L., and Albert A. Hudimac, "The Propagation of Sound in a Wedge Shaped Shallow Water Duct," Rep. NOLTR 70-235, Naval Ordnance Laboratory, Silver Spring, MD (1970).

Brekhovskikh, Leonid M., *Waves in Layered Media* (Academic, New York, 1980).

Brekhovskikh, Leonid M., and O. A. Godin, *Acoustics of Layered Media I* (Springer-Verlag, Berlin, 1990).

Brekhovskikh, Leonid M., and Yury Lysanov, *Fundamentals of Ocean Acoustics* (Springer-Verlag, Berlin, 1991).

Brigham, E. Oran, *The Fast Fourier Transform* (Prentice-Hall, Englewood Cliffs, NJ, 1988).

Bucker, Homer P., "Sound Propagation in a Channel with Lossy Boundaries," J. Acoust. Soc. Am. **48**, 1187-1194 (1970).

Burdic, William S., *Underwater Acoustic System Analysis* (Prentice-Hall, Englewood Cliffs, NJ, 1984).

Chen, Chen-Tung, and Frank J. Millero, "Speed of Sound in Seawater at High Pressures," J. Acoust. Soc. Am. **62**, 1129-1135 (1977).

Choudhary, S., and Leopold B. Felsen, "Asymptotic Theory of Ducted Propagation," J. Acoust. Soc. Am. **63**, 661-666 (1978).

Churchill, Ruel V., and James W. Brown, *Complex Variables and Applications* (McGraw-Hill, New York, 1984).

Clay, Clarence S., and Herman Medwin, *Acoustical Oceanography* (Wiley, New York, 1977).

Cornuelle, B., C. Wunsch, D. Behringer, T. Birdsall, M. Brown, R. Heinmiller, R. Knox, K. Metzger, W. Munk, J. Spiesberger, R. Spindel, D. Webb, and P. Worcester, "Tomographic Maps of the Ocean Mesoscale-1: Pure Acoustics," J. Phys. Oceanogr. **15**, 133-152 (1985).

Cornyn, John J., "GRASS: A Digital Computer Ray Tracing and Transmission Loss Prediction System," Reps. 7621 and 7642, Naval Research Laboratory, Washington, DC (1973).

Del Grosso, Vincent A., "New Equation for the Speed of Sound in Natural Waters (with Comparisons to Other Equations)," J. Acoust. Soc. Am. **56**, 1084-1091 (1974).

DeSanto, John A., Ed., *Ocean Acoustics* (Springer-Verlag, Berlin, 1979).

DeSanto, John A., "Boundary Value Problems for Scalar Waves- Part I," Rep. CWP-078, Colorado School of Mines, Golden, Colorado (1989).

Desaubies, Yves, Albert Tarantola, and J. Zinn-Justin, Eds., *Oceanographic and Geophysical Tomography* (Elsevier, Amsterdam, 1990).

DiNapoli, Frederick R., and Roy L. Deavenport, "Theoretical and Numerical Green's Function Field Solution in a Plane Multilayered System," J. Acoust. Soc. Am. **67**, 92-105 (1980).

Dostoyevsky, Fyodor, *The Brothers Karamazov* (Random House, New York, 1977), translated by Constance Garnett, originally published 1879-1880.

Eisler, Thomas J., "An Introduction to Green's Functions," Rep. 69-6, Institute of Ocean Science and Engineering, Catholic University of America, Washington, DC (1969).

Erdelyi, A., *Asymptotic Expansions* (Dover, New York, 1956).

Ewing, W. Maurice, and John L. Worzel "Long-Range Sound Transmission," in *Propagation of Sound in the Ocean*, Geol. Soc. Am. Mem. **27** (1948).

Ewing, W. Maurice, Wenceslas S. Jardetzky, and Frank Press, *Elastic Waves in Layered Media* (McGraw-Hill, New York, 1957).

Felsen, Leopold B., and Nathan Marcuvitz, *Radiation and Scattering of Waves* (Prentice-Hall, Englewood Cliffs, NJ, 1973).

Felsen, Leopold B., "Progressing and Oscillatory Waves for Hybrid Synthesis of Source Excited Propagation and Diffraction," IEEE Trans. Ant. Propag. **AP-32**, 775-796 (1984).

Feynman, Richard P., and A. R. Hibbs, *Quantum Mechanics and Path Integrals* (McGraw-Hill, New York, 1965).

Flatte, Stanley M., Ed., *Sound Transmission Through a Fluctuating Ocean* (Cambridge University, New York, 1979).

Frisk, George V., and Herbert Überall, "Creeping Waves and Lateral Waves in Acoustic Scattering by Large Elastic Cylinders," J. Acoust. Soc. Am. **59**, 46-54 (1976).

Frisk, George V., Alan V. Oppenheim, and David R. Martinez, "A Technique for Measuring the Plane Wave Reflection Coefficient of the Ocean Bottom," J. Acoust. Soc. Am. **68**, 602-612 (1980).

Frisk, George V., and James F. Lynch, "Shallow Water Waveguide Characterization Using the Hankel Transform," J. Acoust. Soc. Am. **76**, 205-216 (1984).

Frisk, George V., James F. Lynch, and Subramaniam D. Rajan, "Determination of Compressional Wave Speed Profiles Using Modal Inverse Techniques in a Range-Dependent Environment in Nantucket Sound," J. Acoust. Soc. Am. **86**, 1928-1939 (1989).

Frisk, George V., "Inverse Methods in Ocean Bottom Acoustics," in *Oceanographic and Geophysical Tomography*, edited by Yves Desaubies, Albert Tarantola, and J. Zinn-Justin (Elsevier, Amsterdam, 1990).

Gradshteyn, I. S., and I. M. Ryzhik, *Tables of Integrals, Series, and Products* (Academic, New York, 1965).

Grant, F. S., and G. F. West, *Interpretation Theory in Applied Geophysics* (McGraw-Hill, New York, 1965).

Graves, R. D., Anton Nagl, Herbert Überall, and G. L. Zarur, "Range-Dependent Normal Modes in Underwater Sound Propagation: Application to the Wedge-Shaped Ocean," J. Acoust. Soc. Am. **58**, 1171-1177 (1975).

Hamilton, Edwin L., "Geoacoustic Modeling of the Sea Floor," J. Acoust. Soc. Am. **68**, 1313-1340 (1980).

Hovem, Jens M., Michael D. Richardson, and Robert D. Stoll, Eds., *Shear Waves in Marine Sediments* (Kluwer, Dordrecht, 1991).

Howe, Bruce M., Peter F. Worcester, and Robert C. Spindel, "Ocean Acoustic Tomography: Mesoscale Velocity," J. Geophys. Res. **92**, 3785-3805 (1987).

Jackson, John David, *Classical Electrodynamics* (Wiley, New York, 1975).

Jensen, Finn B., William A. Kuperman, Michael B. Porter, and Henrik Schmidt, *Computational Ocean Acoustics* (American Institute of Physics, New York, 1993).

Jones, R. Michael, J. P. Riley, and T. M. Georges, "HARPO: A Versatile Three-Dimensional Hamiltonian Ray-Tracing Program for Acoustic Waves in an Ocean with Irregular Bottom," National Oceanic and Atmospheric Administration Environmental Research Laboratories Report, Wave Propagation Laboratory, Boulder, CO (1986).

Kamel, Aladin H., and Leopold B. Felsen, "On the Ray Equivalent of a Group of Modes," J. Acoust. Soc. Am. **71**, 1445-1452 (1982).

Kawahara, Hiroshi, and George V. Frisk, "A Canonical Ocean Bottom Sound Velocity Profile," J. Acoust. Soc. Am. **76**, 1254-1257 (1984).

Keller, Joseph B., and John S. Papadakis, Eds., *Wave Propagation and Underwater Acoustics* (Springer-Verlag, Berlin, 1977).

Kinsler, Lawrence E., Austin R. Frey, Alan B. Coppens, and James V. Sanders, *Fundamentals of Acoustics* (Wiley, New York, 1982).

Kuperman, William A., and Finn B. Jensen, Eds., *Bottom-Interacting Ocean Acoustics* (Plenum, New York, 1980).

Landau, Lev Davidovich, and E. M. Lifshitz, *Fluid Mechanics* (Pergamon, New York, 1987).

Leroy, C. C., "Development of Simple Equations for Accurate and More Realistic Calculation of the Speed of Sound in Seawater," J. Acoust. Soc. Am. **46**, 216-226 (1969).

Lindsay, R. Bruce, "The Story of Acoustics," J. Acoust. Soc. Am. **39**, 629-644 (1966).

Lindsay, R. Bruce, Ed., *Acoustics: Historical and Philosophical Development* (Dowden, Hutchinson and Ross, Stroudsburg, PA, 1972).

Mackenzie, Kenneth V., "Discussion of Sea Water Sound Speed Determinations," J. Acoust. Soc. Am. **70**, 801-806 (1981).

Mackenzie, Kenneth V., "Nine Term Equation for Sound Speed in the Oceans," J. Acoust. Soc. Am. **70**, 807-812 (1981).

Merzbacher, Eugen, *Quantum Mechanics* (Wiley, New York, 1970).

Milder, D. M., "Ray and Wave Invariants for SOFAR Channel Propagation," J. Acoust. Soc. Am. **46**, 1259-1263 (1969).

Morse, Philip M., and Herman Feshbach, *Methods of Theoretical Physics* (McGraw-Hill, New York, 1953).

Morse, Philip M., and K. Uno Ingard, *Theoretical Acoustics* (Princeton University, Princeton, NJ, 1986).

Munk, Walter, and Carl Wunsch, "Ocean Acoustic Tomography: A Scheme for Large Scale Monitoring," Deep-Sea Research **26**, 123-161 (1979).

Newhall, Arthur E., James F. Lynch, C. S. Chiu, and J. R. Daugherty, "Improvements in Three-Dimensional Raytracing Codes for Underwater Acoustics," in *Computational Acoustics - Volume I*, edited by D. Lee, A. Cakmak, and R. Vichnevetsky (Elsevier, Amsterdam, 1990).

Officer, Charles B., *Introduction to the Theory of Sound Transmission* (McGraw-Hill, New York, 1958).

Ogilvy, J. A., *Theory of Wave Scattering from Random Rough Surfaces* (Adam Hilger, Bristol, 1991).

Pekeris, C. L., "Theory of Propagation of Explosive Sound in Shallow Water," in *Propagation of Sound in the Ocean*, Geol. Soc. Am. Mem. **27** (1948).

Pierce, Allan D., "Extension of the Method of Normal Modes to Sound Propagation in an Almost-Stratified Medium," J. Acoust. Soc. Am. **37**, 19-27 (1965).

Pierce, Allan D., *Acoustics: An Introduction to Its Physical Principles and Applications* (American Institute of Physics, New York, 1989).

Porter, Michael B., and Edward L. Reiss, "A Numerical Method for Bottom Interacting Ocean Acoustic Normal Modes," J. Acoust. Soc. Am. **77**, 1760-1767 (1985).

Rajan, Subramaniam D., George V. Frisk, and James F. Lynch, "On the Determination of Modal Attenuation Coefficients and Compressional Wave Attenuation Profiles in a Range-Dependent Environment in Nantucket Sound," IEEE J. Ocean. Eng. **OE-17**, 118-128 (1992).

Rayleigh, Lord (John William Strutt), *The Theory of Sound* (Dover, New York, 1945), originally published 1877-78.

Roberts, B. G., "Horizontal-Gradient Acoustical Ray Trace Program - TRIMAIN," Rep. 7827, Naval Research Laboratory, Washington, DC (1974).

Sachs, David A., and A. Silbiger, "Focusing and Refraction of Harmonic Sound and Transient Pulses in Stratified Media," J. Acoust. Soc. Am. **49**, 824-840 (1971).

Sachs, David A., "Sound Propagation in Shadow Zones," J. Acoust. Soc. Am. **51**, 1091-1097 (1972).

Schiff, Leonard I., *Quantum Mechanics* (McGraw-Hill, New York, 1968).

Schmidt, Henrik, and Finn B. Jensen, "A Full Wave Solution for Propagation in Multilayered Viscoelastic Media with Application to Gaussian Beam Reflection at Fluid-Solid Interfaces," J. Acoust. Soc. Am. **77**, 813-825 (1985).

Sommerfeld, Arnold, *Partial Differential Equations in Physics* (Academic, New York, 1949).

Spiesberger, John L., and Kurt Metzger "A New Algorithm for Sound Speed in Seawater," J. Acoust. Soc. Am. **89**, 2677-2688 (1991).

Spitzer, Lyman, Jr. (Director, Sonar Analysis Group), *Physics of Sound in the Sea* (Department of the Navy, Headquarters Naval Material Command, Washington, DC, 1969), originally published as Summary Technical Report of Division 6, NDRC Vol. 8, Washington, DC, 1946.

Stickler, David C., "Reflected and Lateral Waves for the Sommerfeld Model," J. Acoust. Soc. Am. **60**, 1061-1070 (1976).

Stickler, David C., "Negative Bottom Loss, Critical-Angle Shift, and the Interpretation of the Bottom Reflection Coefficient," J. Acoust. Soc. Am. **61**, 707-710 (1977).

Stickler, David C., and Egbert Ammicht, "Uniform Asymptotic Evaluation of the Continuous Spectrum Contribution for the Pekeris Model," J. Acoust. Soc. Am. **67**, 2018-2024 (1980).

Stoll, Robert D., "Marine Sediment Acoustics," J. Acoust. Soc. Am. **77**, 1789-1799 (1985).

Stoll, Robert D., *Sediment Acoustics* (Springer-Verlag, New York, 1989).

Stratton, Julius A., *Electromagnetic Theory* (McGraw-Hill, New York, 1941).

Tamir, T., and Leopold B. Felsen, "On Lateral Waves in Slab Configurations and Their Relation to Other Wave Types," IEEE Trans. Ant. Propag. **AP-13**, 410-422 (1965).

Tindle, Christopher T., A. P. Stamp, and K. M. Guthrie, "Virtual Modes and the Surface Boundary Condition in Underwater Acoustics," J. Sound Vib. **49**, 231-240 (1976).

Tindle, Christopher T., and David E. Weston, "Connection of Acoustic Beam Displacement, Cycle Distances, and Attenuations for Rays and Normal Modes," J. Acoust. Soc. Am. **67**, 1614-1622 (1980).

Tindle, Christopher T., "Ray Calculations with Beam Displacement," J. Acoust. Soc. Am. **73**, 1581-1586 (1983).

Tolstoy, Ivan, *Wave Propagation* (McGraw-Hill, New York, 1973).

Tolstoy, Ivan, and Clarence S. Clay, *Ocean Acoustics: Theory and Experiment in Underwater Sound* (American Institute of Physics, New York, 1987).

Tyras, George, *Radiation and Propagation of Electromagnetic Waves* (Academic, New York, 1969).

Überall, Herbert, "Surface Waves in Acoustics," in *Physical Acoustics*, W. P. Mason and R. N. Thurston, Eds. (Academic, New York, 1973), Vol. 10.

Urick, Robert J., *Principles of Underwater Sound* (McGraw-Hill, New York, 1975).

Weinberg, Henry, and Robert Burridge, "Horizontal Ray Theory for Ocean Acoustics," J. Acoust. Soc. Am. **55**, 63-79 (1974).

Williams, Arthur O., Jr., "Pseudoresonances and Virtual Modes in Underwater Sound Propagation," J. Acoust. Soc. Am. **64**, 1487-1491 (1978).

Wilson, W., "Speed of Sound in Sea Water as a Function of Temperature, Pressure, and Salinity," J. Acoust. Soc. Am. **32**, 641-644 (1960).

Index

A

Absorption (intrinsic) 54-55, 166-167
Acoustic Brewster angle 44-45
Acoustic wave equation 5-6
Adiabatic condition 13
Adiabatic mode/WKB theory 269-274
Air-water interface 42-43
Airy function 265
Airy phase 154, 166
Ampere's law 8
Angle of intromission 44-45
Asymptotic analysis
  evaluation of Hankel transform result for point source in single minimum profile *see* Ray theory
  leaky mode decomposition 141-145, 158-162
  modified method of stationary phase *see* Stationary phase method
  point source in homogeneous fluid layer with penetrable boundaries 176-186
  relationship to ray theory 238-239
  spherical wave reflection from homogeneous fluid half-space *see* Spherical wave reflection
Asymptotic expansion 96, 238-239
  uniform 96, 104, 163, 259
Asymptotic form 22, 238
Attenuation
  intrinsic 54-55, 166-167
  modal 166-167

B

Beam displacement 156, 258
Bessel function 22-23, 89, 280
  orthonormality relation 199
  recursion relation 199
  Wronskian 164
Bessel's equation 22, 116, 280
Bimodal interference pattern 123-124
Bottom acoustic parameters 43-44, 206
  sand 50
  silty-clay 45
Bottom loss 54
Boundary condition 32
  Cauchy *see* impedance
  Dirichlet *see* soft
  hard 32-33, 62, 70, 72-73, 80-83, 114, 116-117, 174, 275
  homogeneous 62-63, 83, 171, 231-232
  impedance 33, 62-63, 114, 171
  mixed *see* impedance
  Neumann *see* hard
  periodic 114
  pressure-release *see* soft
  rigid *see* hard
  soft 32, 62, 68, 72-73, 82-83, 106, 114, 116-117, 130, 145-147, 174, 181-182, 187, 275
  Sommerfeld radiation condition 33-34, 38, 61, 83, 86, 109, 114, 161, 280
Boundary wave *see* Lateral wave
Bound states 110
Branch cuts
  for Pekeris waveguide 187-191, 193-198
  for spherical wave reflection from homogeneous fluid half-space 92-104, 108-109, 181
Branch line integral *see* Branch cuts
Brewster angle (acoustic) 44-45

C

Carrier frequency 121

Cauchy boundary condition *see* Boundary condition
Cauchy's residue theorem 143-145, 190-191
Causality 108
Caustic 263-264
Characteristic equation *see* Eigenvalue equation
Characteristic impedance 25
Closure relation 114-115, 271
Coherent propagation 54
Completeness relation 114-115
Complex eigenvalues *see* Eigenvalue equation
Conservation of energy 53-54, 221-222
Continuity conditions 38
Continuity equation 13
Continuous spectrum *see* Modal spectrum, Normal modes, *and* Pekeris waveguide
Continuum
  modal *see* Modal continuum, Modal spectrum, Normal modes, *and* Pekeris waveguide
  SOFAR arrivals 251-253
Convergence zone 263-264
Coupling, mode 274
Critical angle 45-51, 56-57, 97-104, 146-147, 163, 181
Critical point 92-93, 101, 142-143
Cutoff frequency
  modal 120, 150-152, 274
Cycle distance 125-126, 156, 184, 192, 200, 257-258
Cylindrical wave 23

D

Delta function *see* Dirac delta function
Density
  in air 42-43
  in ocean 42-43
  in seabed 45, 50
Depth-dependent Green's function *see* Green's function
Diffracted field 104, 179-181, 186, 259
Dipole source (Lloyd mirror) 77-80
Dirac delta function 60-61, 63-64, 66, 85-86, 107, 111-116, 169-170, 250-251, 271, 279
Dirichlet boundary condition *see* Boundary condition
Discrete spectrum *see* Modal spectrum, Normal modes, *and* Pekeris waveguide
Dispersion
  geometric 119
  intrinsic 119
  modal 122, 153

E

Eigenangle 125
Eigenfunction 113-114, *see also* Normal modes
Eigenfunction expansion method *see* Normal modes
Eigenray 226-231, 236-238
Eigenvalue equation 113, 115, *see also* Normal modes
  for homogeneous fluid layer bounded by horizontally stratified media 129-130
  for homogeneous fluid layer with soft top and higher velocity bottom 146-147, 188-189
  for homogeneous fluid layer with soft top and lower velocity bottom 130-131
  for single minimum profile 254
Eikonal equation 214, 216, 224, 238
EJP cuts 93, 188-189
Elastic scalar potential 12-13
Elastic vector potential 12-13
Elastic wave equation 11-13
Electromagnetic scalar potential 9-10
Electromagnetic vector potential 9-10
Electromagnetic wave equation 7-11
Electromagnetics 7-11
Endpoint method *see* Green's function construction

Energy conservation 53-54, 221-222
Energy flux density 26-27, *see also* Intensity
  plane wave 27
  spherical wave 27-28
Equation of state 13
Euler's equation 13
Evanescent field 50

F

Factorial function 23
Faraday's law 8
Fast Field program 172
Fast Fourier transform 172
Fermat's principle 214
Fourier-Bessel series 199
Fourier decomposition
  for one-dimensional wave equation 18-19
  for Schrodinger wave equation 28
  for three-dimensional wave equation 19-20
Fourier transform method *see* Green's function construction
Fourier transform operator 5, 86, 276
Free-space Green's function *see* Green's function

G

Gamma function 23
Gauge transformation 10
Gauss' law 8
Geometrical acoustics approximation 96, 104, 156, 179-180, 184, 186, 238-239, 258
Green's function 61
  depth-dependent 86, 170-171, 190, 231-233
  free-space 61, 68, 105-108, 164, 273
  general solution of boundary value problem with sources 64-67
  properties 61-64
  recipe for solving problems 67-68
Green's function construction
  eigenfunction expansion method *see* Normal modes
  endpoint method 83-84
  Fourier transform method 85-86
  Hankel transform method *see* Hankel transform
  hybrid Fourier transform/endpoint method 85-89
  hybrid Fourier transform/normal mode/endpoint method
    wedge with impenetrable boundaries 275-280
  hybrid Hankel transform/endpoint method *see* Hankel transform
  method of images 68, 174-176, 226
    homogeneous fluid layer with impenetrable boundaries 80-83
    Lloyd mirror effect 73-80
    plane boundary with Dirichlet conditions 68-70
    plane boundary with Neumann conditions 70-72
    quadrant with Dirichlet conditions 72-73
    wedge with pressure-release boundaries 105-106
  normal mode method *see* Normal modes
Green's theorem 65
Ground wave propagation 154
Group velocity 120-121
  modal 121-122, 152-154

H

Hankel function 22-23, 90-91, 106-107, 116, 164, 280
Hankel transform 89-90, 198-199
  hybrid Hankel transform/endpoint method 169-172
  point source in homogeneous fluid layer bounded by horizontally stratified media 172-173
    geometrical acoustics decomposition 173-174

alternative decomposition 181-186
penetrable boundaries 176-181
soft top, hard bottom 174-176
normal mode decomposition 186-187
Pekeris waveguide 187-198
point source in medium with single minimum profile 231-238
Hard boundary condition *see* Boundary condition
Head wave 104
Helmholtz equation 5-6, 16, 18, 19, 21, 24, 54, 206, 238
inhomogeneous 60, 84-85, 250
Historical overview 2-5
Homogeneous boundary condition *see* Boundary condition
Homogeneous differential equation 83
Homogeneous wave equation 110
Horizontal slowness 210
Horizontal wavenumber 89
Hydrostatic pressure 203-206

I

Impedance
characteristic 25
normal specific acoustic 40-41
plane wave 41
specific acoustic 25
plane wave 25
spherical wave 25-26
Impedance boundary condition *see* Boundary condition
Impenetrable boundary 37
Improper modes *see* Normal modes *and* Pekeris waveguide
Improper Sturm-Liouville problem *see* Sturm-Liouville theory
Index of refraction 40, 202
Inhomogeneous Helmholtz equation *see* Helmholtz equation
Inhomogeneous plane wave 48-50, 57, 88
Inhomogeneous wave equation *see* Wave equation
Initial condition 32
Intensity 26-27
Lloyd mirror 75
modal 123-124, 155
plane wave 27
ray acoustic *see* Ray theory
spherical wave 27-28
transmitted plane wave 46-47, 49-50
Interface wave *see* Lateral wave
Interference wavelength 123-126, 155-156, 257
Intrinsic absorption 54-55
Intrinsic attenuation 54-55
Intromission angle 44-45
Inversion for bottom properties 172
Irrotational fluid flow 23-24

K

Kronecker delta 114

L

Lateral wave 46-50, 98-104
Laterally varying environment *see* Range-dependent environment
Leaky modes *see* Improper modes
Limiting ray 260
Line source 23, 106-107, 116
Linear acoustic equations 13-15
Lloyd mirror effect 73-80, 124, 136
Lorentz condition 10

M

Maxwell's equations 7-8
Mean-square pressure 27-28, 30
Meromorphic function 141
Method of images *see* Green's function construction
Mittag-Leffler expansion 141, 161

Mixed boundary condition *see* Boundary condition
Modal attenuation 166-167
Modal continuum 126-128, 274, *see also* Modal spectrum
Modal spectrum 126-128, 146-147, 163, 191, 198
Mode coupling 274

N

Neumann boundary condition *see* Boundary condition
Neumann function 22-23
Nodes
  of modal eigenfunctions 118-119, 149-150
  of standing waves 18-19
Normal modes 111-116, *see also* Eigenvalue equation, Hankel transform, Modal continuum, *and* Modal spectrum
  adiabatic mode/WKB theory 269-274
  coupling 274
  homogeneous fluid layer with soft top and hard bottom 116-118
    cutoff frequency 120
    cycle distance 125-126
    dispersion curves 122
    group velocity 121-122
    intensity 123-124
    interference wavelength 123-126
    nodes of eigenfunctions 118-119
    phase velocity 119-122
  homogeneous fluid layer with soft top and higher velocity bottom *see* Pekeris waveguide
  homogeneous fluid layer with soft top and lower velocity bottom 130-136
    improper modes 136-145
  in WKB approximation 253-258
Normal specific acoustic impedance *see* Impedance

O

One-dimensional plane wave 18-19
One-dimensional standing wave 18-19
One-dimensional wave equation 17-19
Orthonormality 114, 148, 199

P

Particle displacement 11
Particle speed 25, 26
Particle velocity 25, 33, 40-41
Pekeris waveguide 145-147, 188-198, *see also* Normal modes
  improper modes 157-162, 188-190
  proper modes 148-149, 188-193
    cutoff frequency 150-152
    cycle distance 155-156
    dispersion curves 153
    group velocity 152-154
    intensity 155
    interference wavelength 155-156
    nodes of eigenfunctions 149-150
    phase velocity 152-153
  total field 162-163, 198
Perfect foci 261-262
Periodic boundary condition *see* Boundary condition
Perturbation theory 138-141, 158-160, 166-167
Phase velocity 17
  modal 119-122, 152-153
Physical oceanography 253
Physical sheet *see* Riemann sheet
Pillbox function 198-199
Plane wave
  impedance 25
  inhomogeneous 49-50, 57, 88
  intensity 27
  one-dimensional 18-19
  three-dimensional 19-20
Plane wave ray/WKB theory *see* WKB theory
Plane wave reflection
  from hard boundary 37
  from homogeneous fluid half-space 37-41

higher velocity, more dense half-space 45-51
lower velocity, less dense half-space 41-43
lower velocity, more dense half-space 43-45
from homogeneous fluid layer overlying horizontally stratified medium 51-53
from horizontally stratified medium 34-37
from soft boundary 37
Point source 21, *see also* Hankel transform, Normal modes, Ray theory, *and* Spherical wave reflection
Poisson sum formula 156-157, 165
Poles
for improper modes 138-145, 158-162
for point source in single minimum profile 253-254
for proper modes 186-187
Pekeris waveguide 187-190
for spherical wave reflection from homogeneous fluid half-space 92-93, 109
Potential
elastic scalar 12-13
elastic vector 12-13
electromagnetic scalar 9-10
electromagnetic vector 9-10
quantum mechanical 6-7
velocity 23-24
Poynting vector 27
Pressure-release boundary condition *see* Boundary condition
Proper modes *see* Normal modes *and* Pekeris waveguide
Proper Sturm-Liouville problem *see* Sturm-Liouville theory
Pulse propagation
in homogeneous medium 107-108
in SOFAR channel 239-253

Q

Quantum mechanics 6-7
bound states 110
Quantum mechanical potential 6-7

R

Range-dependent environment 269-280
Ray bundle 221-225
Ray displacement 156, 186
Ray intensity *see* Ray theory
linear gradient 225
Ray invariant 210
Ray parameter 210
Ray phase *see* Ray theory
canonical profiles 265-266
linear gradient 220-221
Ray range *see* Ray theory
canonical profiles 265-266
linear gradient 218-219
Ray series *see* Ray theory
Ray theory
criterion for validity 210-212
deficiencies 258-259
caustics 263-264
convergence zones 263-264
perfect foci 261-262
reflection coefficients and turning points 264-265
shadow zones 260
in three dimensions for point source in horizontally stratified medium 214-215
ray acoustic intensity 221-225
ray phase 219-221
ray range 217-219
ray travel time *see* ray phase
Snell's law 215-217
in three dimensions for unbounded medium with no sources 212-214
in two dimensions *see* WKB theory
point source in medium with single minimum profile 226
asymptotic evaluation of Hankel transform representation 231-238
ray series solution 226-231

SOFAR propagation in symmetric bilinear profile *see* SOFAR propagation
Ray tracing 214, 226-227
Ray trajectory *see* Ray tracing
Ray travel time *see* Ray theory
canonical profiles 265-266
linear gradient 220-221
Rayleigh reflection coefficient 40-41, 52, 55-56, 90, 130, 135
Reciprocity 64, 200
Reduced wave equation 5-6, *see also* Helmholtz equation
Reflection coefficient 86-88, 90-92, 96-104, 130-131, 134-135, 146-147, 171-192, 232, 264-265, *see also* Plane wave reflection
Refracted arrival 46
Residue series:
for improper modes 141-145, 160-162
for Pekeris waveguide 191-193
for point source in single minimum profile 255-258
Resonances *see* Poles
Riemann sheet 93-94, 97-99, 188-191
Rigid boundary condition *see* Boundary condition

S

Saddle points 92
for point source in homogeneous fluid layer with penetrable boundaries 178-186
for point source in single minimum profile 235-238
for spherical wave reflection from homogeneous fluid half-space 93-101
Salinity 203-204
Sandy bottom parameters 50, 206
Schrodinger wave equation *see* Wave equation
Seismics 11-13
Shadow zones 260, 263
Silty-clay bottom parameters 45, 206
Skip distance *see* Cycle distance
Snell's law 39-40, 48, 210, 215-217, 219-221, 224, 243, 258-260
SOFAR channel 4, 204-205, 239-240, 252-253
SOFAR propagation 239-242
ray amplitude 247
ray phase and travel time 244-247
ray range, inital angle, and turning depth 242-244
time-dependent behavior of delta function pulse 248-253
Soft boundary condition *see* Boundary condition
Sommerfeld radiation condition *see* Boundary condition
Sound channel axis *see* SOFAR channel
Sound velocity 15
in air 42-43
in ocean 42-43, 203-206
in seabed 45, 50, 206
Sound velocity profile
Arctic 263
canonical 206, 265-266
hyperbolic cosine 261-262
linear gradient 218-221, 225
North Atlantic 204-205
single minimum 226-258
symmetric bilinear 239-253
Source
line 23, 106-107, 116
point *see* Point source
term in wave equation 60
Specific acoustic impedance *see* Impedance
Specular reflection 35-36, 96, 98, 104, 179-180, *see also* Method of images
Speed of sound 3, *see also* Sound velocity
Spherical spreading 28
Spherical wave 20-21, *see also* Point source
impedance 25-26
intensity 27-28
Spherical wave reflection
from homogeneous fluid half-space 89-104
higher velocity half-space 96-104

lower velocity half-space 93-96
from horizontally stratified medium 85-89
Standard wave equation *see* Wave equation
Standing wave *see* Normal modes
one-dimensional 18-19
three-dimensional 20
Stationary phase method 94-104, 177-186, 231-239
Stationary points *see* Saddle points
Sturm-Liouville theory 3, 113-115, 128
eigenvalue equation 113, 270-271
improper Sturm-Liouville problem 130-131, 146
proper Sturm-Liouville problem 114-115, 131-136, 157, 271, 278
Surface wave *see* Lateral wave

T

Temperature 203-205
Thermocline 204-205
Three-dimensional plane wave 19-20
Three-dimensional standing wave 20
Three-dimensional wave equation 19-21
Tomographic inversion methods 253
Total internal reflection 45-51, 96, 146-148, 181, 188, 212
Total transmission 44-45
Total wavenumber 6
Trace wave *see* Lateral wave
Transmission coefficient 38
Transport equation 213, 238
Trapped modes *see* Normal modes *and* Pekeris waveguide
Tunneling 57-58
Turning point 212, 227, 230-232, 234, 240, 243-244, 247, 254-255, 258, 261, 263-265
Two-dimensional wave equation 21-23

U

Uniform asymptotic expansion 96, 104, 163, 259
Unphysical sheet *see* Riemann sheet

V

Velocity potential 23-24
Vertical wavenumber 86
Virtual modes *see* Improper modes

W

Water wave propagation 154
Wave 17-18
cylindrical 23
plane 18-20
spherical 20-21
tunneling 57
Wave equation
acoustic 5-6
elastic 11-13
electromagnetic 7-11
homogeneous 110
inhomogeneous 60, 110-113, 169, 249-250, 269-270, 275
one-dimensional 17-19
reduced 5-6, *see also* Helmholtz equation
Schrodinger 6-7, 28, 163-164
standard 5, 7, 9, 12, 24
three-dimensional 19-21
two-dimensional 21-23
Wave packet 121
Wave train 121
Wave vector 19-20
Wavefront 19-20, 214
continuity along boundary 35-36, 39
Wavelength 18
Wavenumber
horizontal 89
total 6
vertical 86
Wedge propagation 105-106, 275-280

WKB theory 3, *see also* Ray theory
adiabatic mode/WKB theory 269-274
asymptotic evaluation of Hankel transform result for point source in single minimum profile *see* Ray theory
in two dimensions for unbounded horizontally stratified medium with no sources 206-212
relationship to normal modes *see* Normal modes
Wronskian 84, 164, 170-171, 233, 279-280